CLASSICAL MECHANICS AND ELECTRODYNAMICS

Second Edition

CLASSICAL MECHANICS AND ELECTRODYNAMICS

Second Edition

Jon Magne Leinaas
University of Oslo, Norway

World Scientific

NEW JERSEY • LONDON • SINGAPORE • BEIJING • SHANGHAI • HONG KONG • TAIPEI • CHENNAI

Published by

World Scientific Publishing Co. Pte. Ltd.
5 Toh Tuck Link, Singapore 596224
USA office: 27 Warren Street, Suite 401-402, Hackensack, NJ 07601
UK office: 57 Shelton Street, Covent Garden, London WC2H 9HE

Library of Congress Control Number: 2024945456

British Library Cataloguing-in-Publication Data
A catalogue record for this book is available from the British Library.

CLASSICAL MECHANICS AND ELECTRODYNAMICS
Second Edition

ISBN 978-981-12-9844-8 (hardcover)
ISBN 978-981-98-0074-2 (paperback)
ISBN 978-981-12-9845-5 (ebook for institutions)
ISBN 978-981-12-9846-2 (ebook for individuals)

For any available supplementary material, please visit
https://www.worldscientific.com/worldscibooks/10.1142/13992#t=suppl

Desk Editor: Muhammad Ihsan Putra

Typeset by Stallion Press
Email: enquiries@stallionpress.com

Preface to the Second Edition

After six years, a new part has been added to the book: Part IV. Its name is *Classical Field Theory*, showing that there are close connections between the new part and earlier parts of the book. Thus, Hamilton's principle and Lagrange's equations belong to both the cases. There are however differences, with Part I being restricted to the motion of particles and Part IV having higher dimensions with the presence of classical fields. Another thing to note is that in Part III, there are Maxwell fields, but with no Lagrange formulation. However, with the new part, this is changed, with Lagrange formulations included for both Klein–Gordon and Maxwell fields. Noether's theorem is central in the use of the Lagrangian on these fields.

In the Preface in 2018, I referred to the well-organized course Prof. Erik Eriksen (2026–2017) had given with similar subjects several years before. Also, this time I will refer to him, since I have used some of his course on *Klassisk teoretisk fysikk*. This part of the classical field shows the central content of Noether's theorem, with applications that have the meaning to understand the theorem better.

I will further thank Prof. Are Raklev for reading and making comments on the new part of the book.

Jon Magne Leinaas

University of Oslo, 2024

Preface to the First Edition

Even if physics, as taught in our universities, is much focussed on modern aspects of this discipline, the basic principles of classical physics remain an important part of the physicist's education. Thus, classical mechanics and classical electrodynamics still form much of the fundament for the modern theoretical and experimental developments in these fields, also when quantum effects are of importance. Classical physics can be given an attractive, analytic form, where important physical consequences are derived from a limited set of basic physics principles. This analytic approach, which is applied here, gives a valuable contribution to the student's understanding of theoretical methods in physics, and to his or her abilities to work in the field. It complements in many ways the approach met in introductory physics courses, where the focus is on a basic understanding and description of physical phenomena, with only a gradual development of the theoretical framework.

The present text is based on a physics course I have given for several years at the University of Oslo. It is aimed at Bachelors students in their last or Masters students in their first year. The text consists of three parts, where Part 1 gives an introduction to Analytical Mechanics, which is the mechanics of Newton expressed in a general and more abstract form. The basic element here is the Lagrangian, which determines the dynamics of the system. With this as the starting point, Lagrange's and Hamilton's equations are derived and applied to different types of systems. A geometric point of view is shown to be useful for understanding the Lagrangian dynamics, and the dynamical evolution is illustrated as motion in the phase space of the system. Part 1 also includes a discussion of Hamilton's principle and its correspondence to more general variational problems.

Part 2 gives an introduction to Special Relativity, with derivation of physical consequences of the Lorentz transformations. The lack of absolute simultaneity is shown to lie behind the fundamental effects of length contraction and time dilation, and the twin paradox is discussed in some detail. Minkowski diagrams are much used to illustrate the effects. The relativistic four-vector notation is introduced, and relativistic covariance is shown to be useful for finding the correct relativistic extension of non-relativistic formulas. This is made use of in the description of relativistic kinematics and dynamics. As a particular application, the correct relativistic expressions for conservation of energy and momentum in collision processes are derived, and a relativistic Lagrange formulation for charged particles in an electromagnetic field is found.

Part 3 is devoted to the study of basic elements of Classical Electrodynamics. Maxwell's equations are taken as the starting point, first summarized in their non-covariant formulation and then re-formulated in relativistic, covariant form. The simplification obtained by introducing the electromagnetic potentials is discussed, and the relativistic transformation properties of fields and currents are studied. The main aim of Part 3 is to study methods for finding solutions to Maxwell's equations with stationary and with time dependent sources. The wave solutions for the free fields are first discussed, with focus on the polarization effects. To find solutions with non-vanishing sources the technique of multipole expansion is applied. Special attention is given to the radiation phenomena, with its links to the relativistic retardation effect.

With both classical mechanics and electrodynamics included in the text, each of the two subjects is obviously not treated with the same depth as when classical mechanics and classical electromagnetism are treated in separate courses. However, the idea is that, at the level of education that it is aimed, such a broad course in theoretical physics will be attractive to a rather wide group of students. The aim of this text is to discuss rather explicitly, with detailed derivations, the subjects that are included. Several examples are included, and all the chapters end with a set of exercises. A set of appendices are also included at the end of the text, where several useful formulas are found, as well as comments on Dirac's delta function and the Levi-Civita symbol.

My teaching of the course and of writing the present text, has been much inspired by my own meeting with the subject in my student days. Back then, an early version of the same course was taught by Prof. Erik Eriksen here, and his clear and well organized lecturing of the subject made his course on Theoretical Physics much valued by myself and many of my co-students.

I would like to thank Prof. Are Raklev, who has now taken over the teaching of the course, for his help with corrections and for his suggestions to improve the presentation in several places of the book.

Jon Magne Leinaas
University of Oslo, 2018

Contents

PART 1
Analytical Mechanics

Introduction

The form of classical mechanics we shall discuss here is often called *analytical mechanics*. It is essentially the same as the mechanics of Isaac Newton (1642–1727), but brought into a more abstract form. The analytical formulation of mechanics was developed in the 18th and 19th century by several physicists, but two of these had a particularly strong influence on the development of the field, namely Joseph Louis Lagrange (1736–1813) and William Rowan Hamilton (1805–1865). The mathematical formulation given to mechanics by these two, and developed further by others, is generally admired for its formal beauty. Although this formalism was developed a long time ago, it is still a basic element of modern theoretical physics and has influenced much the later theories of relativity and quantum mechanics.

Lagrange and Hamilton gave two different, but equivalent, formulations of mechanics. In principle we may make a choice between the two, but instead it is common to study both these formulations as two sides of the analytic approach to mechanics. This is because they have different useful properties, and it is advantageous to be able to choose between these methods the one which fits best the problem to be solved. One should note, however, a certain limitation in both these formulations of mechanics, since they in the standard form assume the forces to be conservative. Thus, mechanical systems that involve friction and dissipation are generally not handled by these formulations of mechanics. We refer to systems that can be handled by the Lagrangian and Hamiltonian formalisms to be *Hamiltonian systems.*

In Newtonian mechanics *force* and *acceleration* are central concepts, and in modern terminology we often refer to this as a *vector formulation* of mechanics. Lagrangian and Hamiltonian mechanics are different since force is not a central concept, and *potential* and *kinetic energy* instead are functions that determine the dynamics. In some sense they can be viewed as extensions of the usual formulation of *statics*, where a typical problem is to find the minimum of a potential. As a curious difference the *Lagrangian*, which is the function that regulates the dynamics in Lagrange's formulation, is the *difference* between kinetic and potential energy, while the *Hamiltonian*, which is the basic dynamical function in Hamilton's formulation, is usually the *sum* of kinetic and potential energy.

The Hamiltonian and Lagrangian formulations are generally more easy to apply to composite systems than the Newtonian formulation is. The main problem is to identify the physical degrees of freedom of the mechanical system, to choose a corresponding set of independent variables and to express the kinetic and potential energies in terms of these. The dynamical equations, or equations of motion, are then derived in a straight forward way as differential equations determined either by the Lagrangian or the Hamiltonian. Newtonian mechanics on the other hand expresses the dynamics as motion in three-dimensional space, and students who have struggled with the vector equations of a composite mechanical system know that such a vector analysis is not always simple. However, as is generally common when a higher level of abstraction is used, there is something to gain and something to lose. A well formulated abstract theory may introduce sharper tools for analyzing a physical system, but often at the expense of more intuitive physical interpretation. That is the case also for analytical mechanics, and the vector formulation of Newtonian mechanics is often indispensable for the physical interpretation of the theory.

In the following we shall derive the basic equations of the Lagrangian and Hamiltonian mechanics from Newtonian mechanics. In this derivation there are certain complications, like the distinction between *virtual* and *physical* displacements, but application of the derived formalism does not depend on these intermediate steps. The typical problem of using the Lagrangian or Hamiltonian formalism is based on a simple standardized algorithm with the following steps: First determine the degrees of freedom of the mechanical system and choose a set of independent coordinates, one for each degree of freedom. Next find the Lagrangian or Hamiltonian expressed in terms of the coordinates and their time derivatives and formulate the dynamics either as Lagrange's or Hamilton's equations. The final problem, which is the purely mathematical part, is then to solve the corresponding differential equations with the given initial conditions.

Hamilton's principle introduces a different view of the dynamics of Hamiltonian systems. The time evolution of the system is here expressed as solutions of a variational problem, rather than of a set of differential equations. The basic object in this description is the *action*, which is the time integral of the Lagrangian along paths which may be taken by the system. As will be shown, there is a simple relation between the two types of formulations, and this relation can be used to reformulate other types of variational problems in the form of differential equations. We illustrate this by some examples.

Chapter 1

Generalized coordinates

1.1 Physical constraints and independent variables

In the description of mechanical systems we often meet constraints, which means that the motion of one part of the system strictly follows the motion of another part. In the vector analysis of such a system there will be unknown forces associated with the constraints, and a part of the analysis of the system consists in eliminating the unknown forces by applying the constraint relations. One of the main simplifications of the Lagrangian and Hamiltonian formulations is that the dynamics is expressed in variables that from the beginning take these constraints into account. These independent variables are known as *generalized coordinates*, and they are usually different from the Cartesian coordinates of the system. The number of generalized coordinates corresponds to the number of degrees of freedom of the system, which is equal to the remaining number of variables after all the constraint relations have been imposed.

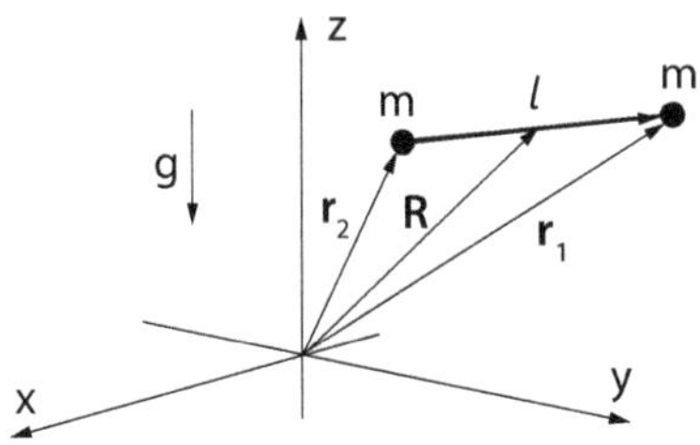

Fig. 1.1 Two small bodies connected by a rigid rod.

As a simple example let us consider two small bodies of equal mass m, attached to the endpoints of a thin rigid, massless rod of length l, and moving in the earth's gravitational field, as shown in Fig. 13.1. In a vector analysis of the system we write the following equations,

$$\begin{aligned} m\ddot{\mathbf{r}}_1 &= m\mathbf{g} + \mathbf{f}\,, \\ m\ddot{\mathbf{r}}_2 &= m\mathbf{g} - \mathbf{f}\,, \\ |\mathbf{r}_1 - \mathbf{r}_2| &= l\,, \end{aligned} \tag{1.1}$$

where we here, and in the following, use the standard notations for time derivatives, $\dot{\mathbf{r}} = \frac{d\mathbf{r}}{dt}, \ddot{\mathbf{r}} = \frac{d^2\mathbf{r}}{dt^2}$. The two first equations are Newton's second law applied to particle 1 and particle 2, with $\mathbf{f}$ as the force from the rod on particle 1, and $\mathbf{g}$ as the gravitational acceleration. The third equation is the constraint equation which expresses that the length of the rod is fixed. The number of degrees of freedom d is easy to find,

$$d = 3 + 3 - 1 = 5\,, \tag{1.2}$$

where each of the two vector equations gives the contribution 3, corresponding to the three components of each of the two vectors $\mathbf{r}_1$ and $\mathbf{r}_2$. The constraint equation, however, removes one degree of freedom, since this equation can be used to express one of the variables in terms of the others, thus reducing the number of *independent* coordinates. As a set of generalized coordinates corresponding to these 5 degrees of freedom we may choose the center of mass vector $\mathbf{R} = (X, Y, Z)$ and the two angles (ϕ, θ), which determine the direction of the rod in space. Expressed in terms of these independent coordinates the position vectors of the endpoints of the rod are

$$\begin{aligned} \mathbf{r}_1 &= \left(X + \frac{l}{2}\sin\theta\cos\phi\right)\mathbf{i} + \left(Y + \frac{l}{2}\sin\theta\sin\phi\right)\mathbf{j} + \left(Z + \frac{l}{2}\cos\theta\right)\mathbf{k}\,, \\ \mathbf{r}_2 &= \left(X - \frac{l}{2}\sin\theta\cos\phi\right)\mathbf{i} + \left(Y - \frac{l}{2}\sin\theta\sin\phi\right)\mathbf{j} + \left(Z - \frac{l}{2}\cos\theta\right)\mathbf{k}\,, \end{aligned} \tag{1.3}$$

with $\mathbf{i}$, $\mathbf{j}$ and $\mathbf{k}$ as unit vectors along the three Cartesian coordinate axes. The corresponding expressions for the kinetic energy T and the potential energy V are

$$\begin{aligned} T &= \frac{1}{2}m(\dot{\mathbf{r}}_1^2 + \dot{\mathbf{r}}_2^2) = m(\dot{X}^2 + \dot{Y}^2 + \dot{Z}^2) + \frac{1}{4}ml^2(\dot{\theta}^2 + \sin^2\theta\,\dot{\phi}^2)\,, \\ V &= mg(z_1 + z_2) = 2mgZ\,, \end{aligned} \tag{1.4}$$

with the z-axis chosen in the vertical direction. These functions, which depend only on the 5 generalized coordinates are the input functions in

Lagrange's and Hamilton's equations, and the elimination of constraint relations means that the unknown constraint force **f** does not appear in the equations.

Let us now make a more general formulation of the transition from Cartesian to generalized coordinates. Following the above example we assume that a general mechanical system can be viewed as composed of a number of small bodies with masses $m_i, i = 1, 2, ..., N$, and position vectors $\mathbf{r}_i, i = 1, 2, ..., N$. We assume that these cannot all move independently, due to a set of constraints, which can be expressed as functional dependences between the coordinates,

$$f_j(\mathbf{r}_1, \mathbf{r}_2, ..., \mathbf{r}_N; t) = 0 \quad j = 1, 2, ..., M\,. \tag{1.5}$$

One should note that such a dependence between the coordinates is not the most general possible. The constraints may, for example, also depend on velocities. However, the possibility of *time dependent* constraints is included in the expression. Constraints that can be written in the form (1.5) are called *holonomic*, and in the following we will restrict the discussion to constraints of this type.

The number of variables of the system is $3N$, since for each particle there are three variables corresponding to the components of the position vector $\mathbf{r}_i$, but the number of *independent* variables is smaller, since each constraint equation reduces the number of independent variables by 1. The number of degrees of freedom of the system is therefore

$$d = 3N - M\,, \tag{1.6}$$

with M as the number of (independent) constraint equations. d then equals the number of generalized coordinates that is needed to give a complete description of the system, and we denote in the following such a set of coordinates by $\{q_k, k = 1, 2, ..., d\}$. Without specifying the constraints we cannot give explicit expressions for the generalized coordinates, but that is not needed for the general discussion. What is needed is to realize that when the constraints are imposed, the $3N$ Cartesian coordinates can in principle be written as functions of the smaller number of generalized coordinates,

$$\mathbf{r}_i = \mathbf{r}_i(q_1, q_2, ..., q_d; t)\,, \quad i = 1, 2, ..., N\,. \tag{1.7}$$

The time dependence in the relation between the Cartesian and generalized coordinates reflects the possibility that the constraints may be time dependent. For convenience we will use the notation $q = \{q_1, q_2, ..., q_d\}$ for the whole set, so that (1.7) gets the more compact form

$$\mathbf{r}_i = \mathbf{r}_i(q, t)\,, \quad i = 1, 2, ..., N\,. \tag{1.8}$$

Note that the set of generalized coordinates can be chosen in many different ways, and often the coordinates will not all have the same physical dimension. For example some of them may have dimension of length, like the center of mass coordinates in the example above, and others may be dimensionless, like the angles in the same example. Below are some further examples of constraints and generalized coordinates.

Examples

A planar pendulum

We consider a small body with mass m attached to a thin, massless, rigid rod of length l. The rod can oscillate freely about one endpoint, as shown in Fig. 1.2a. We assume the motion to be limited to a two-dimensional plane. There are two Cartesian coordinates in this case, corresponding to the components of the position vector for the small massive body, $\mathbf{r} = x\mathbf{i} + y\mathbf{j}$. The coordinates are restricted by one constraint equation, $f(\mathbf{r}) = |\mathbf{r}| - l = 0$. The number of degrees of freedom is $d = 2 - 1 = 1$, and therefore one generalized coordinate is needed to describe the motion of the system. A natural choice for the generalized coordinate is the angle θ indicated in the figure. Expressed in terms of the generalized coordinate the position vector of the small body is

$$\mathbf{r}(\theta) = l(\sin\theta\mathbf{i} - \cos\theta\mathbf{j})\,, \tag{1.9}$$

and it is straight forward to check that the constraint is satisfied when $\mathbf{r}$ is written in this form. We may further use this expression to find the kinetic and potential energies expressed in terms of the generalized coordinate,

$$\begin{aligned} T(\dot{\theta}) &= \frac{1}{2} m l^2 \dot{\theta}^2\,, \\ V(\theta) &= -mgl\cos\theta\,. \end{aligned} \tag{1.10}$$

A double pendulum

A slightly more complicated case is given by the double pendulum shown in Fig. 1.2b. If we use the same step by step analysis of this system, we start by specifying the Cartesian coordinates of the two massive bodies, $\mathbf{r}_1 = x_1\mathbf{i} + y_1\mathbf{j}$ and $\mathbf{r}_2 = x_2\mathbf{i} + y_2\mathbf{j}$. There are 4 such coordinates, x_1, y_1, x_2

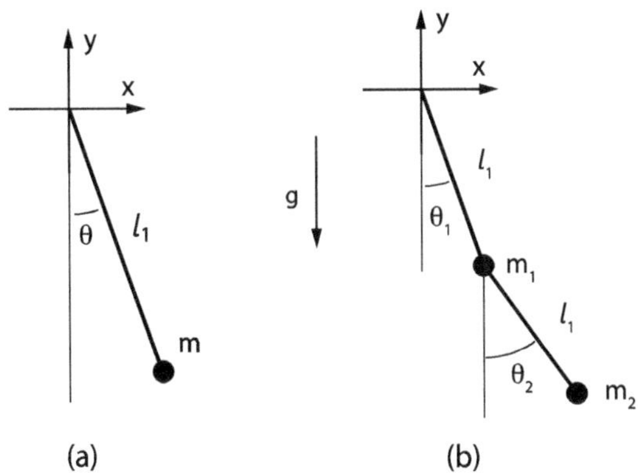

Fig. 1.2 A planar pendulum (a) and a double pendulum (b).

and y_2. However they are not all independent due to the two constraints, $f_1(\mathbf{r}_1) = |\mathbf{r}_1| - l_1 = 0$ and $f_2(\mathbf{r}_1, \mathbf{r}_2) = |\mathbf{r}_1 - \mathbf{r}_2| - l_2 = 0$. The number of degrees of freedom is therefore $d = 4 - 2 = 2$, and a natural choice for the two generalized coordinates is the angles θ_1 and θ_2. The Cartesian coordinates are now expressed in terms of the generalized coordinates as

$$\begin{aligned}\mathbf{r}_1(\theta_1) &= l_1(\sin\theta_1\mathbf{i} - \cos\theta_1\mathbf{j}),\\ \mathbf{r}_2(\theta_1, \theta_2) &= l_1(\sin\theta_1\mathbf{i} - \cos\theta_1\mathbf{j}) + l_2(\sin\theta_2\mathbf{i} - \cos\theta_2\mathbf{j}). \qquad (1.11)\end{aligned}$$

This gives for the kinetic energy the expression,

$$\begin{aligned}T(\theta_1, \theta_2, \dot{\theta}_1, \dot{\theta}_1) &= \frac{1}{2}(m_1\dot{\mathbf{r}_1}^2 + m_2\dot{\mathbf{r}_2}^2)\\ &= \frac{1}{2}(m_1 + m_2)l_1^2\dot{\theta_1}^2 + \frac{1}{2}m_2 l_2^2\dot{\theta_2}^2 + m_2 l_1 l_2\cos(\theta_1 - \theta_2)\dot{\theta}_1\dot{\theta}_2,\end{aligned} \qquad (1.12)$$

and for the potential energy,

$$\begin{aligned}V(\theta_1, \theta_2) &= m_1 g y_1 + m_2 g y_2\\ &= -(m_1 + m_2)g l_1\cos\theta_1 - m_2 g l_2\cos\theta_2. \qquad (1.13)\end{aligned}$$

Rigid body

As a third example we consider a three-dimensional rigid body. We may think of this as being composed of a large number N of small parts, each associated with a position vector $\mathbf{r}_k, k = 1, 2, ..., N$. These vectors are not independent since the distance between any pair of the small parts is fixed.

This corresponds to a set of constraints, $|\mathbf{r}_k - \mathbf{r}_l| = d_{kl}$ with d_{kl} fixed. However, to count the number of independent constraints for the N parts is not so straight forward, and in this case it is therefore easier to find the number of degrees of freedom by a direct argument.

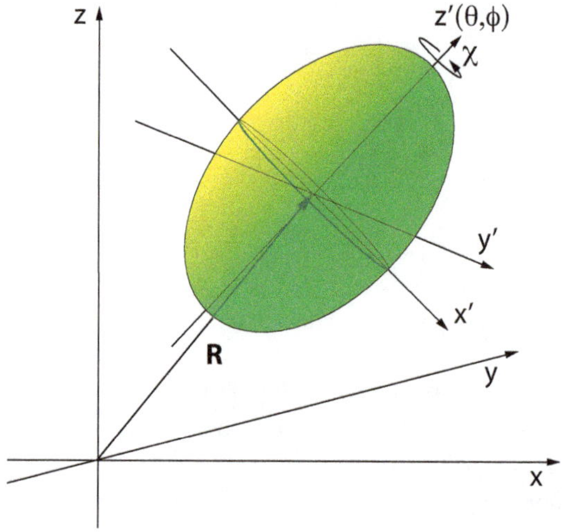

Fig. 1.3 The six degrees of freedom of a rigid body. Three of these are described by the components of the center-of-mass vector $\mathbf{R}$, and three of them by the angular coordinates (θ, ϕ, χ) which determine the orientation of the body in three-dimensional space. The orientation is here illustrated by the rotation of a fixed frame (x, y, z) into a body-fixed frame (x', y', z').

The Cartesian components of the center-of-mass vector obviously define a set of independent variables,

$$\mathbf{R} = X\mathbf{i} + Y\mathbf{j} + Z\mathbf{k}\,. \tag{1.14}$$

When these coordinates are fixed, there is a further freedom to rotate the body about the center of mass. To specify the orientation of the body after performing this rotation, three coordinates are needed. This is easily seen by specifying the orientation of the body in terms of the directions of the axes of an imagined body-fixed orthogonal frame (see Fig. 1.3). For these axes, denoted by (x', y', z'), we see that two angles (ϕ, θ) are needed to fix the orientation of the z' axis, while the orientation of the remaining two axes may be fixed by a rotation angle χ in the x', y' plane. A complete set of generalized coordinates may thus be chosen as

$$q = \{X, Y, Z, \phi, \theta, \chi\}\,. \tag{1.15}$$

The number of degrees of freedom of a three-dimensional rigid body is consequently 6.

Time dependent constraint

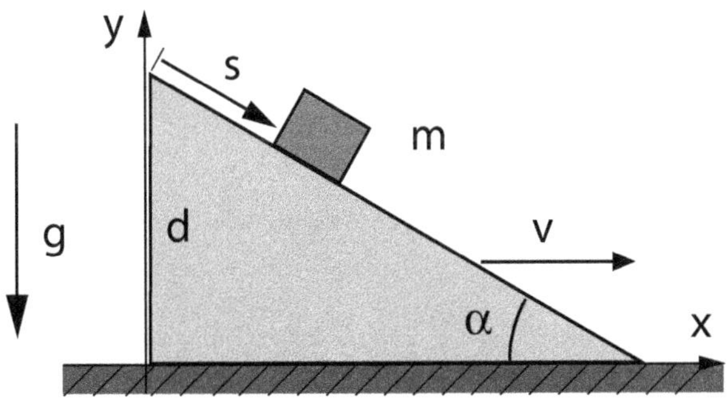

Fig. 1.4 A body is sliding on an inclined plane while the plane moves with constant velocity in the horizontal direction.

We consider a small body with mass m sliding on an inclined plane, shown in Fig. 1.4, and assume the motion to be restricted to the two-dimensional x, y-plane shown in the figure. The angle of inclination is α, and d is the value of the vertical coordinate y for $x = 0$. We first consider the case where the inclined plane is at rest ($v = 0$). With x and y as the two Cartesian coordinates of the body, there is one constraint equation

$$y = d - x \tan \alpha \,, \tag{1.16}$$

and therefore one degree of freedom for the moving body. As generalized coordinate we may conveniently choose the distance s along the plane. The position vector, expressed as a function of this generalized coordinate, is simply

$$\mathbf{r}(s) = s \cos \alpha \, \mathbf{i} + (d - s \sin \alpha) \, \mathbf{j} \,. \tag{1.17}$$

Let us next assume the inclined plane to be moving with constant velocity v in the x-direction. The number of degrees of freedom is still one, but the constraint equation is now time dependent

$$y = d - (x - vt) \tan \alpha \,. \tag{1.18}$$

We may use the same generalized coordinate s as in the time independent case, and find that the position vector $\mathbf{r}$ now depends on both the generalized coordinate s and on time t

$$\mathbf{r}(s,t) = (s\cos\alpha + vt)\,\mathbf{i} + (d - s\sin\alpha)\,\mathbf{j}\,. \tag{1.19}$$

The kinetic and potential energies, when expressed in terms of the generalized coordinate s, then take the forms

$$\begin{aligned} T &= \frac{1}{2}m\dot{\mathbf{r}}^2 = \frac{1}{2}m(\dot{s}^2 + 2v\dot{s}\cos\alpha + v^2)\,, \\ V &= mgy = mg(d - s\sin\alpha)\,. \end{aligned} \tag{1.20}$$

In the general discussion to follow we will assume that the constraints may be time dependent, since this possibility can readily be taken care of by the formalism.

Non-holonomic constraint

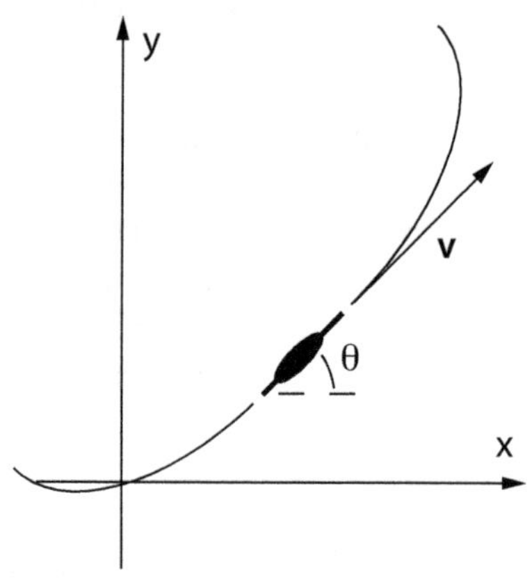

Fig. 1.5 Example of a non-holonomic constraint. The velocity $\mathbf{v}$ of a skate moving on ice is related to the direction of the skate, here indicated by the angle θ. However there is no direct relation between the position coordinates (x, y) and the angle θ.

Even if, in the analysis to follow, we shall restrict the constraints to be holonomic, it may be of interest to consider a simple example of a non-holonomic constraint. Let us study the motion of one of the skates of a person who is skating on ice. As coordinates for the skate we may choose the two Cartesian components of the position vector $\mathbf{r} = x\mathbf{i} + y\mathbf{j}$ together with the angle θ that determines the orientation of the skate. There is no functional relation between these three coordinates, since for an arbitrary position $\mathbf{r}$ the skate can have any angle θ. However, under normal skating there is a constraint on the motion, since the direction of the velocity will be the same as the direction of the skate. This we may write as

$$\dot{\mathbf{r}} = v(\cos\theta\mathbf{i} + \sin\theta\mathbf{j}), \tag{1.21}$$

which gives the following relation

$$\dot{y} = \dot{x}\tan\theta. \tag{1.22}$$

This is a non-holonomic constraint, since it is not a functional relation between coordinates alone, but between velocities and coordinates. Such a relation cannot simply be used to reduce the number of variables, but should be treated in a somewhat different way.

1.2 The configuration space

To sum up what we have discussed so far: A three-dimensional mechanical system, which is composed of N small parts, and which is subject to M holonomic constraints, has a number of degrees of freedom $d = 3N - M$. For each degree of freedom an independent generalized coordinate q_i can be chosen, so that the time evolution is fully described by the time dependence of the set of generalized coordinates

$$q = \{q_1, q_2, ..., q_d\}. \tag{1.23}$$

The set q can be interpreted as a set of coordinates on a d dimensional space (a *manifold*[1]), which is referred to as the *configuration space* of the system. Each point q corresponds to a possible configuration of the composite system, which specifies the positions of all the parts of the system in accordance with the constraints imposed on the system.

In the Lagrangian formulation the time evolution in the configuration space is governed by the Lagrangian, which is a function of the generalized coordinates q, of their velocities $\dot{q}$, and possibly of time t. The normal form of the Lagrangian is given as the difference between the kinetic and potential energy

$$L(q, \dot{q}, t) = T(q, \dot{q}, t) - V(q, t). \tag{1.24}$$

[1]Mathematically a manifold is a topological space which locally is Euclidean. As example, a surface in three-dimensional Euclidean space will be a manifold.

In the following we shall derive the equations of motion expressed in terms of the Lagrangian. In this derivation we begin with the vector formulation of Newton's second law, applied to each part of the system, and show how this can be reformulated in terms of the generalized coordinates.

For the discussion to follow it may be of interest to give a *geometrical* representation of the constraints and generalized coordinates. Again we assume the system to be composed of N parts, the position of each part being specified by a three-dimensional vector, $\mathbf{r}_k, k = 1, 2, ..., N$. Together these vectors can be thought of as a vector in a $3N$-dimensional space,

$$\mathbf{R} = \{\mathbf{r}_1, \mathbf{r}_2, ..., \mathbf{r}_N\} = \{x_1, y_1, z_1; ...; x_N, y_N, z_N\}\,, \tag{1.25}$$

which is a Cartesian product of N copies of 3-dimensional, physical space, where each copy corresponds to one of the parts of the composite system. When the vector $\mathbf{R}$ is specified, this implies that the positions of all parts of the system are specified.

The constraints impose restrictions on the positions of the parts, which can be expressed through the functional dependence of $\mathbf{R}$ on the generalized coordinates,

$$\mathbf{R} = \mathbf{R}(q_1, q_2, ..., q_d; t)\,. \tag{1.26}$$

When the generalized coordinates q are varied, the vector $\mathbf{R}$ will trace out a surface of dimension d in the $3N$-dimensional vector space.[2] This surface, where the constraints are satisfied, represents the configuration space of the system, and the set of generalized coordinates are coordinates on this surface, as schematically shown in the figure. Note that the configuration space will in general *not* be a vector space like the $3N$-dimensional space.

Consider first the constraints to be *time independent*, with the *d*-dimensional surface as a fixed surface in the $3N$-dimensional vector space, and assume that we turn on the time evolution, so that the coordinates become time dependent, $q = q(t)$. The composite position vector $\mathbf{R}$ then describes the time evolution of the system in the form of a curve in $\mathbb{R}^{3N}$, which is constrained to the d-dimensional surface,

$$\mathbf{R}(t) = \mathbf{R}(q(t))\,, \tag{1.27}$$

and the velocity vector $\mathbf{V} = \dot{\mathbf{R}}$ is a tangent vector to the surface, as shown in the figure.

If the constraints instead are *time dependent*, and the surface therefore changes with time, the velocity vector $\mathbf{V}$ will in general no longer be a

[2]Such a higher-dimensional surface is often referred to as a *hypersurface*.

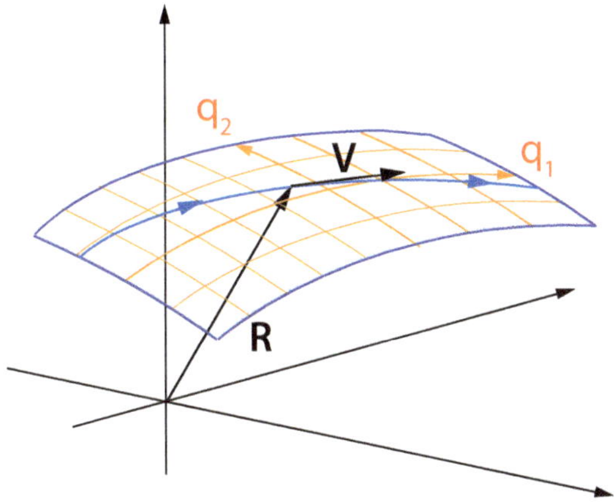

Fig. 1.6 Geometrical representation of the configuration space as a surface in the $3N$ dimensional vector space defined by the Cartesian coordinates of the N small parts of the physical system. The points $\mathbf{R}$ that are confined to the surface are those that satisfy the constraints and the generalized coordinates define a coordinate system that covers the surface (orange lines). The time evolution of the system describes a curve $\mathbf{R}(t)$ on the surface (blue line). If the surface is time independent, the velocity $\mathbf{V} = \dot{\mathbf{R}}$ defines at all times a tangent vector to the surface.

tangent vector to the surface at any given moment, due to the motion of the surface itself. However, for the discussion to follow it is convenient to introduce a type of (infinitesimal) displacement, usually written as $q \to q + \delta q$. It corresponds to a situation where we "freeze" the surface during the displacement. The corresponding displacement vector $\delta\mathbf{R}$ is a tangent vector to the surface. When the constraints are time dependent, such a displacement can obviously not correspond to a physical motion of the system, since the displacement takes place at a fixed time. For this reason one refers to this type of change of position as a *virtual* displacement.

1.3 Virtual displacements

We again express the position vectors of each part of the system as functions of the generalized coordinates,

$$\mathbf{r}_k = \mathbf{r}_k(q_1, q_2, ..., q_d; t)\,, \tag{1.28}$$

where we have included the possibility of time dependent constraints. We refer to this as an *explicit* time dependence, since it does not come from the change of the coordinates q during motion of the system. A general displacement of the positions, which satisfies the constraints, can then be decomposed in a contribution from the change of general coordinates q at fixed t and a contribution from change of t with fixed q,

$$d\mathbf{r}_k = \sum_{j=1}^{d} \frac{\partial \mathbf{r}_k}{\partial q_j} dq_j + \frac{\partial \mathbf{r}_k}{\partial t} dt \,. \tag{1.29}$$

In particular, if we consider the dynamical evolution of the system, the velocities can be expressed as

$$\dot{\mathbf{r}}_k = \sum_{j=1}^{d} \frac{\partial \mathbf{r}_k}{\partial q_j} \dot{q}_j + \frac{\partial \mathbf{r}_k}{\partial t} \,. \tag{1.30}$$

The motion in part comes from the time evolution of the generalized coordinates, $q = q(t)$, and in part from the explicit time dependence of the position vectors.

Note that in the above expression for the velocity, we distinguish between the two types of time derivatives, referred to as *explicit* time derivative,

$$\frac{\partial}{\partial t} \,, \tag{1.31}$$

and *total* time derivative

$$\frac{d}{dt} = \sum_{j=1} \dot{q}_j \frac{\partial}{\partial q_j} + \frac{\partial}{\partial t} \,. \tag{1.32}$$

The first one is simply the partial derivative with respect to t, with the generalized coordinates q kept fixed. The total time derivative, on the other hand, is meaningful only when we consider a particular time evolution, or path in configuration space, expressed by time dependent coordinates $q = q(t)$. It acts on variables that are defined on such a path in configuration space.

A virtual displacement of the positions $\mathbf{r}_k$ corresponds to a displacement δq_i at *fixed time* t,

$$\delta\mathbf{r}_k = \sum_{j=1}^{d} \frac{\partial \mathbf{r}_k}{\partial q_j} \delta q_j \,. \tag{1.33}$$

With no contribution from the explicit time dependence, it will in general not correspond to a real, physical displacement. Instead it can be viewed as an imagined displacement, which measures the functional dependence of the position vectors $\mathbf{r}_k$ on the generalized coordinates q.

1.4 Applied forces and constraint forces

The total force acting on part k of the system can be thought of as consisting of two parts,

$$\mathbf{F}_k = \mathbf{F}_k^a + \mathbf{f}_k\,, \tag{1.34}$$

where $\mathbf{f}_k$ is the generally unknown *constraint force*, and $\mathbf{F}_k^a$ is the so-called *applied force*. The constraint forces can be regarded as a response to the applied forces caused by the presence of constraints.

As a simple example, consider a body sliding on an inclined plane under the action of the gravitational force. The forces acting on the body are the gravitational force, the normal force, which acts on the body perpendicular to the tilted plane, and finally the friction force acting parallel to the plane. The normal force is counteracting the perpendicular component of the gravitational force, thus preventing any motion of the body in this direction. We identify the normal force as the constraint force, while the other forces are considered as applied forces.

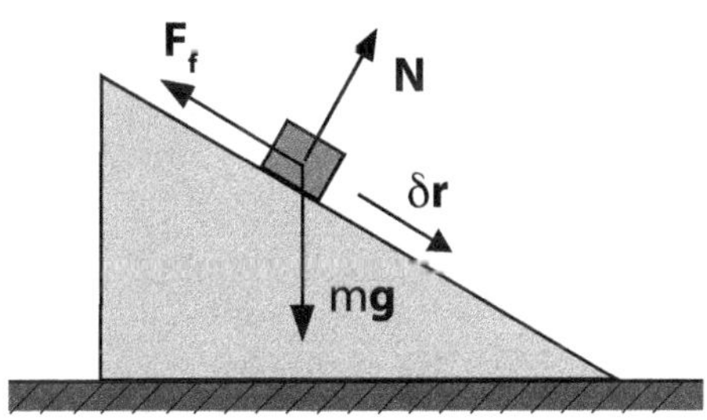

Fig. 1.7 A body on an inclined plane. The applied forces are the force of gravity and the friction. The normal force is a constraint force. It neutralizes the normal components of the other forces, which would otherwise create motion in conflict with the constraints. The direction of virtual displacements $\delta\mathbf{r}$ is along the inclined plane. This is so even if the plane itself is moving since a *virtual* displacement is an imagined displacement at fixed time.

We assume now that a general constraint force is similar to the normal force, in the sense of being orthogonal to any *virtual* displacement of the system. We write this condition as

$$\mathbf{f}\cdot\delta\mathbf{R} = 0\,, \tag{1.35}$$

where we have introduced a $3N$-dimensional vector $\mathbf{f} = (\mathbf{f}_1, \mathbf{f}_2, ..., \mathbf{f}_N)$ for the constraint forces, similar to the $3N$-dimensional position vector $\mathbf{R} =$

$(\mathbf{r}_1, \mathbf{r}_2, ..., \mathbf{r}_N)$. Thus $\mathbf{f}$ specifies the forces acting on all the N parts of the system. The vector $\mathbf{f}$ acts perpendicular to the surface defined by the constraints, and it can be viewed as a reaction to other (applied) forces that have components perpendicular to the surface. Since $\mathbf{f}$ has vanishing components along the surface, it will not affect the motion of the system in these directions and it can therefore be eliminated by changing from Cartesian to generalized coordinates. This we shall exploit in the following.

The orthogonality condition (1.35) can be re-written in terms of three-dimensional vectors as

$$\sum_k \mathbf{f}_k \cdot \delta\mathbf{r}_k = 0\,, \tag{1.36}$$

and we note that the expression can be interpreted as the *work* performed by the constraint forces under the displacement $\delta\mathbf{r}_k$. Thus, *the work performed by the constraint forces under any virtual displacement vanishes.* One should note that this does *not* mean that the work done by a constraint force under the time evolution will always vanish, since the *real* displacement $d\mathbf{r}_k$ may have a component along the constraint force if the constraint is time dependent.

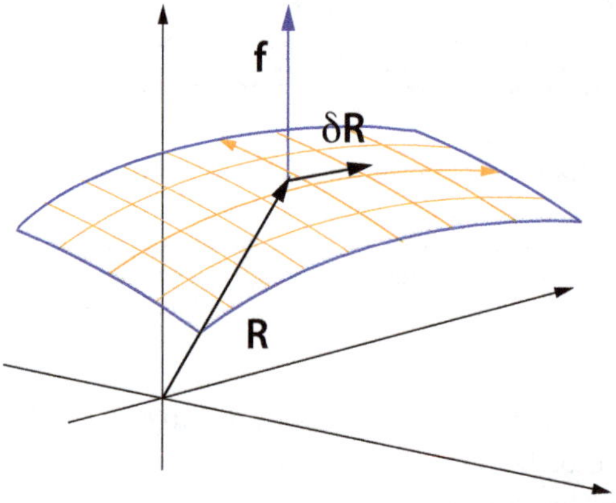

Fig. 1.8 The constraint force $\mathbf{f}$ is a force that is perpendicular to the virtual displacements, and therefore to the surface that defines the configuration space.

1.5 Equilibrium conditions and the principle of virtual work

We assume that a mechanical system is in the state of equilibrium. This means that there is a balance between the forces acting on each part of the system so that there is no motion,

$$\mathbf{F}_k^a + \mathbf{f}_k = 0\,, \quad k = 1, 2, ..., N\,. \tag{1.37}$$

Since the virtual work performed by the constraint forces always vanishes (see (1.36)), the virtual work done by the applied forces will in a situation of equilibrium also vanish,

$$\sum_k \mathbf{F}_k^a \cdot \delta \mathbf{r}_k = 0\,. \tag{1.38}$$

This form of the condition for static equilibrium is often referred to as the *principle of virtual work*.

This condition can be re-expressed in terms of the 3N-dimensional vectors as

$$\mathbf{F}^a \cdot \delta \mathbf{R} = 0\,. \tag{1.39}$$

Geometrically this means that in a point of equilibrium on the d-dimensional surface in $\mathbb{R}^{3N}$, the applied force has to be orthogonal to the surface. This seems easy to understand: If the applied force has a non-vanishing component along the surface this will induce a motion of the system in that direction. That cannot happen at a point of equilibrium.

Let us reconsider the virtual work and re-express it in terms of the generalized coordinates. We have

$$\begin{aligned}
\delta W &= \sum_k \mathbf{F}_k \cdot \delta \mathbf{r}_k \\
&= \sum_k \mathbf{F}_k^a \cdot \delta \mathbf{r}_k \\
&= \sum_k \sum_j \mathbf{F}_k^a \cdot \frac{\partial \mathbf{r}_k}{\partial q_j} \delta q_j \\
&= \sum_j \mathcal{F}_j \delta q_j\,,
\end{aligned} \tag{1.40}$$

where, at the last step we have introduced the *generalized force*, defined by

$$\mathcal{F}_j = \sum_k \mathbf{F}_k^a \cdot \frac{\partial \mathbf{r}_k}{\partial q_j}\,. \tag{1.41}$$

We note that the generalized force depends only on the applied forces, not on the constraint forces. At equilibrium the virtual work δW should vanish for *any* virtual displacement δq, and since all the coordinates q_i are independent, that means that the coefficients of δq_i have all to vanish,

$$\mathcal{F}_j = 0\,, \quad j = 1, 2, ..., d \quad \text{(equilibrium condition)}\,. \tag{1.42}$$

Thus at equilibrium all the generalized forces have to vanish. Note that the same conclusion cannot be drawn about the applied forces, since the coefficients of $\mathbf{r}_k$ may not all be independent due to the constraints.

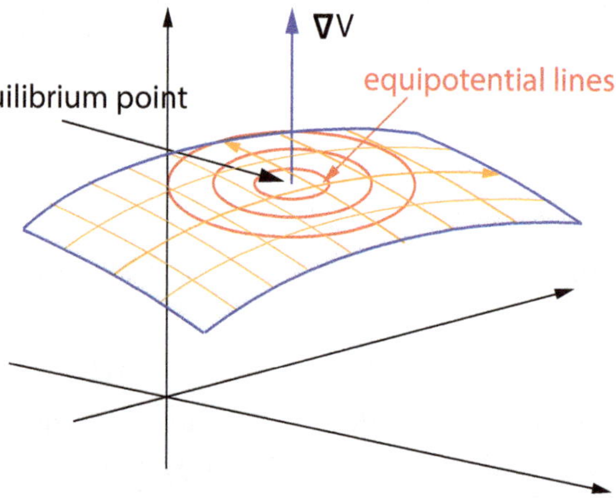

Fig. 1.9 Equilibrium point. At this point the derivatives of the potential with respect to the generalized coordinates vanish and the gradient of the potential is perpendicular to the surface defined by the configuration space.

In the special cases where the applied forces can be derived from a potential $V(\mathbf{r}_1, \mathbf{r}_2, ...)$,

$$\mathbf{F}_k^a = -\boldsymbol{\nabla}_k V\,; \tag{1.43}$$

with $\boldsymbol{\nabla}_k$ being the gradient with respect to the coordinates $\mathbf{r}_k$ of part k of the system, the generalized force can be expressed as a gradient in configuration space,

$$\mathcal{F}_j = -\sum_k \boldsymbol{\nabla}_k V \cdot \frac{\partial \mathbf{r}_k}{\partial q_j} = -\frac{\partial V}{\partial q_j}\,. \tag{1.44}$$

The equilibrium condition is then simply

$$\frac{\partial V}{\partial q_j} = 0\,, \quad j = 1, 2, ..., d \quad \text{(equilibrium condition)}\,, \tag{1.45}$$

which means that the potential has a local minimum, or more generally is stationary, at points of equilibrium in the d-dimensional configuration space of the system.

Example

The configuration space of the pendulum

We illustrate here the concepts that have been introduced in this section, using the planar pendulum as a simple example. The situation is illustrated in Fig. 1.10. We focus on the pendulum bob as the physical body, while the (massless) pendulum rod is only considered as causing the constraint on the motion of the pendulum bob. The pendulum is allowed to make full rotations.

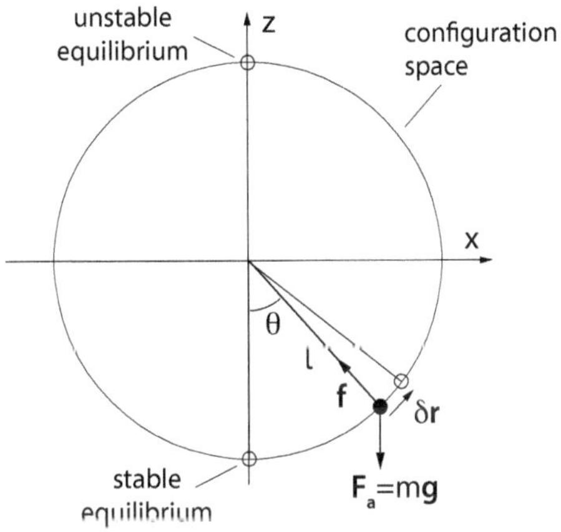

Fig. 1.10 The circle as configuration space of the pendulum. $\mathbf{F}_a$ is the applied force, which is the force of gravity acting on the pendulum bob, and $\mathbf{f}$ is the constraint force, which is the force acting from the pendulum rod on the bob. $\delta\mathbf{r}$ denotes a virtual displacement of the bob, which is directed along the circle.

With the pendulum bob constrained to move on a circle, the circle can be regarded as defining the configuration space of the system. The angle θ, which determines the angular position on the circle is the natural generalized coordinate of the system, and the position vector in the plane depends on θ as $\mathbf{r}(\theta) = l(\sin\theta\mathbf{i} - \cos\theta\mathbf{k})$, with l as the radius of the circle, $\mathbf{i}$ as a unit vector in the horizontal direction, and $\mathbf{k}$ as a unit vector in the vertical direction. A virtual displacement of the pendulum bob is then given by

$$\delta\mathbf{r} = \frac{\partial\mathbf{r}}{\partial\theta}\delta\theta = l(\cos\theta\mathbf{i} + \sin\theta\mathbf{k})\delta\theta\,, \tag{1.46}$$

which clearly is a vector directed tangentially to the circle. In the present case, with a time independent constraint, the same expression is valid also for a real displacement of the pendulum bob. However, even if the constraint is time dependent, for example due to rotation of the pendulum plane, the expression for the virtual displacement is the same, since it is defined as a displacement with fixed time coordinate.

The constraint force satisfies the condition $\mathbf{f} \cdot \delta\mathbf{r} = 0$, and is therefore of the form

$$\mathbf{f} = f(-\sin\theta\mathbf{i} + \cos\theta\mathbf{k}), \tag{1.47}$$

where the strength f of the force is defined implicitly by the fact that it constrains the pendulum bob to move on the circle. The applied force, in this case, is the force of gravity which acts on the pendulum bob,

$$\mathbf{F}_a = m\mathbf{g} = -mg\mathbf{k}, \tag{1.48}$$

and the corresponding generalized force is then

$$\begin{aligned}\mathcal{F}_\theta &= \mathbf{F}_a \cdot \frac{\partial \mathbf{r}}{\partial \theta} \\ &= m\mathbf{g} \cdot (\cos\theta\mathbf{i} + \sin\theta\mathbf{k})l \\ &= -mgl\sin\theta\,.\end{aligned} \tag{1.49}$$

It can be identified as the torque of the applied force, with direction orthogonal to the pendulum plane,

$$\boldsymbol{\tau} = \mathbf{r} \times m\mathbf{g} = mgx\mathbf{j} = mgl\sin\theta\mathbf{j}\,. \tag{1.50}$$

The generalized force can also be identified as the derivative of the potential energy of the pendulum with respect to the generalized coordinate

$$\mathcal{F}_\theta = -\frac{dV}{d\theta} = -\frac{d}{d\theta}(mgz) = -mgl\sin\theta\,. \tag{1.51}$$

The equilibrium points of the pendulum are points where the generalized force vanishes, and equivalently where the potential function $V(\theta)$ is stationary,

$$\mathcal{F}_\theta = 0 \quad \Rightarrow \quad \theta = 0, \pi\,, \quad \text{equilibrium condition}\,. \tag{1.52}$$

If the pendulum has zero velocity at such a point, the total force acting on the pendulum bob vanishes, and the constraint force thereby precisely counteracts the gravity force, with $\mathbf{f} = -\mathbf{F}_a = mg\mathbf{k}$.

1.6 Exercises

Problem 1.1

Figure 1.11 shows four different mechanical systems: (a) A pendulum attached to a block, which in turn is attached to a spring, (b) A pendulum attached to a vertical disk, which rotates with a fixed angular frequency ω, (c) A rigid rod, which can tilt without sliding on the top of cylinder, while the cylinder can roll on a horizontal plane, (d) A rotating top which moves on a horizontal floor. In all cases specify the number of degrees of freedom

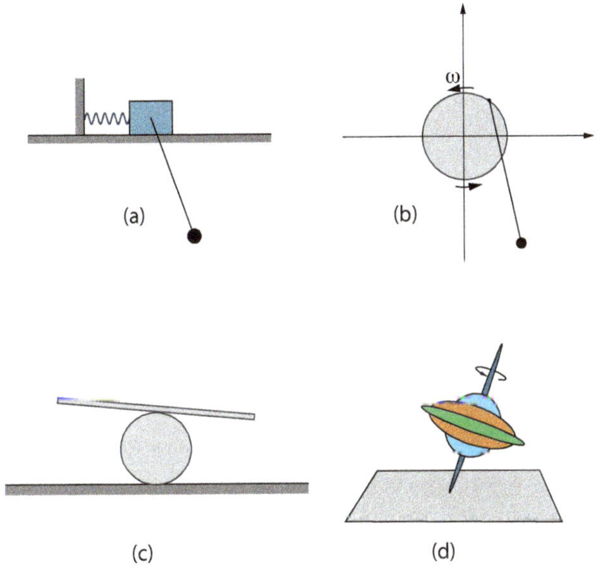

Fig. 1.11 Choosing generalized coordinates.

and choose an appropriate set of generalized coordinates.

Problem 1.2

An Atwood's machine (Fig. 1.12) consists of three weights, with masses $m_1 = 4m$, $m_2 = 2m$ and $m_3 = m$, which move vertically, and two rotating pulleys, which we treat as massless. The ropes, which we also consider as massless, have fixed (vertical) lengths l_1 and l_2.

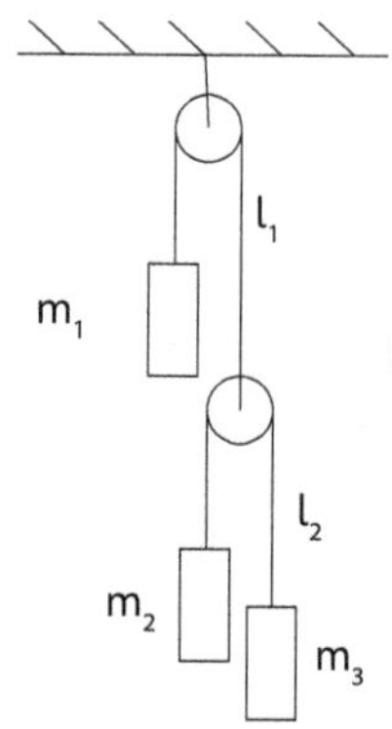

Fig. 1.12 An Atwood's machine.

Explain why the number of degrees of freedom of the system is 2 and choose a corresponding set of generalized coordinates. Find the potential and kinetic energies of the system expressed as functions of the generalized coordinates and their time derivatives.

Problem 1.3

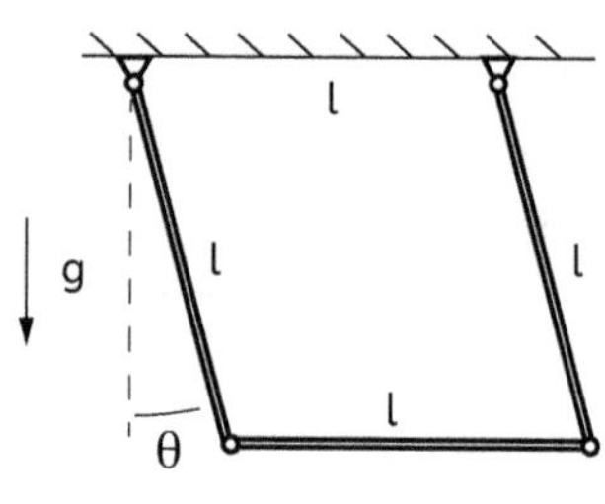

Fig. 1.13 A system of moving rods.

Three identical rods of mass m and length l (Fig. 1.13) are connected by frictionless joints, with the distance between the points of suspension being equal to the length of the rods. The rods move in the vertical plane. Explain why the system has only one degree of freedom, and choose the angle θ as generalized coordinate.

Show that the Lagrangian, defined as the difference between the kinetic and potential energy, $L = T - V$, gets the following form as function of θ

and $\dot{\theta}$,

$$L(\theta, \dot{\theta}) = \frac{5}{6} m l^2 \dot{\theta}^2 + 2mgl \cos\theta \, . \tag{1.53}$$

As a reminder, the moment of inertia about the endpoint of a rod of length l is $I = \frac{1}{3} m l^2$.

Problem 1.4

A particle with mass m moves in three-dimensional space under the influence of a constraint. The constraint is expressed by the following relation between the Cartesian coordinates (x, y, z) of the particle,

$$e^{-(x^2+y^2)} + z = 0 \, . \tag{1.54}$$

a) Explain why the number of degrees of freedom of the particle is 2. Use x and y as generalized coordinates and find the expression for the position vector $\mathbf{r}$ of the particle in terms of x and y.

b) A virtual displacement is a change in the position of the particle $\mathbf{r} \to \mathbf{r} + \delta\mathbf{r}$ which is caused by an infinitesimal change in the generalized coordinates, $x \to x + \delta x$ and $y \to y + \delta y$. Find $\delta\mathbf{r}$ expressed in terms of δx and δy.

c) The constraint can be interpreted as a restriction for the particle to move on a two-dimensional surface in three-dimensional space. Any virtual displacement $\delta\mathbf{r}$ is a tangent vector to the surface, while the constraint force $\mathbf{f}$, which acts on the particle, is perpendicular to the surface. Use this to determine $\mathbf{f}$, up to a position-dependent normalization factor, as a function of x and y.

d) Make a drawing of a section through the surface for $y = 0$. Indicate in the drawing the direction of the two vectors $\mathbf{f}$ and $\delta\mathbf{r}$ for a chosen point on the surface.

Problem 1.5

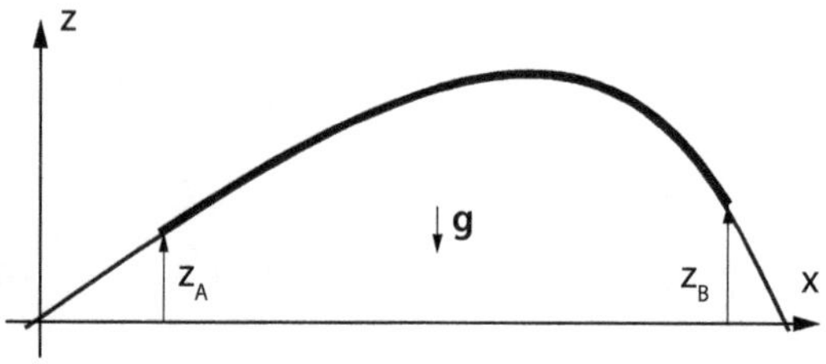

Fig. 1.14 Virtual work on a flexible chain.

A flexible chain can move without friction on a smooth surface, as shown in the figure. The chain has a constant (linear) mass density μ. The vertical heights of the endpoints are z_A and z_B. Use the principle of virtual work to find how z_A and z_B are related when the chain is at static equilibrium.

Chapter 2

Lagrange's equations

2.1 D'Alembert's principle and Lagrange's equations

We examine now how the description of the equilibrium condition discussed in the previous section can be extended to non-equilibrium situations. The equilibrium condition is then replaced by Newton's second law, in the form

$$m_k \ddot{\mathbf{r}}_k = \mathbf{F}_k = \mathbf{F}_k^a + \mathbf{f}_k\,, \tag{2.1}$$

and for a virtual displacement this implies

$$\sum_k (\mathbf{F}_k^a - m_k \ddot{\mathbf{r}}_k) \cdot \delta \mathbf{r}_k = 0\,, \tag{2.2}$$

which is referred to as *D'Alembert's principle.*[1] The important point is, like in the equilibrium case, that by introducing the virtual displacements in the equation one eliminates the (unknown) constraint forces. The expression is in fact similar to the equilibrium condition, although the "force", which appears in this expression, $\mathbf{F}_k^a - m_k \ddot{\mathbf{r}}_k$, is not simply a function of the positions $\mathbf{r}_k$, but also of the accelerations. Nevertheless, the method used to express the equilibrium condition in terms of the generalized coordinates can be generalized to the dynamical case, and that leads to Lagrange's equations. In order to show this we have to rewrite the expressions.

The first part of Eq. (2.2) is easy to handle and we write it as

$$\sum_k \mathbf{F}_k^a \cdot \delta \mathbf{r}_k = \sum_j \mathcal{F}_j \delta q_j\,, \tag{2.3}$$

with $\mathcal{F}_j$ as the generalized forces. Also the second part can be expressed in terms of variations in the generalized coordinates and we rewrite

[1] Jean le Rond d'Alembert (1717–1783) was a French mathematician, physicist and philosopher. He was co-editor with Denis Diderot of the *Encyclopédie*.

D'Alembert's principle as

$$\sum_j \left(\mathcal{F}_j - \sum_k m_k \ddot{\mathbf{r}}_k \cdot \frac{\partial \mathbf{r}_k}{\partial q_j} \right) \delta q_j = 0 \,. \tag{2.4}$$

Since this should vanish for arbitrary virtual displacements, the coefficients of δq_j have to vanish, and this gives

$$\sum_k m_k \ddot{\mathbf{r}}_k \cdot \frac{\partial \mathbf{r}_k}{\partial q_j} = \mathcal{F}_j \,, \quad j = 1, 2, ..., d \,. \tag{2.5}$$

The objective is now to re-express the left hand side in terms of the generalized coordinates and their velocities.

To proceed we split the acceleration term in two parts,

$$\sum_k m_k \ddot{\mathbf{r}}_k \cdot \frac{\partial \mathbf{r}_k}{\partial q_j} = \frac{d}{dt} \left(\sum_k m_k \dot{\mathbf{r}}_k \cdot \frac{\partial \mathbf{r}_k}{\partial q_j} \right) - \sum_k m_k \dot{\mathbf{r}}_k \cdot \frac{d}{dt} \left(\frac{\partial \mathbf{r}_k}{\partial q_j} \right) , \tag{2.6}$$

and examine each of these separately. To rewrite the first term we first note how the velocity vector depends on the generalized coordinates and their velocities,

$$\dot{\mathbf{r}}_k \equiv \frac{d}{dt} \mathbf{r}_k = \sum_j \frac{\partial \mathbf{r}_k}{\partial q_j} \dot{q}_j + \frac{\partial \mathbf{r}_k}{\partial t} \,. \tag{2.7}$$

The expression shows that whereas the position vector depends only on the generalized coordinates, and possibly on time (when there is explicit time dependence),

$$\mathbf{r}_k = \mathbf{r}_k(q, t) \,, \tag{2.8}$$

that is not the case for the velocity vector $\dot{\mathbf{r}}_k$, which depends also on the time derivatives $\dot{q}_j$. At this point we make an extension of the number of independent variables in our description. We simply consider the generalized velocities $\dot{q}_j$ as variables that are independent of the generalized coordinates q_j. Is that meaningful? Yes, as long as we consider all possible motions of the system, we know that specifying the positions will not also determine the velocities. So, assuming the positions to be specified, if we change the velocities that means that we change from one possible motion of the system to another.

In the following we shall therefore consider all coordinates $q = \{q_1, q_2, ..., q_d\}$, all velocities $\dot{q} = \{\dot{q}_1, \dot{q}_2, ..., \dot{q}_d\}$, and time t to be independent variables. Of course, when we consider a particular time evolution, $q = q(t)$, then both q_j and $\dot{q}_j$ become dependent on t. So the challenge is not to mix these two views, the first one when all 2d+1 variables are treated as independent, and the second one when all of them are considered as time dependent functions. However, the idea is not much more complicated than with the usual space and time coordinates (x, y, z, t). In general they can be considered as independent variables, but when applied to the motion of a particle, the space coordinates of the particle become dependent on time, $x = x(t)$ etc. As already discussed these two views are captured in the difference between the partial derivative with respect to time, $\frac{\partial}{\partial t}$ and the total derivative $\frac{d}{dt}$. The latter we may now write as

$$\frac{d}{dt} = \sum_j \left(\ddot{q}_j \frac{\partial}{\partial \dot{q}_j} + \dot{q}_j \frac{\partial}{\partial q_j} \right) + \frac{\partial}{\partial t}, \tag{2.9}$$

since we have introduced $\dot{q}_j$ as independent variables.

From Eq. (2.7) we deduce the following relation between partial derivatives of velocities and positions

$$\frac{\partial \dot{\mathbf{r}}_k}{\partial \dot{q}_j} = \frac{\partial \mathbf{r}_k}{\partial q_j}, \tag{2.10}$$

which gives

$$\dot{\mathbf{r}}_k \cdot \frac{\partial \mathbf{r}_k}{\partial q_j} = \dot{\mathbf{r}}_k \cdot \frac{\partial \dot{\mathbf{r}}_k}{\partial \dot{q}_j} = \frac{1}{2} \frac{\partial}{\partial \dot{q}_j} \dot{\mathbf{r}}_k^2 . \tag{2.11}$$

This further gives

$$\sum_k m_k \frac{d}{dt} \left(\dot{\mathbf{r}}_k \cdot \frac{\partial \mathbf{r}_k}{\partial q_j} \right) = \frac{d}{dt} \left[\frac{\partial}{\partial \dot{q}_j} \left(\sum_k \frac{1}{2} m_k \dot{\mathbf{r}}_k^2 \right) \right] = \frac{d}{dt} \left(\frac{\partial T}{\partial \dot{q}_j} \right), \tag{2.12}$$

with T as the kinetic energy of the system. This expression simplifies the first term on the right-hand side of Eq. (2.6).

The second term we also re-write, and we use now the following identity

$$\begin{aligned} \frac{d}{dt} \left(\frac{\partial \mathbf{r}_k}{\partial q_j} \right) &= \sum_l \frac{\partial^2 \mathbf{r}_k}{\partial q_l \partial q_j} \dot{q}_l + \frac{\partial}{\partial t} \left(\frac{\partial \mathbf{r}_k}{\partial q_j} \right) \\ &= \frac{\partial}{\partial q_j} \left(\sum_l \frac{\partial \mathbf{r}_k}{\partial q_l} \dot{q}_l + \frac{\partial \mathbf{r}_k}{\partial t} \right) \\ &= \frac{\partial \dot{\mathbf{r}}_k}{\partial q_j}, \end{aligned} \tag{2.13}$$

which shows that the order of differentiations $\frac{d}{dt}$ and $\frac{\partial}{\partial q_j}$ can be interchanged. This gives

$$\begin{aligned} \sum_k m_k \dot{\mathbf{r}}_k \cdot \frac{d}{dt} \left(\frac{\partial \mathbf{r}_k}{\partial q_j} \right) &= \sum_k m_k \dot{\mathbf{r}}_k \cdot \frac{\partial \dot{\mathbf{r}}_k}{\partial q_j} \\ &= \frac{\partial}{\partial q_j} \left(\sum_k \frac{1}{2} m_k \dot{\mathbf{r}}_k^2 \right) \\ &= \frac{\partial T}{\partial q_j}. \end{aligned} \tag{2.14}$$

We have then shown that both terms in Eq. (2.6) can be expressed in terms of partial derivatives of the kinetic energy.

By collecting terms, the equation of motion now can be written as

$$\frac{d}{dt}\left(\frac{\partial T}{\partial \dot{q}_j}\right) - \frac{\partial T}{\partial q_j} = \mathcal{F}_j\,, \quad j = 1, 2, ..., d\,. \tag{2.15}$$

In this form the position vectors $\mathbf{r}_k$ have been eliminated from the equation, which only makes reference to the generalized coordinates and their velocities. The equation we have arrived at can be regarded as a reformulation of Newton's second law. It does not have the usual vector form, but is instead expressed in terms of the generalized coordinates and their velocities.

We will make a further modification of the equations of motion based on the assumption that the applied forces can be derived from a potential. For the generalized forces $\mathcal{F}_j$ this means

$$\mathcal{F}_j = -\frac{\partial V}{\partial q_j}\,, \tag{2.16}$$

and the dynamical equations can then be written as

$$\frac{d}{dt}\left(\frac{\partial T}{\partial \dot{q}_j}\right) - \frac{\partial T}{\partial q_j} = -\frac{\partial V}{\partial q_j}\,, \quad j = 1, 2, ..., d\,. \tag{2.17}$$

We further note that since the potential only depends on the coordinates q_j and not on the velocities $\dot{q}_j$ the equations can be written as

$$\frac{d}{dt}\left(\frac{\partial (T-V)}{\partial \dot{q}_j}\right) - \frac{\partial (T-V)}{\partial q_j} = 0\,, \quad j = 1, 2, ..., d\,. \tag{2.18}$$

This motivates the introduction of the *Lagrangian*, defined by $L = T - V$. In terms of this new function the dynamical equations can be written in a compact form, known as Lagrange's equations,

$$\frac{d}{dt}\left(\frac{\partial L}{\partial \dot{q}_j}\right) - \frac{\partial L}{\partial q_j} = 0\,, \quad j = 1, 2, ..., d\,. \tag{2.19}$$

Lagrange's equations give a simple and elegant description of the time evolution of the system. The dynamics is specified by a single, scalar function — the Lagrangian — and the dynamical equations have a form that shows a similarity with the equation which determines the equilibrium in a static problem. In that case the coordinate dependent potential is the relevant scalar function. In the present case it is the Lagrangian, which depends on velocities as well as coordinates. It may in addition depend explicitly on time, in the following way

$$L(q, \dot{q}, t) = T(q, \dot{q}, t) - V(q, t)\,, \tag{2.20}$$

where explicit time dependence appears if the Cartesian coordinates are expressed as time dependent functions of the generalized coordinates (in most cases due to time dependent constraints).

Note that the potential is assumed to depend only on coordinates, not on velocities, but the formalism has a natural extension to velocity dependent potentials. Such an extension is particularly relevant to the description of charged particles in electromagnetic fields, where the magnetic force depends on the velocity of the particles. We will later show in detail how a Lagrangian can be designed for such a system.

Lagrange's equations motivate a general, systematic way to analyze a mechanical system which satisfies the general condition (2.16). It consists of the following steps

(1) Choose a set of generalized coordinates $q = \{q_1, q_2, ..., q_d\}$, which fits the system to be analyzed, one coordinate for each degree of freedom.
(2) Find the potential energy V and the kinetic energy T expressed as functions of coordinates q, velocities $\dot{q}$ and possibly time t.
(3) Write down the set of Lagrange's equations, one equation for each generalized coordinate.
(4) Solve the set of Lagrange's equations with the relevant initial conditions.

Other ways to analyze the system, in particular the vector approach of Newtonian mechanics, would usually also, when the unknown forces are eliminated, end up with a set of equations, like in point (4). However, the method outlined above is in many cases more convenient, since it is less dependent on a visual understanding of the action of forces on different parts of the mechanical system.

In the following we illustrate the Lagrangian method by some simple examples.

Examples

Particle in a central potential, planar motion

We consider a point particle of mass m, which moves in a rotationally invariant potential $V(r)$. For simplicity we assume the particle motion to be constrained to a plane (the x, y-plane). We follow the schematic approach outlined above.

1. Since the particle can move freely in the plane, the system has two degrees of freedom and a convenient set of (generalized) coordinates is, due to the rotational invariance, the polar coordinates (r, ϕ), with $r = 0$ as the center of the potential.

2. The potential energy, expressed in these coordinates, is simply the function $V(r)$, while the kinetic energy is $T = \frac{1}{2}m(\dot{r}^2 + r^2\dot{\phi}^2)$, and the Lagrangian is

$$L = T - V = \frac{1}{2}m(\dot{r}^2 + r^2\dot{\phi}^2) - V(r)\,. \tag{2.21}$$

3. There are two Lagrange's equations, corresponding to the two coordinates r and ϕ. The r equation is

$$\frac{d}{dt}\left(\frac{\partial L}{\partial \dot{r}}\right) - \frac{\partial L}{\partial r} = 0 \quad \Rightarrow \quad m\ddot{r} - mr\dot{\phi}^2 + \frac{dV}{dr} = 0\,, \tag{2.22}$$

and the ϕ equation is

$$\frac{d}{dt}\left(\frac{\partial L}{\partial \dot{\phi}}\right) - \frac{\partial L}{\partial \phi} = 0 \quad \Rightarrow \quad \frac{d}{dt}(mr^2\dot{\phi}) = 0\,. \tag{2.23}$$

From the last one follows

$$mr^2\dot{\phi} = \ell\,, \tag{2.24}$$

with ℓ as a constant. The physical interpretation of this constant is the angular momentum of the particle.

4. Equation (2.24) can be used to solve for $\dot{\phi}$, and inserted in (2.22) this gives the following differential equation for $r(t)$,

$$m\ddot{r} - \frac{\ell^2}{mr^3} + \frac{dV}{dr} = 0\,. \tag{2.25}$$

To proceed one should solve this equation with given initial conditions, but since we are less focussed on solving the equation of motion than on illustrating the use of the Lagrangian formalism, we stop the analysis of the system at this point.

For the case discussed here Newton's second law, in vector form, would soon lead to the same equation of motion, with a change from Cartesian to polar coordinates. The main difference between the two approaches is then that with the vector formulation this change of coordinates is made *after* the (vector) equation of motion has been established, whereas in Lagrange's formulation the choice of coordinates is done *before* Lagrange's equations are established.

Atwood's machine

We consider the composite system illustrated in Fig. 2.1. Two bodies with masses m_1 and m_2 are connected by a cord with fixed length, which is suspended over a pulley. We assume that the two bodies move only vertically, and that the cord rolls over the pulley without sliding. The pulley has a moment of inertia I. We will establish the Lagrange equation for the composite system.

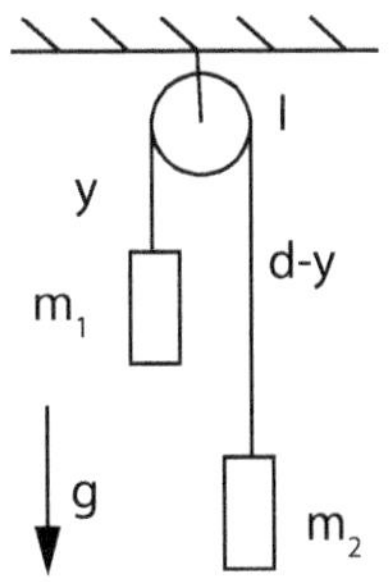

Fig. 2.1 Atwood's machine with two weights.

1. The system has only one degree of freedom, and we may use the length of the cord on the left-hand side of the pulley, denoted y, as the corresponding (generalized) coordinate. This coordinate measures the (negative) height of the mass m_1 relative to the position of the pulley. The corresponding position of the mass m_2 is $d - y$, with d as the sum of the two parts of the cord on both sides of the pulley. With R as the radius of the pulley, the angular coordinate ϕ of the pulley is related to the coordinate y by $y = R\phi$.

2. The potential energy, expressed as a function of y is

$$V = -m_1 gy - m_2(d-y)g = (m_2 - m_1)gy - m_2 d\,, \tag{2.26}$$

where the last term is an unimportant constant. For the kinetic energy we find the expression

$$T = \frac{1}{2}m_1\dot{y}^2 + \frac{1}{2}m_2\dot{y}^2 + \frac{1}{2}I\dot{\phi}^2 = \frac{1}{2}\left(m_1 + m_2 + \frac{I}{R^2}\right)\dot{y}^2\,. \tag{2.27}$$

This gives the following Lagrangian

$$L = \frac{1}{2}\left(m_1 + m_2 + \frac{I}{R^2}\right)\dot{y}^2 + (m_1 - m_2)gy + m_2 d\,. \tag{2.28}$$

It is the functional dependence of L on y and $\dot{y}$ that is interesting, since it is the partial derivatives of L with respect to these two variables that enter into Lagrange's equations.

3. The partial derivatives of the Lagrangian, with respect to coordinate and velocity, are

$$\frac{\partial L}{\partial y} = (m_1 - m_2)g\,, \quad \frac{\partial L}{\partial \dot{y}} = \left(m_1 + m_2 + \frac{I}{R^2}\right)\dot{y}\,, \tag{2.29}$$

and for the Lagrange equation this gives

$$\begin{aligned} &\frac{d}{dt}\left(\frac{\partial L}{\partial \dot{y}}\right) - \frac{\partial L}{\partial y} = 0 \\ \Rightarrow \quad &\left(m_1 + m_2 + \frac{I}{R^2}\right)\ddot{y} + (m_2 - m_1)g = 0 \\ \Rightarrow \quad &\ddot{y} = \frac{m_1 - m_2}{m_1 + m_2 + \frac{I}{R^2}}\, g\,. \end{aligned} \tag{2.30}$$

This equation shows that the weights move with constant acceleration, and with specified initial data the solution is easy to find.

Pendulum with accelerated point of suspension

As previously discussed, the Lagrangian formulation applies also to situations with explicit time dependence, and we consider here a particular example of this kind. A pendulum performs oscillations in the x, y-plane, with y as the vertical direction. The pendulum bob has mass m and the pendulum rod, with fixed length l, is considered as massless. It is suspended in a point A which moves with constant acceleration in the x-direction, so that the coordinates of this point are

$$x_A = \frac{1}{2}at^2\,, \quad y_A = 0\,, \tag{2.31}$$

with a as the (constant) acceleration. We will establish the equation of motion of the pendulum.

1. The pendulum bob moves in a plane with a fixed distance to the point of suspension. This means that the system has one degree of freedom, and we choose the angle θ between the pendulum rod and the vertical direction as generalized coordinate. Expressed in terms of θ the Cartesian coordinates of the pendulum bob are

$$\begin{aligned} x &= x_A + l\sin\theta = l\sin\theta + \frac{1}{2}at^2\,, \\ y &= -l\cos\theta\,, \end{aligned} \tag{2.32}$$

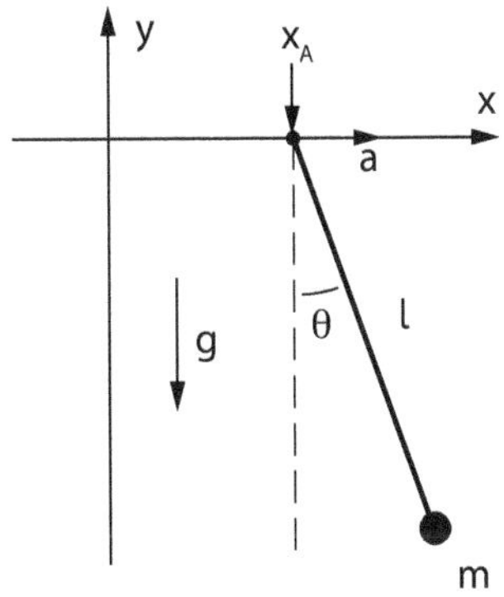

Fig. 2.2 Pendulum with accelerated point of suspension.

with the corresponding velocities,

$$\dot{x} = l\dot{\theta}\cos\theta + at\,,$$
$$\dot{y} = l\dot{\theta}\sin\theta\,. \tag{2.33}$$

2. The potential energy is

$$V = -mgl\cos\theta\,, \tag{2.34}$$

and the kinetic energy is

$$\begin{aligned} T &= \frac{1}{2}m(\dot{x}^2 + \dot{y}^2) \\ &= \frac{1}{2}m(l^2\dot{\theta}^2 + 2atl\dot{\theta}\cos\theta + a^2t^2)\,. \end{aligned} \tag{2.35}$$

This gives the following expression for the Lagrangian,

$$L = \frac{1}{2}m(l^2\dot{\theta}^2 + 2atl\dot{\theta}\cos\theta + a^2t^2) + mgl\cos\theta\,. \tag{2.36}$$

As expected it depends on the generalized coordinate θ, its velocity $\dot{\theta}$, and also explicitly on time t. The explicit time dependence follows from the (externally determined) motion of the point of suspension.

3. Lagrange's equation,

$$\frac{d}{dt}\left(\frac{\partial L}{\partial \dot{\theta}}\right) - \frac{\partial L}{\partial \theta} = 0\,, \tag{2.37}$$

can be expressed as a differential equation in the variable θ by evaluating the partial derivatives of L with respect to θ and $\dot{\theta}$,

$$\frac{\partial L}{\partial \theta} = -ma\,t\,l\dot{\theta}\sin\theta - mg\,l\sin\theta\,,$$
$$\frac{\partial L}{\partial \dot{\theta}} = ml^2\dot{\theta} + ma\,t\,l\cos\theta\,. \tag{2.38}$$

This gives

$$ml^2\ddot{\theta} + ml(g\sin\theta + a\cos\theta) = 0. \tag{2.39}$$

We note that the term which is linear in $\dot{\theta}$ disappears from the equation. It is convenient to re-write the equation by introducing a fixed angle θ_0, defined by

$$g = \sqrt{g^2 + a^2}\cos\theta_0\,, \quad a = -\sqrt{g^2 + a^2}\sin\theta_0\,. \tag{2.40}$$

The equation of motion is then

$$ml^2\ddot{\theta} + ml\sqrt{g^2 + a^2}\sin(\theta - \theta_0) = 0\,, \tag{2.41}$$

and we recognize this as a standard pendulum equation, however, with equilibrium position for the rotated direction $\theta = \theta_0$ rather than for the vertical direction $\theta = 0$, and with a stronger effective acceleration of gravity $\sqrt{g^2 + a^2}$.

Again we leave out the last point, which is to solve this equation with given boundary conditions. We only note that the form of the equation of motion is in fact what we should expect from general reasoning. If we consider the motion in an accelerated reference frame, which follows the motion of the point of suspension A, we eliminate the explicit time dependence caused by the motion of the point A. However, in such an accelerated frame there will be a fictitious gravitational force caused by the acceleration. The corresponding acceleration of gravity is a and the direction is opposite of the direction of acceleration, which means in the negative x-direction. In this frame the effective gravitational force therefore has two components, the true gravitational force in the negative y-direction and the fictitious gravitational force in the negative x-direction. The effective acceleration of gravity is therefore $\sqrt{g^2 + a^2}$ and the direction is given by the angle θ_0. The equation of motion (2.41) can therefore be interpreted as describing the pendulum oscillations in an accelerated reference frame, which follows the motion of the suspension point x_A.

2.2 Small oscillations about an equilibrium point

Consider a system described by a Lagrangian $L(q, \dot{q})$, for simplicity with only one generalized coordinate q, and with no explicit time dependence. Since the Lagrangian is generally quadratic in velocities, it can be written in the form

$$L = \frac{1}{2}a(q)\dot{q}^2 + b(q)\dot{q} + c(q)\,. \tag{2.42}$$

This gives

$$\frac{\partial L}{\partial \dot{q}} = a\dot{q} + b\,, \tag{2.43}$$

and

$$\frac{\partial L}{\partial q} = \frac{1}{2}a'\dot{q}^2 + b'\dot{q} + c'\,, \tag{2.44}$$

with $a' = da/dq$ etc. Lagrange's equation is then

$$\frac{d}{dt}\left(\frac{\partial L}{\partial \dot{q}}\right) - \frac{\partial L}{\partial q} = a\ddot{q} - \frac{1}{2}a'\dot{q}^2 - c' = 0\,. \tag{2.45}$$

Note that the b term in the Lagrangian gives no contribution to the equation.

Assume that q_0 is an equilibrium point. This means that $q = q_0$ is solution of the equation of motion with $\dot{q} = \ddot{q} = 0$, and therefore $c'(q_0) = 0$. To study the motion close to equilibrium we write $q = q_0 + \eta$, which gives $\dot{q} = \dot{\eta}$ and $\ddot{q} = \ddot{\eta}$, and expand in the deviation η from equilibrium,

$$\begin{aligned} a(q) &= a_0 + a_1\eta + \frac{1}{2}a_2\eta^2 \ldots\,, \quad c(q) = c_0 + \frac{1}{2}c_2\eta^2 + \ldots \\ \Rightarrow \quad & a'(q) = a_1 + a_2\eta\ldots\,, \quad c'(q) = c_2\eta + \ldots\,. \end{aligned} \tag{2.46}$$

We assume η to be so small that we can simplify Eq.(2.45) by keeping only terms that are first order in η and its time derivatives. This gives

$$a_0\ddot{\eta} - c_2\eta = 0\,. \tag{2.47}$$

If $c_2/a_0 < 0$, the point q_0 is a stable equilibrium point. The equation then takes the form of a harmonic oscillator equation,

$$\ddot{\eta} + \omega^2\eta = 0\,, \tag{2.48}$$

with $\omega = \sqrt{|c_2/a_0|}$. If instead $c_2/a_0 > 0$, the point q_0 is an unstable equilibrium point, and the equation will show an exponential growth of η with time.

As a simple example, we take the Lagrangian of a particle moving in a potential $V(x)$,

$$L = \frac{1}{2}m\dot{x}^2 - V(x)\,. \tag{2.49}$$

The corresponding equation of motion is

$$\ddot{x} = -\frac{1}{m}V'(x)\,, \tag{2.50}$$

where V' is the derivative of V with respect to x. We assume $x = x_0$ to be a point of stable equilibrium, with $\eta = x - x_0$ as the deviation from equilibrium. Expansion of the potential about x_0 gives the linearized equation

$$\ddot{\eta} + \frac{1}{m} V''(x_0)\eta = 0\,, \tag{2.51}$$

where we have used that $V'(x_0) = 0$ at the equilibrium point. The equation is of the form (2.48), with $\omega^2 = V''(x_0)/m$, which has real solutions for ω provided $V''(x_0)$ is positive. This is indeed the condition for the equilibrium to be stable.

As a second example let us take the following equation of motion, taken from one of the exercises,

$$\left(1 - \frac{1}{2}\cos^2\theta\right)\ddot{\theta} + \frac{1}{2}\sin\theta\cos\theta\,\dot{\theta}^2 + \frac{g}{d}\sin\theta = 0\,. \tag{2.52}$$

Clearly $\theta = 0$ is an equilibrium point, and we therefore linearize the equation in the variable θ. Let us do this here in two steps. First we note that, when expanded to first order in θ, the trigonometric functions are approximated by $\cos\theta \approx 1$, $\sin\theta \approx \theta$. Inserted in the equation this gives

$$\frac{1}{2}\ddot{\theta} + \frac{1}{2}\theta\dot{\theta}^2 + \frac{g}{d}\theta = 0\,. \tag{2.53}$$

However, the equation still contains a higher order term, and when this term is excluded we finally get the linear equation

$$\ddot{\theta} + \frac{2g}{d}\theta = 0\,, \tag{2.54}$$

which has the form of a harmonic oscillator equation with angular frequency $\omega = \sqrt{2g/d}$.

If we consider a problem with several variables, an expansion about an equilibrium point $q_0 = \{q_{01}, q_{02}, ..., q_{0n}\}$ can be performed in a similar way, with $q_i = q_{0i} + \eta_i$. This leads to a set of equations of the form

$$\sum_j (a_{ij}\ddot{\eta}_j + b_{ij}\dot{\eta} + c_{ij}\eta_j) = 0\,, \quad i = 1, 2, ..., n\,, \tag{2.55}$$

with a_{ij}, b_{ij} and c_{ij} as constants. This is a coupled set of harmonic oscillator equations, if q_0 is a point of stable equilibrium. By matrix transformations which mix the variables η_i, the equations can be separated into n uncoupled harmonic oscillator equations, each corresponding to what is known as a *normal mode* of the coupled harmonic oscillators.

Note that if the "small oscillation approximation" is performed on the Lagrangian L, rather than on the equations of motion, this means that L should include second order terms in the small quantity. This is because Lagrange's equations are determined by the partial derivatives of L with respect to the variables η_j and their time derivatives, and the quadratic terms in L therefore determine the linear terms in the equations of motion.

2.3 Symmetries and constants of motion

There is in physics a general and interesting connection between *symmetries* of a physical system and constants of motion. Well known examples of this kind are the relations between rotational symmetry and spin conservation and between translational symmetry and conservation of linear momentum. The Lagrangian formulation of classical mechanics gives a convenient way to *derive* constants of motion from symmetries in a direct way. A general form of this connection was shown in field theory by Emmy Noether (*Noether's theorem*) in 1918.[2] In a simpler form it is valid also for systems with a discrete set of variables, as we discuss here. One of the important consequences of finding constants of motion is that they can be used to reduce the number of variables in the problem. And even if the equations of motion cannot be fully solved, the conserved quantities may give important partial information about the motion of the system.

Before discussing this connection between symmetries and constants of motion, it may be of interest with some general comments about symmetries in physics. *Symmetry* may have somewhat different meanings depending on whether we consider a static or a dynamical situation. A body is symmetric under rotations if it looks identical when viewed from rotated positions. Similarly a crystal is symmetric under a discrete group of rotations, translations and reflections, when the lattice structure is invariant under these transformations. These are static situations, where the symmetry transformations leave unchanged the body or structure that we consider.[3]

In a dynamical situation we refer to certain transformations as symmetries *when they leave the equations of motion invariant* rather than physical bodies or structures. In general the equations of motion take different forms depending on the coordinates we use, but in some cases a change of coordinates will introduce no change in the form of the equations. A well-known example is the case of inertial frames, where Newton's second law has the same form whether we use the coordinates of one inertial reference frame or another. It is this form of symmetry that is of interest for the further discussion.

Let us describe the time evolution of system by the set of coordinates $q = \{q_i, i = 1, 2, ..., d\}$, where d is the number of degrees of freedom of

[2]Amalie Emmy Noether (1882–1935) was a German mathematician known for her landmark contributions to abstract algebra and theoretical physics.

[3]The symmetries we consider are often restricted to space (or space-time) transformations, but more general types of symmetry transformations may be considered, which may involve the change of particle properties like charge and intrinsic spin.

a system. A particular solution of the equations of motion we denote by $q = q(t)$. A coordinate transformation is a mapping

$$q \to q' = q'(q, t)\,, \tag{2.56}$$

where we may regard the new set of coordinates q' as a function of the old set q (and possibly of time t). This transformation is a symmetry transformation of the system if any solution $q(t)$ of the equation of motions *is mapped into a new solution* $q'(t)$ of the same equations of motion.

We shall here focus on symmetries that follow from *invariance of the Lagrangian* under coordinate transformations, in the sense

$$L(q', \dot{q}', t) = L(q, \dot{q}, t)\,. \tag{2.57}$$

This equation means that under a change of variables $q \to q'$ the Lagrangian will have the same *functional dependence* of the new and old variables. Since the Lagrangian determines the form of the equations of motion, this implies that the time evolution of the system, described by coordinates $q(t)$ and by coordinates $q'(t)$ will satisfy the same equations of motion, and the mapping $q \to q'$ is thus a symmetry transformation of the system.

If the Lagrangian is invariant under a set of *continuous* coordinate transformations, then it follows from Lagrange's equations that there are one or more constants of motion associated with these transformations. This we will, in the following, show to be the case quite generally, but first we will discuss the special case where the Lagrangian has a *cyclic coordinate.*

Cyclic coordinates

We consider a Lagrangian of the general form

$$L = L(q, \dot{q}, t)\,, \tag{2.58}$$

with $q = (q_1, q_2, ..., q_d)$ as the set of generalized coordinates. We further assume that the Lagrangian is independent of one of the coordinates, say q_1. This means

$$\frac{\partial L}{\partial q_1} = 0\,, \tag{2.59}$$

and we refer to q_1 as a *cyclic coordinate.* From Lagrange's equation then follows

$$\frac{d}{dt}\left(\frac{\partial L}{\partial \dot{q}_1}\right) = 0\,. \tag{2.60}$$

This means that the physical variable

$$p_1 \equiv \frac{\partial L}{\partial \dot{q}_1}\,, \tag{2.61}$$

which we refer to as the *conjugate momentum*[4] to the coordinate q_1, is a constant of motion. Thus, for every cyclic coordinate there is a constant of motion.

The presence of a cyclic coordinate can be used to reduce the number of independent variables from d to $d-1$. The coordinate q_1 is already eliminated, since it does not appear in the Lagrangian, but $\dot{q}_1$ is generally present. However also this can be eliminated by using the fact that p_1 is a constant of motion. Let us write this condition in the following way,

$$\frac{\partial L}{\partial \dot{q}_1} = p_1(q_2, ..., q_d; \dot{q}_1, \dot{q}_2, ..., \dot{q}_d; t) = k\,, \tag{2.62}$$

with k as a constant. In this equation we have written explicitly the functional dependence of p_1 on all coordinates and velocities. Since q_1 is cyclic it does not appear in the expression. The equation can be solved for $\dot{q}_1$,

$$\dot{q}_1 = f(q_2, ...q_d; \dot{q}_2, ...\dot{q}_d; k; t)\,, \tag{2.63}$$

with the function f as the unspecified solution. In this way both q_1 and $\dot{q}_1$ are eliminated as basic variables, and the number of independent equations of motion is reduced from d to $d-1$. Note, however, that the $d-1$ equations will not only depend on the $d-1$ remaining coordinates and their velocities, but also on the constant of motion k. The value of this constant is determined by the initial conditions.

Example

Point particle moving under influence of gravity on the surface of a sphere

We consider a point particle of mass m, which moves without friction on the surface of a sphere, under the influence of gravity. The gravitational field is assumed to point in the negative z-direction. This system has two degrees of freedom, since the three Cartesian coordinates (x, y, z) of the particle are subject to one constraint equation $r = \sqrt{x^2 + y^2 + z^2} = const.$

[4]The conjugate momentum is also referred to as *generalized momentum* or *canonical momentum*.

As generalized coordinates we chose the polar angles (ϕ, θ), so that the Cartesian coordinates are

$$\begin{aligned} x &= r\cos\phi\sin\theta\,, \\ y &= r\sin\phi\sin\theta\,, \\ z &= r\cos\theta\,, \end{aligned} \tag{2.64}$$

with r as a constant. The corresponding velocities are

$$\begin{aligned} \dot{x} &= r(\cos\phi\cos\theta\,\dot{\theta} - \sin\phi\sin\theta\,\dot{\phi})\,, \\ \dot{y} &= r(\sin\phi\cos\theta\,\dot{\theta} + \cos\phi\sin\theta\,\dot{\phi})\,, \\ \dot{z} &= -r\sin\theta\,\dot{\theta}\,. \end{aligned} \tag{2.65}$$

The potential energy is

$$V = mgz = mgr\cos\theta\,, \tag{2.66}$$

with g as the acceleration of gravitation, and the kinetic energy is

$$T = \frac{1}{2}m(\dot{x}^2 + \dot{y}^2 + \dot{z}^2) = \frac{1}{2}mr^2(\dot{\theta}^2 + \sin^2\theta\dot{\phi}^2)\,. \tag{2.67}$$

This gives the following expression for the Lagrangian

$$L = \frac{1}{2}mr^2(\dot{\theta}^2 + \sin^2\theta\dot{\phi}^2) - mgr\cos\theta\,. \tag{2.68}$$

Clearly ϕ is a cyclic coordinate,

$$\frac{\partial L}{\partial \phi} = 0\,, \tag{2.69}$$

and therefore Lagrange's equation for this variable reduces to

$$\frac{\partial L}{\partial \dot{\phi}} = mr^2\dot{\phi}\sin^2\theta = \ell\,, \tag{2.70}$$

with ℓ as a constant. Lagrange's equation for the variable θ is

$$mr^2\ddot{\theta} - mr^2\dot{\phi}^2\sin\theta\cos\theta - mgr\sin\theta = 0\,. \tag{2.71}$$

To eliminate the variable ϕ from the equation, we express, by use of (2.70), $\dot{\phi}$ in terms of the constant of motion ℓ,

$$\dot{\phi} = \frac{\ell}{mr^2\sin^2\theta}\,. \tag{2.72}$$

Inserted in (2.71) this gives

$$mr^2\ddot{\theta} - \frac{\ell^2\cos\theta}{mr^2\sin^3\theta} - mgr\sin\theta = 0\,. \tag{2.73}$$

This illustrates the general discussion of cyclic coordinates. In the present case the elimination of the coordinate ϕ has reduced the equations of motion to one, and the only remaining trace of the coordinate ϕ is the appearance of the conserved quantity ℓ in the equation.

In this example the relation between symmetries and conserved quantities is clear. Thus, the independence of ϕ means that the Lagrangian is invariant under rotations about the z axis and at the same time the cyclic coordinate gives rise to a conserved quantity ℓ. It is straight forward to show that this constant has the physical interpretation as the z-component of the orbital angular momentum of the particle.

Symmetries of the Lagrangian

The existence of a cyclic coordinate will always imply the presence of a continuous symmetry in the form given by Eq. (14.46). Thus, the fact that q_1 is cyclic means that the Lagrangian is invariant under the transformation

$$\begin{aligned} q_1 &\to q_1' = q_1 + a\,, \\ q_i &\to q_i' = q_i\,, \quad i \neq 1\,, \end{aligned} \tag{2.74}$$

where a is a constant. The transformation has the form of *translation* in the cyclic coordinate, and since a can take arbitrarily small values, it is referred to as a continuous transformation. This shows that there is a continuous symmetry transformation associated with each cyclic coordinate. In the previous example the symmetry transformation corresponds to rotations about the z axis.

We shall now discuss more generally how invariance of the Lagrangian under a continuous coordinate transformation is related to the presence of a constant of motion. In the general case there may be no cyclic coordinate corresponding to this symmetry. We consider then a continuous *time independent* coordinate transformation

$$q \to q' = q'(q)\,, \tag{2.75}$$

where we may assume the change in the coordinates to be arbitrarily small. We write this as $q_i' = q_i + \delta q_i$, and assume in the following that terms which are higher order in δq_i can be neglected. By expansion to first order in δq, we have the identity

$$L(q', \dot{q}') = L(\dot{q}, q) + \sum_k \left(\frac{\partial L}{\partial q_k} \delta q_k + \frac{\partial L}{\partial \dot{q}_k} \delta \dot{q}_k \right)\,, \tag{2.76}$$

and invariance of the Lagrangian thus implies

$$\sum_k \left(\frac{\partial L}{\partial q_k} \delta q_k + \frac{\partial L}{\partial \dot{q}_k} \delta \dot{q}_k \right) = 0 \,. \tag{2.77}$$

This we may rewrite as

$$\sum_k \left[\frac{\partial L}{\partial q_k} \delta q_k - \frac{d}{dt}\left(\frac{\partial L}{\partial \dot{q}_k} \right) \delta q_k + \frac{d}{dt}\left(\frac{\partial L}{\partial \dot{q}_k} \delta q_k \right) \right] = 0 \,. \tag{2.78}$$

We will now apply this identity to a solution $q(t)$ of Lagrange's equations. The two first terms then will cancel, and this gives

$$\frac{d}{dt}\left(\sum_k \frac{\partial L}{\partial \dot{q}_k} \delta q_k \right) = 0 \,. \tag{2.79}$$

Therefore the following quantity is a constant of motion,

$$\delta K = \sum_k \frac{\partial L}{\partial \dot{q}_k} \delta q_k \,. \tag{2.80}$$

With δq_k as an infinitesimal change of the coordinates, it can be written as

$$\delta q_k = \epsilon J_k \,, \tag{2.81}$$

where J_k is a *finite* parameter characteristic for the transformation, while ϵ is time independent and infinitesimal. The parameter ϵ can be omitted and that gives the following expression for the finite (non-infinitesimal) constant of motion associated with the symmetry,

$$K = \sum_k \frac{\partial L}{\partial \dot{q}_k} J_k \,. \tag{2.82}$$

To summarize, if we can identify a symmetry of the system, expressed as invariance of the Lagrangian under a coordinate transformation, we can use the above expression to derive a conserved quantity corresponding to this symmetry. Clearly there is one constant of motion for each independent continuous symmetry of the Lagrangian.

Example

Particle in rotationally invariant potential

In order to illustrate the general discussion we examine a rotationally invariant system with kinetic and potential energies

$$T = \frac{1}{2} m \dot{\mathbf{r}}^2 \,, \quad V = V(r) \,, \tag{2.83}$$

which give the following Lagrangian in Cartesian coordinates,

$$L = \frac{1}{2}m(\dot{x}^2 + \dot{y}^2 + \dot{z}^2) - V(\sqrt{x^2 + y^2 + z^2}\,)\,, \tag{2.84}$$

and in polar coordinates,

$$L = \frac{1}{2}m(\dot{r}^2 + r^2\dot{\theta}^2 + r^2\sin^2\theta\,\dot{\phi}^2) - V(r). \tag{2.85}$$

The system is obviously symmetric under all rotations about the origin (the center of the potential), but we note that expressed in Cartesian coordinates there is no cyclic coordinate corresponding to these symmetries. In polar coordinates there is one cyclic coordinate, ϕ. The corresponding conserved quantity is the conjugate momentum,

$$p_\phi = \frac{\partial L}{\partial \dot{\phi}} = mr^2\sin^2\theta\,\dot{\phi}\,, \tag{2.86}$$

and the physical interpretation of p_ϕ is the z-component of the angular momentum,

$$(m\mathbf{r}\times\dot{\mathbf{r}}\,)_z = m(x\dot{y} - y\dot{x}) = mr^2\sin^2\theta\,\dot{\phi}\,. \tag{2.87}$$

Clearly also the other components of the angular momentum are conserved, but there are not cyclic coordinates corresponding to all the three components.

We use the expression derived in the last section to find the conserved quantities associated with the rotational symmetry. First we note that an infinitesimal rotation can be expressed in the form

$$\mathbf{r} \to \mathbf{r}\,' = \mathbf{r} + \delta\boldsymbol{\alpha}\times\mathbf{r} \tag{2.88}$$

or

$$\delta\mathbf{r} = \delta\boldsymbol{\alpha}\times\mathbf{r}\,, \tag{2.89}$$

where the direction of the vector $\delta\boldsymbol{\alpha}$ specifies the direction of the axis of rotation and the absolute value $\delta\alpha$ specifies the angle of rotation.

We can explicitly verify that to first order in $\delta\boldsymbol{\alpha}$ the transformation (2.88) leaves $\mathbf{r}^2$ unchanged, and since the velocity $\dot{\mathbf{r}}$ transforms in the same way (by time derivative of (2.88)) also $\dot{\mathbf{r}}^2$ is invariant under the transformation. Consequently, the Lagrangian is invariant under the infinitesimal rotations (2.88), which are therefore symmetry transformations of the system.

By use of the expression (2.79) we find the following expression for the conserved quantity associated with the symmetry transformation,

$$K = \sum_{k=1}^{3}\frac{\partial L}{\partial \dot{x}_k}\delta x_k = m\dot{\mathbf{r}}\cdot\delta\mathbf{r} = m(\mathbf{r}\times\dot{\mathbf{r}})\cdot\delta\boldsymbol{\alpha}\,. \tag{2.90}$$

Since this quantity is conserved for arbitrary values of the constant vector $\delta\boldsymbol{\alpha}$, we conclude that the vector quantity

$$\boldsymbol{\ell} = m\mathbf{r} \times \dot{\mathbf{r}} \tag{2.91}$$

is conserved. This demonstrates that the general expression we have found for a constant of motion reproduces, as expected, the angular momentum as a constant of motion when the particle moves in a rotationally invariant potential.

Time invariance and energy conservation

We consider a Lagrangian $L = L(q, \dot{q})$ which has no explicit time dependence, so that

$$\frac{\partial L}{\partial t} = 0\,. \tag{2.92}$$

This functional independence of t we note to be similar to the functional independence of q_1, when this is a cyclic coordinate. Time is certainly *not* a coordinate in the same sense as q_i, and in particular there is no conjugate momentum to t. Nevertheless, there is a conserved quantity that can be derived from the time independence of L. To show this we consider the total time derivative of L, evaluated for a path $q(t)$, which satisfies the equations of motion. The total time derivative picks up contributions both from the explicit time dependence of L, and from its dynamical time dependence, which comes from the time dependence of the generalized coordinates $q_i(t)$,

$$\frac{dL}{dt} = \sum_i \left(\frac{\partial L}{\partial \dot{q}_i}\ddot{q}_i + \frac{\partial L}{\partial q_i}\dot{q}_i \right) + \frac{\partial L}{\partial t}\,. \tag{2.93}$$

We rewrite this equation and make use of the fact that the time dependence of q_i is determined by Lagrange's equation,

$$\begin{aligned}\frac{dL}{dt} &= \sum_i \frac{d}{dt}\left(\frac{\partial L}{\partial \dot{q}_i}\dot{q}_i \right) - \sum_i \left[\frac{d}{dt}\left(\frac{\partial L}{\partial \dot{q}_i}\right) - \frac{\partial L}{\partial q_i} \right] \dot{q}_i + \frac{\partial L}{\partial t} \\ &= \sum_i \frac{d}{dt}\left(\frac{\partial L}{\partial \dot{q}_i}\dot{q}_i \right) + \frac{\partial L}{\partial t}\,.\end{aligned} \tag{2.94}$$

This shows that the following quantity

$$H = \sum_i \frac{\partial L}{\partial \dot{q}_i}\dot{q}_i - L\,, \tag{2.95}$$

which is called the *Hamiltonian* of the system, satisfies the equation

$$\frac{dH}{dt} = -\frac{\partial L}{\partial t}\,. \tag{2.96}$$

This means that if L has no explicit time dependence, $\frac{\partial L}{\partial t} = 0$, then H is time independent under the evolution of the system, and is therefore a constant of motion.

When the Lagrangian has the standard form $L = T - V$, and when the constraints are *time independent*, the Hamiltonian corresponds to the sum of kinetic and potential energy, $H = T + V$. To show this we note that with time independent constraints we have the following relation,

$$\dot{\mathbf{r}}_k = \sum_i \frac{\partial \mathbf{r}_k}{\partial q_i} \dot{q}_i \,, \tag{2.97}$$

which implies that the kinetic energy is quadratic in the generalized velocity $\dot{q}$, and the Lagrangian therefore has the form

$$L = \frac{1}{2} \sum_{ij} g_{ij}(q)\, \dot{q}_i \dot{q}_j - V(q) \,, \tag{2.98}$$

with

$$g_{ij}(q) = \sum_k \frac{\partial \mathbf{r}_k}{\partial q_i} \cdot \frac{\partial \mathbf{r}_k}{\partial q_j} \,. \tag{2.99}$$

This gives

$$\frac{\partial L}{\partial \dot{q}_i} = \sum_j g_{ij}(q)\, \dot{q}_j \,, \tag{2.100}$$

and therefore

$$\begin{aligned} H &= \sum_j g_{ij}(q)\, \dot{q}_i \dot{q}_j - \left(\frac{1}{2} \sum_{ij} g_{ij}(q)\, \dot{q}_i \dot{q}_j - V(q) \right) \\ &= \frac{1}{2} \sum_{ij} g_{ij}(q)\, \dot{q}_i \dot{q}_j + V(q) \\ &= T + V \,. \end{aligned} \tag{2.101}$$

Thus, in this case the conserved quantity H is identical to the total energy $T + V$ of the system.

Note that with *time dependent* constraints we have

$$\dot{\mathbf{r}}_k = \sum_i \frac{\partial \mathbf{r}_k}{\partial q_i} \dot{q}_i + \frac{\partial \mathbf{r}}{\partial t} \,, \tag{2.102}$$

where the last term gives rise to new terms in the kinetic energy, which now takes the form

$$T = \frac{1}{2} \sum_{ij} g_{ij}(q,t)\, \dot{q}_i \dot{q}_j + \sum_i h_i(q,t) \dot{q}_i + f(q,t) \,. \tag{2.103}$$

The additional terms lead to an expression for the Hamiltonian that is in general different from $T + V$. One should note that even if the constraints are time *dependent*, the Lagrangian may in some cases be time *independent*, provided the functions g_{ij}, h_i, f and V are all independent of time. In that case the Hamiltonian H is a constant of motion, but it is *not* identical to the total energy of the system.

In a similar way as the Lagrangian L is the fundamental quantity in Lagrange's description of the dynamics, the Hamiltonian H is the fundamental quantity in Hamilton's description. We shall study that in some detail in Chapter 3.

2.4 Generalizing the formalism

Adding a total time derivative

A change of the Lagrangian,

$$L(q, \dot{q}, t) \to L'(q, \dot{q}, t), \tag{2.104}$$

will usually lead to a change in the corresponding equations of motions, but not always. Let us consider a change given by

$$L'(q, \dot{q}, t) = L(q, \dot{q}, t) + \frac{d}{dt} f(q, t), \tag{2.105}$$

where $f(q, t)$ is a differentiable function of the coordinates q_i, but is independent of the velocities $\dot{q}_i$. The additional term, which has the form of a total time derivative, does not change the (Lagrange) equations of motion, as we can easily demonstrate. We define the additional term as

$$g(q, \dot{q}, t) \equiv \frac{d}{dt} f = \sum_i \frac{\partial f}{\partial q_i} \dot{q}_i + \frac{\partial f}{\partial t}, \tag{2.106}$$

and consider the contribution to the Lagrange equation from this additional term,

$$\frac{d}{dt}\left(\frac{\partial L'}{\partial \dot{q}_i}\right) - \frac{\partial L'}{\partial q_i} = \frac{d}{dt}\left(\frac{\partial L}{\partial \dot{q}_i}\right) - \frac{\partial L}{\partial q_i} + \frac{d}{dt}\left(\frac{\partial g}{\partial \dot{q}_i}\right) - \frac{\partial g}{\partial q_i}. \tag{2.107}$$

We have

$$\frac{\partial g}{\partial q_i} = \sum_m \frac{\partial^2 f}{\partial q_i \partial q_m} \dot{q}_m + \frac{\partial^2 f}{\partial q_i \partial t}, \tag{2.108}$$

and

$$\frac{d}{dt}\left(\frac{\partial g}{\partial \dot{q}_i}\right) = \frac{d}{dt}\left(\frac{\partial f}{\partial q_i}\right) = \sum_m \frac{\partial^2 f}{\partial q_m \partial q_i} \dot{q}_m + \frac{\partial^2 f}{\partial t \partial q_i}, \tag{2.109}$$

and provided the function f is well behaved, so the order of differentiations can be interchanged, these two expressions are equal. This means that the contribution to Lagrange's equation vanishes,

$$\frac{d}{dt}\left(\frac{\partial g}{\partial \dot{q}_i}\right) - \frac{\partial g}{\partial q_i} = 0\,. \tag{2.110}$$

Therefore, two Lagrangians which differ by a total time derivative, as in (2.105), are equivalent in the sense that they give rise to the same equations of motion. In particular, if the Lagrangian is given by the standard expression $L = T - V$, this implies an equally valid Lagrangian for the same system, is obtained by adding (or subtracting) a total time derivative to the expression $T - V$. This observation is sometimes useful in order to simplify the expression for the Lagrangian.

One should also note, that even if a symmetry of a physical system will often correspond to *invariance* of the Lagrangian under a given transformation, *invariance up to a total time derivative* would more generally give rise to a symmetry of the equations of motion. Also in this case, when the Lagrangian is invariant up to the addition of a total time derivative, there is a constant of motion corresponding to the symmetry. This can be shown in essentially the same way as we have done for the case of an invariant Lagrangian.

Velocity dependent potentials

Let us return to the equation of motion in the early form Eq. (2.15), which was established before the potential was introduced,

$$\frac{d}{dt}\left(\frac{\partial T}{\partial \dot{q}_j}\right) - \frac{\partial T}{\partial q_j} = \mathcal{F}_j\,, \quad j = 1, 2, ..., d\,. \tag{2.111}$$

We derived Lagrange's equations from this by writing the generalized force as $\mathcal{F}_j = -\frac{\partial V}{\partial q_j}$ with V as a velocity independent function. However, there is an obvious possibility of extending the formalism by assuming the potential to be velocity dependent, written as $U = U(q, \dot{q}, t)$, with the generalized force depending on U as

$$\begin{aligned}\mathcal{F}_i &= \frac{d}{dt}\left(\frac{\partial U}{\partial \dot{q}_i}\right) - \frac{\partial U}{\partial q_i} \\ &= \sum_j \frac{\partial^2 U}{\partial \dot{q}_j \partial \dot{q}_i}\ddot{q}_j + \sum_j \frac{\partial^2 U}{\partial q_j \partial \dot{q}_i}\dot{q}_j + \frac{\partial^2 U}{\partial t\, \partial \dot{q}_i} - \frac{\partial U}{\partial q_i}\,.\end{aligned} \tag{2.112}$$

We note the presence of an acceleration dependent term (depending on $\ddot{q}$) which may not seem natural to include in the generalized force, but if

U depends linearly on the velocity this term is absent. The equation of motion (2.5) can then be written in the standard Lagrangian form, with the Lagrangian defined as

$$L(q, \dot{q}, t) = T(q, \dot{q}, t) - U(q, \dot{q}, t). \tag{2.113}$$

This generalized form of Lagrange's equation has an important application in the description of charged particles in electromagnetic fields, as we shall see. In that case the potential U depends linearly on the velocity and this dependence on the velocity gives rise to the magnetic force that acts on the particles.

2.5 Charged particle in an electromagnetic field

Lagrangian for a charged particle

We consider the motion of a charged particle in an electromagnetic field, and since there are no constraints, the Cartesian coordinates of the particle are used as the generalized coordinates. The equation of motion is

$$m\mathbf{a} = e(\mathbf{E}(\mathbf{r}, t) + \mathbf{v} \times \mathbf{B}(\mathbf{r}, t)) \equiv \mathbf{F}(\mathbf{r}, \mathbf{v}, t), \tag{2.114}$$

where e is the charge of the particle, $\mathbf{E}$ the electric field and $\mathbf{B}$ the magnetic field, $\mathbf{r}$ is the (time dependent) position vector of the particle, and $\mathbf{v} = \dot{\mathbf{r}}$ the velocity. The electromagnetic force $\mathbf{F}$ is generally known as the *Lorentz force.*

Only in the electrostatic case, with $\mathbf{B} = 0$, this equation of motion can be derived from a Lagrangian of the standard form $L = T - V$, with $V = e\phi$ as the electrostatic potential. However, as we shall see, in the general case the force can be expressed in terms of a velocity dependent potential as

$$F_i = \frac{d}{dt}\left(\frac{\partial U}{\partial \dot{x}_i}\right) - \frac{\partial U}{\partial x_i}, \tag{2.115}$$

and therefore the equation of motion can be derived from the Lagrangian $L = T - U$.

In order to show this, we introduce the electromagnetic potentials Φ and $\mathbf{A}$, defined by

$$\mathbf{E} = -\boldsymbol{\nabla}\Phi - \frac{\partial \mathbf{A}}{\partial t}, \quad \mathbf{B} = \boldsymbol{\nabla} \times \mathbf{A}. \tag{2.116}$$

It follows from Maxwell's equations, to be discussed in Chapter 9, that such a representation of $\mathbf{E}$ and $\mathbf{B}$ in terms of potentials can always be made. The corresponding expression for the electromagnetic force is

$$\begin{aligned}\mathbf{F} &= e\left[-\boldsymbol{\nabla}\Phi - \frac{\partial \mathbf{A}}{\partial t} + \mathbf{v}\times(\boldsymbol{\nabla}\times\mathbf{A})\right] \\ &= e\left[-\boldsymbol{\nabla}\Phi - \frac{\partial \mathbf{A}}{\partial t} + \boldsymbol{\nabla}(\mathbf{v}\cdot\mathbf{A}) - (\mathbf{v}\cdot\boldsymbol{\nabla})\mathbf{A}\right], \end{aligned} \tag{2.117}$$

which in component form is

$$\begin{aligned}F_i &= e\left(-\frac{\partial\Phi}{\partial x_i} - \frac{\partial A_i}{\partial t} + \mathbf{v}\cdot\frac{\partial \mathbf{A}}{\partial x_i} - \mathbf{v}\cdot\boldsymbol{\nabla}A_i\right) \\ &= \frac{d}{dt}(-eA_i) - \frac{\partial}{\partial x_i}[e\Phi - e\mathbf{v}\cdot\mathbf{A}]. \end{aligned} \tag{2.118}$$

If the velocity dependent potential U is defined as

$$U = e\Phi - e\mathbf{v}\cdot\mathbf{A}, \tag{2.119}$$

this gives

$$\frac{\partial U}{\partial \dot{x}_i} = -eA_i, \tag{2.120}$$

and, as follows from (2.118), the force $\mathbf{F}$ is then related to U by Eq. (2.115). This implies that the Lagrangian

$$L = T - U = \frac{1}{2}m\dot{\mathbf{r}}^2 - e\Phi(\mathbf{r},t) + e\dot{\mathbf{r}}\cdot\mathbf{A}(\mathbf{r},t) \tag{2.121}$$

will correctly reproduce the equation of motion (2.114).

Let us further examine the form of the conjugate momentum and the Hamiltonian in this case. We have

$$p_i = \frac{\partial L}{\partial \dot{x}_i} = m\dot{x}_i + eA_i, \tag{2.122}$$

which gives

$$m\dot{\mathbf{r}} = \mathbf{p} - e\mathbf{A}. \tag{2.123}$$

This shows that the *canonical momentum* $\mathbf{p}$ in this case is not identical to the *mechanical momentum* $m\mathbf{v}$ of the charged particle. The Hamiltonian is now

$$\begin{aligned}H &= \mathbf{p}\cdot\dot{\mathbf{r}} - L \\ &= \mathbf{v}\cdot(m\mathbf{v} + e\mathbf{A}) - \frac{1}{2}m\mathbf{v}^2 + e\Phi - e\mathbf{v}\cdot\mathbf{A} \\ &= \frac{1}{2}m\mathbf{v}^2 + e\Phi \\ &= \frac{1}{2m}(\mathbf{p} - e\mathbf{A})^2 + e\Phi. \end{aligned} \tag{2.124}$$

We note that this is different from $T+U$, but is identical to the total energy $T+V$, with $V=e\Phi$ as the potential energy of the charge in the electromagnetic field.

According to the previous discussion H should be a constant of motion if the Lagrangian has no explicit time dependence. In the present case this can be related to a more direct argument for conservation of energy in the following way. We first note that time independence of L means that the potentials and therefore the electric and magnetic fields are time independent. The electric part of the force in (2.114) is $\mathbf{F}_e=-e\boldsymbol{\nabla}\Phi$. This is a conservative force. It does not change the total energy of the particle, but only shifts it from the kinetic to the potential part. The magnetic part of the force, $\mathbf{F}_m=e\mathbf{v}\times\mathbf{B}$, acts in a direction perpendicular to the direction of motion, and therefore performs no work on the particle, so the total energy is left unchanged. If the potentials on the other hand are time dependent, the electric force is no longer conservative, $\mathbf{F}_e=-e(\boldsymbol{\nabla}\Phi-\frac{\partial\mathbf{A}}{\partial t})$, and the interaction of the particle with the electric field will change its energy.

There is one point about the Lagrangian that is worth noting. It is not gauge invariant, even if the equation of motion is gauge invariant. A gauge transformation is a modification of the potentials of the form

$$\Phi\to\Phi'=\Phi-\frac{\partial\chi}{\partial t}\,,\qquad \mathbf{A}\to\mathbf{A}'=\mathbf{A}+\boldsymbol{\nabla}\chi\,,\tag{2.125}$$

with $\chi=\chi(\mathbf{r},t)$ as an arbitrary differentiable function of space and time. The fields $\mathbf{E}$ and $\mathbf{B}$ are left unchanged by this transformation, and usually gauge transformations are therefore considered as not corresponding to any physical change. The question is whether the non-invariance of the Lagrangian is consistent with this view. The transformation induces the following change of the Lagrangian

$$L\to L'=L+e\left(\frac{\partial\chi}{\partial t}+\mathbf{v}\cdot\boldsymbol{\nabla}\chi\right)=L+e\frac{d\chi}{dt}\,.\tag{2.126}$$

So we see that the gauge transformation adds a term to the Lagrangian, which can be written as a total time derivative. As already discussed, Lagrangians that differ by a total time derivative are equivalent, so in this sense no essential change is made under the gauge transformation.

Examples

Charged particle in a constant magnetic field

The electromagnetic potentials in this case are

$$\Phi = 0\,, \quad \mathbf{A} = -\frac{1}{2}\mathbf{r}\times\mathbf{B}\,, \tag{2.127}$$

with $\mathbf{B} \equiv B_0\mathbf{k}$ as a constant vector in the z-direction. It is straight forward to check that $\mathbf{B} = \boldsymbol{\nabla}\times\mathbf{A}$, so that $\mathbf{B}$ represents a constant magnetic field. We use the established expression (2.121) for the Lagrangian of a charged particle in an electromagnetic field, which in the present case simplifies to

$$L = \frac{1}{2}m\mathbf{v}^2 + e\mathbf{v}\cdot\mathbf{A} = \frac{1}{2}m(\dot{x}^2+\dot{y}^2+\dot{z}^2) + \frac{1}{2}eB_0(x\dot{y}-y\dot{x}). \tag{2.128}$$

z is clearly a cyclic coordinate, and the conjugate momentum, $p_z = m\dot{z}$, and therefore the z component of the velocity, v_z, is a constant of motion. Since L is time independent, the Hamiltonian, and therefore the kinetic energy, is also a constant of motion. This means that $\mathbf{v}^2$, and therefore $v_x^2 + v_y^2$ is constant.

The motion in the x,y-plane is decoupled from the motion in the z-direction, and is described by the following Lagrange's equations for the x and y coordinates,

$$m\ddot{x} = eB_0\dot{y}\,, \quad m\ddot{y} = -eB_0\dot{x}\,. \tag{2.129}$$

These can be integrated to give

$$m\dot{x} = eB_0(y-y_0)\,, \quad m\dot{y} = -eB_0(x-x_0)\,, \tag{2.130}$$

with x_0 and y_0 as integration constants. From these equations follow that $(x-x_0)^2 + (y-y_0)^2 \equiv R^2$ is a constant of motion. This means that in the x,y-plane the motion is circular, with constant angular velocity $\omega_0 = -eB_0/m$.

Thus, the motion of the charged particle is a combination of a constant drift with velocity v_z in the z-direction and circular motion, with radius R and angular velocity ω_0 in the x,y-plane. The velocity v_z and the radius R are determined by initial conditions, while the angular velocity is determined by the strength of the magnetic field (and the particle properties m and e). Finally, the kinetic energy is determined by these as

$$T = \frac{1}{2}m(R^2\omega_0^2 + v_z^2). \tag{2.131}$$

Motion of a charged particle in an inhomogeneous magnetic field

As discussed in the previous example, a charged particle, which moves in a constant magnetic field, will follow a trajectory which spirals around the magnetic field lines. However, when the field lines are converging, the component of the particle velocity along the field will be reduced until this motion stops, and the particle is reflected back in the opposite direction. A well-known consequence of this effect is the presence of the van Allen radiation belts in the Earth's magnetic field, where charged particles from the Sun are trapped in a continuous motion along the field lines between the two magnetic poles.

In this example we study the phenomenon in a simplified version, where the magnetic field is almost homogeneous (constant), with direction along the z-axis, but with a small inhomogeneous modification, which makes the field lines slowly converge towards the z-axis. We consider the particle motion to be non-relativistic, and we do not include effects of gravity.

The vector potential $\mathbf{A}$, when expressed in cylindrical coordinates, is assumed to take the form

$$A_\rho = 0 \quad A_\phi = \frac{1}{2}\rho B_0 f(z) \quad A_z = 0\,, \tag{2.132}$$

with

$$f(z) = 1 + z^2/d^2\,, \tag{2.133}$$

where d is a parameter with dimension length, which measures how rapidly the magnetic field is changing along the z-direction. The corresponding magnetic field $\mathbf{B} = \boldsymbol{\nabla} \times \mathbf{A}$, has the following components in cylinder coordinates (see Appendix C.3),

$$\begin{aligned} B_\rho &= \frac{1}{\rho}\frac{\partial A_z}{\partial \phi} - \frac{\partial A_\phi}{\partial z} = -\frac{1}{2}\rho B_0 \frac{df}{dz}\,, \\ B_\phi &= \frac{\partial A_\rho}{\partial z} - \frac{\partial A_z}{\partial \rho} = 0\,, \\ B_z &= \frac{1}{\rho}\frac{\partial(\rho A_\phi)}{\partial \rho} - \frac{1}{\rho}\frac{\partial A_\rho}{\partial \phi} = B_0 f(z)\,. \end{aligned} \tag{2.134}$$

In the limit $d \to \infty$ we have $f = 1$, which gives a constant magnetic field oriented in the z-direction.

The Lagrangian of the charged particle moving in the magnetic field is

$$\begin{aligned} L &= \frac{1}{2}m\mathbf{v}^2 + e\mathbf{v}\cdot\mathbf{A} \\ &= \frac{1}{2}m(\dot\rho^2 + \rho^2\dot\phi^2 + \dot z^2) + \frac{1}{2}eB_0\rho^2\dot\phi f(z)\,, \end{aligned} \tag{2.135}$$

with m as the mass and e as the electric charge of the particle. This gives the partial derivatives

$$\frac{\partial L}{\partial \dot{\rho}} = m\dot{\rho}\,, \qquad \frac{\partial L}{\partial \rho} = m\rho\dot{\phi}^2 + eB_0\rho\dot{\phi}f \tag{2.136}$$

and from these Lagrange's equation for the ρ variable,

$$\frac{d}{dt}\frac{\partial L}{\partial \dot{\rho}} - \frac{\partial L}{\partial \rho} = m\ddot{\rho} - m\rho\dot{\phi}^2 - eB_0\rho\dot{\phi}f = 0\,. \tag{2.137}$$

The angular variable ϕ is cyclic, which gives a constant of motion, ℓ_z,

$$\frac{\partial L}{\partial \phi} = 0 \quad \Rightarrow \quad \frac{\partial L}{\partial \dot{\phi}} = m\rho^2\dot{\phi} + \frac{1}{2}eB_0\rho^2 f(z) = \ell_z\,. \tag{2.138}$$

The constant ℓ_z can be interpreted as the conserved angular momentum, corresponding to rotational invariance around the z-axis. We note that in addition to the mechanical angular momentum (the first term), there is an electromagnetic contribution (the second term). For the variable z, the partial derivatives are

$$\frac{\partial L}{\partial \dot{z}} = m\dot{z}\,, \qquad \frac{\partial L}{\partial z} = \frac{1}{2}eB_0\rho^2\dot{\phi}\frac{df}{dz}\,, \tag{2.139}$$

with the corresponding Lagrange's equation

$$\frac{d}{dt}\frac{\partial L}{\partial \dot{z}} - \frac{\partial L}{\partial z} = m\ddot{z} - \frac{1}{2}eB_0\rho^2\dot{\phi}\frac{df}{dz} = 0\,. \tag{2.140}$$

Let us first look at the case with a constant magnetic field, $\mathbf{B} = B_0\,\mathbf{k}$, which corresponds to $f(z) = 1$ (or $d = \infty$). The equations then have simple solutions of the form

$$\rho = \rho_0\,, \quad \dot{\phi} = -eB_0/m \equiv \omega_0\,, \quad \dot{z} = u_0\,, \tag{2.141}$$

which corresponds to the particle spiralling around the magnetic field lines, with $\rho = 0$ as center and ρ_0 as the radius of the circulations. The parameter u_0 is the velocity component of the particle along the field lines, and both ρ_0 and u_0 are constants. This solution is consistent with the result in the previous example, although the center of the circular motion is here linked to the choice of coordinates.

We consider next the case where the non-homogenous contribution to the magnetic field is included, $f \neq 1$. Convenient initial conditions are chosen in the following way. For $t = 0$, ρ, $\dot{\phi}$ and $\dot{z}$ are given as in (15.50), and furthermore $z(0) = \dot{\rho}(0) = 0$. The equations of motion (2.137), (2.138), and (2.140) cannot be solved analytically, but we will make use of an approximation which is valid when the inhomogeneity is small, $|B_\rho| << |B_z|$. This is implemented by assuming the angular frequency $\dot\phi$ to be approximated by

$$\dot\phi \approx -\frac{e}{m}B_z = -\frac{e}{m}B_0 f(z)\,, \tag{2.142}$$

which is satisfied as an equality when $f = 1$. When the approximation is introduced in the equations of motion, these are reduced to

$$\ddot\rho \approx 0\,, \tag{2.143}$$

$$eB_0\rho^2 f(z) \approx -2\ell_z\,, \tag{2.144}$$

$$m\ddot z + \frac{1}{2m}e^2B_0^2\rho^2 f(z)\frac{df}{dz} \approx 0\,. \tag{2.145}$$

We note that if the approximations are replaced with equalities the first two equations are not consistent when z is time dependent. This is the case since (2.143), with the given initial condition, then implies that ρ is constant. However, the discrepancy can be regarded as small, as it disappears when $\frac{df}{dz} \to 0$. When the angular velocity $\dot\phi$ has been fixed by the assumption (2.142), only two equations should be applied to determine $\rho(t)$ and $z(t)$. We therefore disregard the first equation and replace the approximations with equalities in the two remaining equations.

Equation (2.144) implies

$$\rho^2 f(z) = \rho_0^2\,, \tag{2.146}$$

as a consequence of the left-hand side being time independent, and therefore equal to its value at $t = 0$. Making use of this and of the expression for the angular momentum at $t = 0$, $\omega_0 = -eB_0/m$, we find that (2.145) can be rewritten as

$$m\ddot z + \frac{1}{2}m\omega_0^2\rho_0^2\frac{df}{dz} = 0\,. \tag{2.147}$$

Since df/dz is linear in z this is in fact a harmonic oscillator equation, and we rewrite the equation as

$$\ddot z + \left(\frac{\omega_0\rho_0}{d}\right)^2 z = 0\,. \tag{2.148}$$

From this follows that the solution, which satisfies the correct initial condition $\dot z(0) = u_0$, can be written as

$$z(t) = a\sin\left(\frac{u_0}{a}t\right)\,, \quad a \equiv \frac{u_0 d}{\rho_0\omega_0}\,. \tag{2.149}$$

The radial coordinate is then determined by (13.33), as

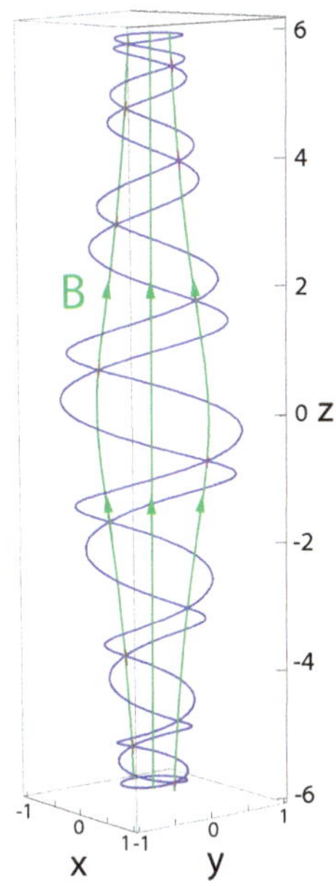

Fig. 2.3 Charged particle trapped in an inhomogeneous magnetic field. The magnetic field (green curves) is of the form given in the text. The particle path (blue curve) is shown for a full period of oscillations in the vertical direction.

$$\rho(t) = \frac{\rho_0}{\sqrt{f(z)}} = \frac{\rho_0}{\sqrt{1 + (a^2/d^2)\sin^2(u_0 t/a)}}\,, \tag{2.150}$$

and, with the angular variable determined by (2.142),

$$\dot{\phi} = -\frac{e}{m}B_0 f(z) = -\frac{e}{m}B_0\sqrt{1 + (a^2/d^2)\sin^2(u_0 t/a)}\,, \tag{2.151}$$

we have the full solution of the problem, valid under the approximation which we have adopted. It describes the motion of the particle as circulation around the magnetic field lines combined with oscillations along the field lines. The motion is illustrated in Fig. 2.3, where the parameters are chosen with the following dimensionless values $d/\rho_0 = 15$ and $a/\rho_0 = 6$.

2.6 Exercises

Problem 2.1

A particle with mass m moves freely in a horizontal plane. There is no constraint on the motion, but in the following we will consider the free motion described in a rotating reference frame. We refer to the Cartesian coordinates of a fixed frame as (x, y) and the coordinates of the rotating frame as (ξ, η). They are related by the standard expressions

$$x = \xi \cos\omega t - \eta \sin\omega t\,,$$
$$y = \xi \sin\omega t + \eta \cos\omega t. \tag{2.152}$$

a) Find the Lagrangian expressed in terms of the coordinates ξ, η and their time derivatives.

b) Find the corresponding equations of motion for the two variables, and identify the Coriolis and centrifugal terms by comparing with the standard expression for Newton's 2nd law in a rotating reference frame. It has the general form

$$m\ddot{\mathbf{r}} = \mathbf{F} - m\boldsymbol{\omega} \times (\boldsymbol{\omega} \times \mathbf{r}) - 2m\boldsymbol{\omega} \times \dot{\mathbf{r}}\,, \tag{2.153}$$

where $\mathbf{F}$ is the force acting on the particle in an inertial frame, $\mathbf{r}$ is the position vector in the rotating reference frame, and $\boldsymbol{\omega}$ is the angular velocity vector of the rotating frame.

Problem 2.2

Two identical rods of mass m and length l are connected to each other with a frictionless joint (Fig. 2.4). The first rod is connected, at one end, to a joint in the ceiling, and at the other end, to the joint at the center of the second rod. The motion takes place in the vertical plane.

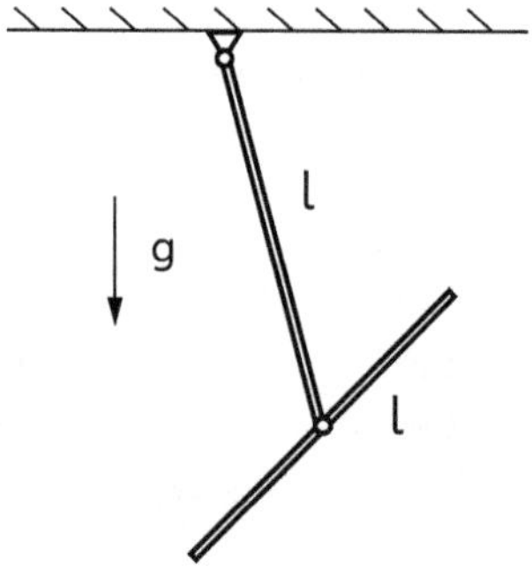

Fig. 2.4 A two-rod problem.

a) Choose suitable generalized coordinates for the system, and find the corresponding Lagrangian.

b) Formulate Lagrange's equations for the system, and find the angular frequency for small oscillations of the upper rod about its equilibrium position.

As a reminder, the moment of inertia of a rod (with even mass distribution) about an endpoint is $I_1 = ml^2/3$ and about the midpoint is $I_2 = ml^2/12$.

Problem 2.3

A small body with mass m moves without friction on a rod (see Fig. 2.5). The rod rotates in the horizontal plane, with constant angular velocity ω about a fixed point, which we assign the radial coordinate $r = 0$.

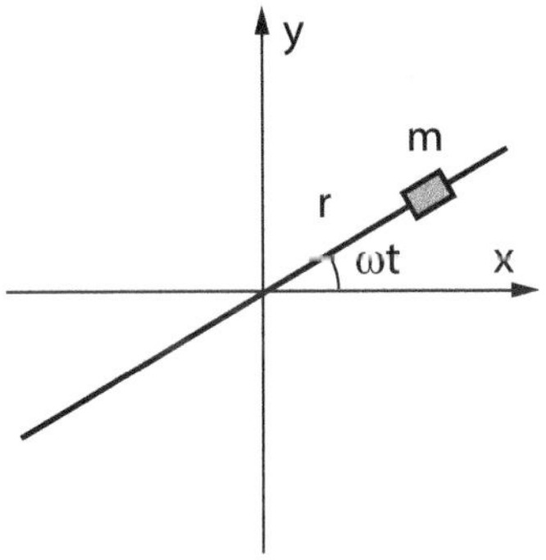

Fig. 2.5 Body sliding on a rotating rod

a) Find Lagrange's equation for the radial coordinate r, and show that, with the initial condition $\dot{r} = 0$ and $r = r_0$ at $t = 0$, the equation has the solution $r = r_0 \cosh \omega t$.

b) Make a plot of the orbit in the x, y-plane, with dimensionless parameter $r_0 = 1$, and with t restricted by $\omega t \lesssim \pi$.

Problem 2.4

A pendulum consists of a rigid rod, which we consider as massless, and a pendulum bob of mass m (see Fig. 2.6). The point of suspension of the pendulum has horizontal coordinate $x = s$ and vertical coordinate $y = 0$.

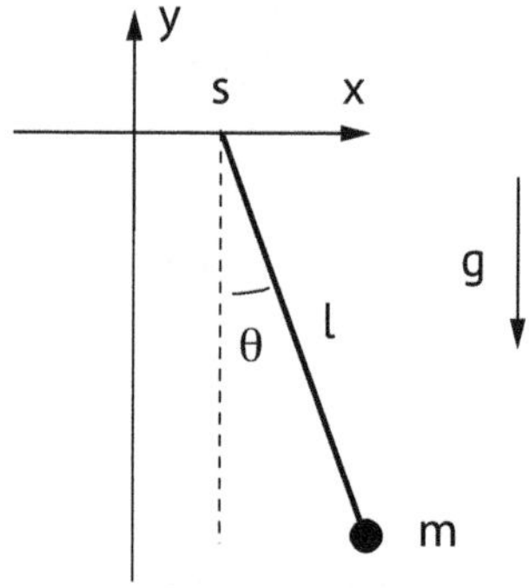

Fig. 2.6 A pendulum problem.

a) Assume first that the point of suspension is kept fixed, with $s = 0$. Use the angle θ as generalized coordinate, find the Lagrangian of the system, and determine the form of Lagrange's equation for the system. Check that it has the standard form of a pendulum equation.

b) The point of suspension is now released so it can move freely in the horizontal direction (x-direction). Use s and θ as generalized coordinates for the system and determine the corresponding set of Lagrange's equations.

c) Show that s can be eliminated to give an equation which only depends on θ and its derivatives. Further show that this equation implies that the vertical motion of the pendulum bob is identical to free fall in the gravitational field (in reality restricted by the length l of the rod).

Problem 2.5

A rigid circular metal hoop rotates with constant angular velocity ω around an axis through the center. A bead with mass m slides without friction along the circle and there is no gravity. Use the angular variable θ as the generalized coordinate (see Fig. 2.7).

a) Find the Lagrangian expressed in terms of θ and $\dot{\theta}$, and derive Lagrange's equation for this variable.

b) Show that the Lagrangian is the same as the Lagrangian of a particle moving in a time independent, periodic potential $V(\theta)$, with two stable and two unstable equilibrium points (within a 2π interval).

c) With θ_0 as one of the stable equilibrium points, introduce a small deviation, $\phi = \theta - \theta_0$, and determine the small-angle form of the equation

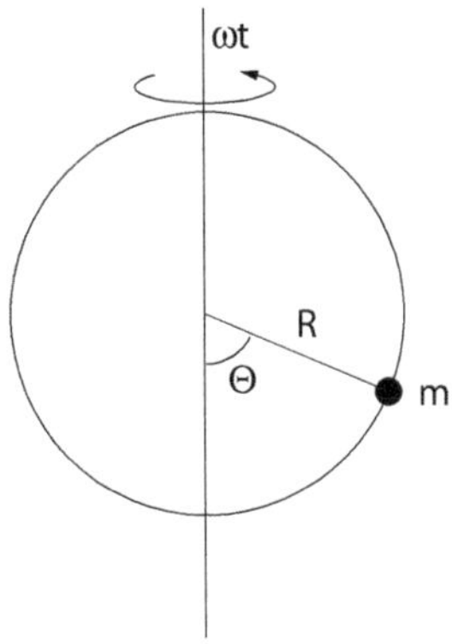

Fig. 2.7 Motion on a rotating hoop.

of motion (keeping only the lowest order in ϕ and its derivative). What is the angular frequency of oscillations about the stationary point?

Problem 2.6

An object with mass m slides without friction on an inclined plane, as shown in Fig. 2.8. The inclined plane is tilted with an angle of 30° relative to the horizontal plane. It is forced to move horizontally with a constant acceleration a. A natural choice of generalized coordinate will be the displacement s of the object along the tilted surface.

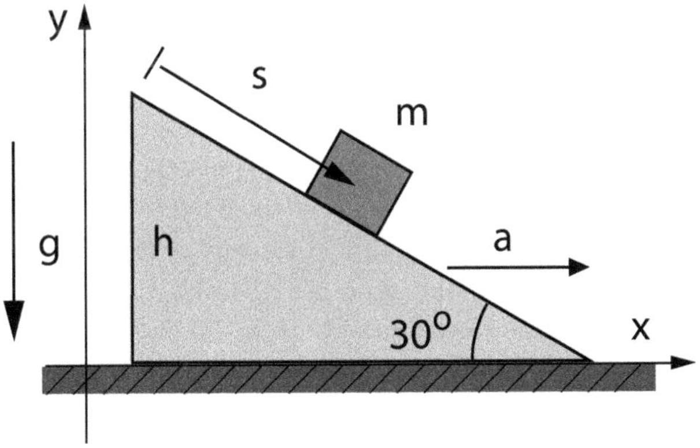

Fig. 2.8 Motion on an inclined plane.

a) Assume first the inclined plane to be at rest, with vanishing velocity and acceleration. Express the Cartesian coordinates x, y of the sliding object as functions of s. Find the Lagrangian expressed in terms of s and $\dot{s}$.

b) Assume next the acceleration a to be constant and non-vanishing. Again find the Cartesian coordinates, and their time derivatives, expressed in terms of s and $\dot{s}$ and determine the Lagrangian.

c) Find Lagrange's equation for the system, and solve the equation under the assumption that the body starts at time $t = 0$ from the top of the inclined plane ($s = 0$) with zero velocity relative to the plane.

Problem 2.7

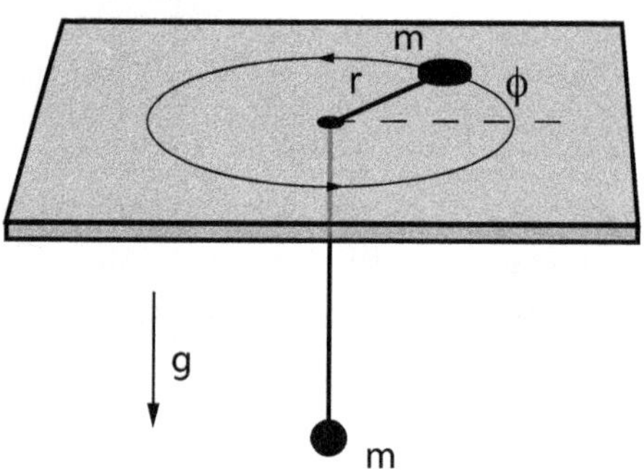

Fig. 2.9 Two bodies connected by a rope through a small hole.

Two bodies with the same mass, m, are connected with a massless rope through a small hole in a horizontal plane (see Fig 2.9). One body is moving on the plane, the other one is hanging at the end of the rope and can move vertically. At all instances the rope is tight and the motion is without friction. The acceleration due to gravity is g.

a) Find Lagrange's equations of motion for the system, expressed in polar coordinates (r, θ) of the body moving on the plane.

b) Reduce the equations of motion to a one-dimensional problem in r and make a qualitative description of the motion.

Problem 2.8

A pendulum is connected to a block, which can slide without friction in a horizontal direction (Fig. 2.10). Assume that all motion takes place in the vertical x, y-plane. The block and the pendulum bob have equal masses m. The pendulum rod, which has length d is considered to be massless. As generalized coordinates in this problem, use s as the x-coordinate of the center of mass of the box, and θ as the angle of the pendulum rod relative to the vertical direction. As initial condition, choose at time $t = 0$ both the block and the pendulum to be at zero velocity, with the pendulum angle being θ_0.

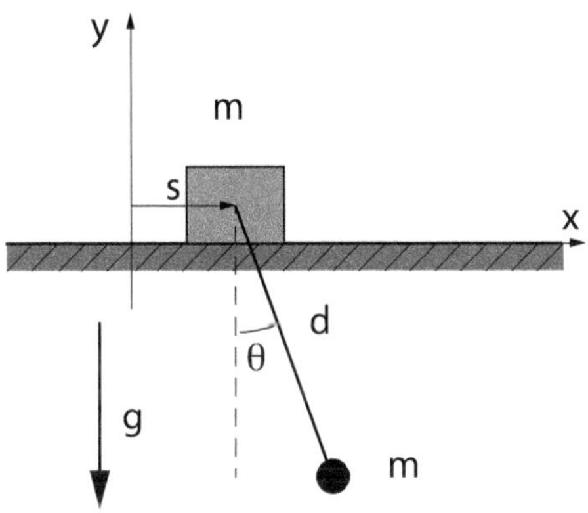

Fig. 2.10 Pendulum attached to a moving block.

a) Find the Lagrangian of the system, expressed in terms of the generalized coordinates and their velocities, and derive Lagrange's equations for these variables.

b) What is meant by s being a cyclic coordinate? Make use of this to reduce the equations to an equation of motion for the variable θ alone, and show that it takes the form

$$\left(1 - \frac{1}{2}\cos^2\theta\right)\ddot{\theta} + \frac{1}{2}\sin\theta\cos\theta\,\dot{\theta}^2 + \frac{g}{d}\sin\theta = 0\,. \tag{2.154}$$

c) With θ_0 sufficiently small, the equation of motion can be simplified by making a "small oscillation" approximation. Show that the equation then gets the form of a harmonic oscillator equation, and determine the angular frequency of the pendulum oscillations.

Problem 2.9

A small body with mass m is constrained to move (without friction) along a spiral-shaped channel, which is etched in a flat, circular disk (Fig. 2.11). The disk rotates in the horizontal plane, with constant angular velocity ω about an axis through the center of the disk. The points on the spiral are characterized by polar coordinates (r, θ), with $r = a\theta$, where a is a constant. The angle variable θ is measured relative to a reference frame which rotates with the disk. In the expression for r the angle θ is chosen to take positive values and is not restricted to be less than 2π. The radius of the disk we refer to as R.

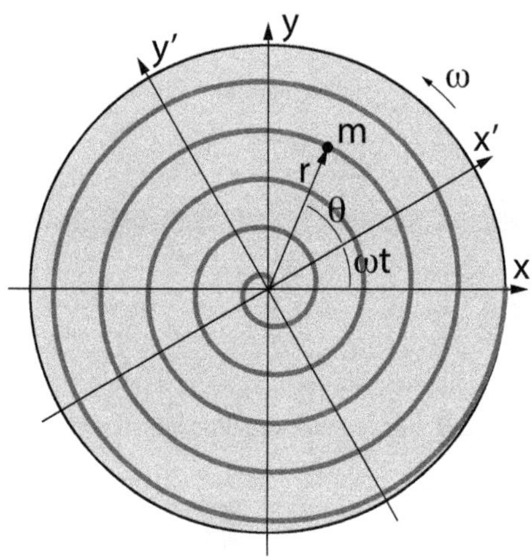

Fig. 2.11 Motion along a spiral-shaped channel.

a) Find the Cartesian coordinates (x, y) of the moving body, measured in the fixed reference frame, expressed in terms of the generalized coordinate θ. Use this to determine the Lagrangian L as a function of $\dot{\theta}$ and θ.

b) Find Lagrange's equation, expressed as a differential equation for θ. Assume the condition $\theta >> 1$ to be satisfied, and show that the equation then can be simplified to

$$\theta\ddot{\theta} + \dot{\theta}^2 - \omega^2 = 0\,. \tag{2.155}$$

c) The initial conditions are $r(0) = r_0$ and $\dot{r}(0) = 0$, with $r_0 >> a$.

Show that Eq.(2.155) has the solution

$$\theta(t) = \sqrt{\omega^2 t^2 + r_0^2/a^2}\,. \tag{2.156}$$

d) Determine the time t, which the body takes to reach the edge of the disk, expressed as a function of the initial coordinate r_0.

Problem 2.10

In this problem we will study the acceleration of charged particles by use of *electromagnetic induction*. This is a method which is used in particle accelerators called *betatrons*. We study the motion of a particle with mass m and charge q in a horizontal plane, which we take as as the x, y-plane. The motion we assume to be non-relativistic.

The particle moves in an electromagnetic field, described by the scalar potential $\Phi = 0$ and a vector potential $\mathbf{A}$, which in cylindrical coordinates has the form (see Appendix C.3),

$$A_r = A_z = 0,\; A_\phi = \frac{1}{2} r B(t)\,, \tag{2.157}$$

with $r = \sqrt{x^2 + y^2}$. These expressions are valid in a rotationally symmetric region, given by $r < R$, where the particle is moving. $B(t)$ we assume to be constant over this region, but varying with time.

a) Determine the magnetic field $\mathbf{B}$ and the induced electric field $\mathbf{E}$, both expressed in cylindrical coordinates.

b) Express the Lagrangian of the charged particle in terms of the polar coordinates (r, ϕ), and find the corresponding Lagrange's equations. Check that the equations are consistent with the equation of motion in vector form, for a charged particle in an electromagnetic field,

$$\frac{d\mathbf{p}}{dt} = q(\mathbf{E} + \mathbf{v} \times \mathbf{B})\,. \tag{2.158}$$

c) Assume that for $t < 0$ that B is time independent, $B = B_0$. The symmetries of the Lagrangian in this case imply that there are two constants of motion. Find expressions for these. Show that Lagrange's equations have solutions corresponding to circular motion with $r = r_0$. Find for such motion the constants of motion expressed in terms of r_0 and B_0. Find also the angular frequency ω_0 of the motion.

d) Assume next that B has B_0 as initial value at $t = 0$, and subsequently increases slowly with time to reach the final value B_1 at time $t = t_1$. The particle moves in a circle with $r = r_0$ at time $t = 0$, and this changes to a circle with radius r_1 at $t = t_1$. One of the two variables that are constants of motion in c), remains constant also when B increases. Which one? Determine the radius r_1 and angular frequency ω_1 at time t_1, expressed in terms of r_0, B_1 and B_0. What is the ratio between the energy $\mathcal{E}_1$ at the final time, and $\mathcal{E}_0$ at the initial time?

Problem 2.11

A particle moves on a parabolic surface given by the equation $z = (\lambda/2)(x^2 + y^2)$ where z is the Cartesian coordinate in the vertical direction, x and y are orthogonal coordinates in the horizontal plane and λ is a constant. The particle has mass m and moves without friction on the surface under influence of gravitation. The gravitational acceleration g acts in the negative z-direction. The particle's position is given by the polar coordinates (r, θ) of the *projection* of the position vector into the x, y-plane.

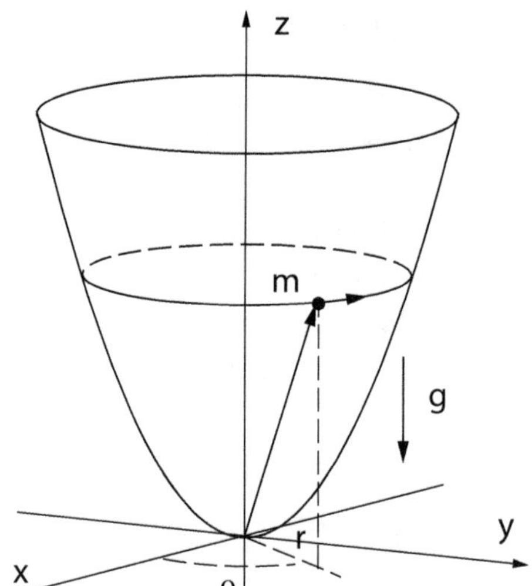

Fig. 2.12 Motion on a parabolic surface.

a) Show that the Lagrangian for this system is

$$L = \frac{1}{2}m[(1 + \lambda^2 r^2)\dot{r}^2 + r^2\dot{\theta}^2 - g\lambda r^2], \tag{2.159}$$

and find Lagrange's equations for the particle.

b) Use the fact that there is a cyclic coordinate to show that the equations can be reduced to a single equation in the radial variable r. What is the condition for the particle to move in a circle with radius $r = r_0$?

c) Assume that the path of the particle deviates little from the circular motion so that $r = r_0 + \rho$, where ρ is small. Show that under this condition the radial equation can be reduced to a harmonic oscillator equation for the small variable ρ and determine the corresponding frequency. Give a qualitative description of the motion of the particle.

Problem 2.12

A small body with mass m and charge q is moving in the horizontal plane (x, y-plane), under influence of a harmonic oscillator potential, $V(r) = \frac{1}{2}m\omega_0^2 r^2$ and a constant magnetic field $\mathbf{B} = B\,\mathbf{k}$, which is directed perpendicular to the plane of the moving particle. The vector potential corresponding to $\mathbf{B}$ can be written as $\mathbf{A} = -\frac{1}{2}\mathbf{r} \times \mathbf{B}$, with $\mathbf{r}$ as the position vector of the particle.

a) With the polar coordinates of the plane (r, ϕ) used as generalized coordinates, show that the Lagrangian takes the form

$$L = \frac{1}{2}m(\dot{r}^2 + r^2(\dot{\phi}^2 - \omega_B\dot{\phi} - \omega_0^2))\,, \tag{2.160}$$

where we have introduced the cyclotron (angular) frequency, $\omega_B = -qB/m$.

b) The polar angle ϕ is cyclic. Explain what that means, and give the expression for the corresponding conserved quantity, which we label ℓ. What is the physical interpretation of the quantity? The form of the Lagrangian implies that there is a second constant of motion. Give the expression and physical interpretation of this quantity.

c) Find the expression for Lagrange's equation for the variable r, and use the fact that ϕ is cyclic to express the equation in the variable r alone.

d) Show that the radial equation has solutions which describe circular motion, and give the radius and angular velocity of the motion as functions of the parameters of the problem. Show also that there is a solution where the particle performs oscillations about the origin, in a direction which rotates with time, and find the oscillation and rotation frequencies. Give a qualitative description of the more general kind of motion, which is described by the equation.

Problem 2.13

In the lobby of the Physics building there is a Foucault pendulum, which is the subject of this exercise. The idea is to use the Lagrange method to study its motion, and in particular to derive the period of rotation of the pendulum due to the effect of Earth's rotation. The length of the pendulum wire is $l = 14m$ and the mass of the brass sphere at the end of the wire is $m = 20kg$. The pendulum is situated at the latitude $60°$ north.

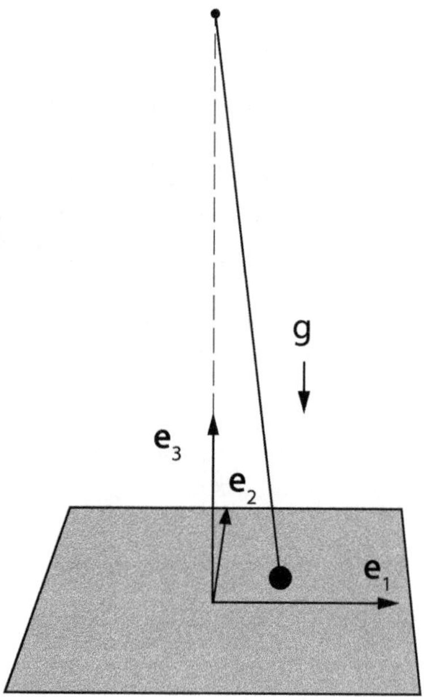

Fig. 2.13 The Foucault pendulum.

We introduce three orthogonal unit vectors $\mathbf{e}_k, k = 1, 2, 3$, which are fixed relative to the building. These are defined with $\mathbf{e}_3$ pointing in the vertical direction, $\mathbf{e}_1$ pointing to the north, and $\mathbf{e}_2$ orthogonal to the two. The three unit vectors are used as the basis vectors of an earth-fixed reference frame S, with the origin of the reference frame taken as the equilibrium position of the pendulum sphere. In addition we introduce $\mathbf{k}$ as a unit vector in the plane spanned by $\mathbf{e}_1$ and $\mathbf{e}_3$, with direction parallel to the Earth's rotational axis. The angle between $\mathbf{e}_3$ and $\mathbf{k}$ we refer to as θ.

The vertical direction is the direction opposite to the *effective* gravitational acceleration **g**, which deviates from the true gravitational acceleration because of the centripetal acceleration of the Physics building, caused by Earth's rotation. However, this effect is small and is not important for the present problem. We shall therefore simply consider the origin of the reference frame S to be non-accelerated, with **g** pointing towards the center of the earth. The fact that the reference frame S rotates with the angular velocity of the earth is however important for the effect to be studied. This rotation can be expressed in terms of the time derivatives of the basis vectors as

$$\dot{\mathbf{e}}_k = \boldsymbol{\omega} \times \mathbf{e}_k \,, \quad k = 1, 2, 3 \,, \tag{2.161}$$

with $\boldsymbol{\omega} = \omega \mathbf{k}$ as the angular velocity of the earth.

The position vector of the pendulum sphere we express in the following as

$$\mathbf{r} = x\mathbf{e}_1 + y\mathbf{e}_2 + z\mathbf{e}_3 \,, \tag{2.162}$$

where the coordinates satisfy the constraint $x^2 + y^2 + (l - z)^2 = l^2$, with l as the length of the pendulum wire.

a) Find the velocity vector $\dot{\mathbf{r}}$ expressed in terms of the basis vectors of S, and use this to derive the kinetic energy T, expressed in terms of the coordinates (x, y, z) and their time derivatives. Use the angle θ as a parameter in the expressions. Terms that are quadratic in ω are very small and may be dropped. Give a comment on why this is a good approximation. Also give the expression for the potential energy V.

b) Make use of x and y as the generalized coordinates of the pendulum, and assume small oscillations $x/l << 1$ and $y/l << 1$. Show that with this assumption the Lagrangian $L(x, y, \dot{x}, \dot{y}, t)$ is approximated by the following expression

$$L = \frac{1}{2}m(\dot{x}^2 + \dot{y}^2) + m\omega\cos\theta(x\dot{y} - y\dot{x}) - \frac{1}{2l}mg(x^2 + y^2) \,. \tag{2.163}$$

c) The form of the Lagrangian suggests that a change to polar coordinates may be convenient, $x = \rho\cos\phi \,,\ y = \rho\sin\phi$. Do that and derive the corresponding Lagrange's equations.

d) Show that the equations have a solution with constant angular velocity $\omega_\phi = \dot{\phi}$, and with oscillations in the radial variable. Determine the angular velocity ω_ϕ and the angular frequency ω_p of the pendulum oscillations.

e) At 12 o'clock on a particular day, the pendulum is performing oscillations in the x, z-plane. At 12 o'clock the next day, the pendulum is swinging in a plane which is rotated by the angle $\Delta\phi$ relative to the original plane of oscillations. Determine the value of $\Delta\phi$.

Chapter 3

Hamiltonian dynamics

3.1 Hamilton's equations

In the Lagrange formulation of mechanics the Lagrangian $L(q, \dot{q}, t)$ controls the dynamical evolution of the physical system. It determines the motion of the system through its partial derivatives with respect to the variables q_i and $\dot{q}_i$. The Hamilton formulation of the dynamics can be viewed as being derived from the Lagrange formulation by a change of the function which controls the dynamics, from the Lagrangian to the Hamiltonian,

$$L(q, \dot{q}, t) \rightarrow H(q, p, t) . \tag{3.1}$$

The transformation from L to H is combined with a change of fundamental variables, from the set of generalized coordinates and velocities $(q, \dot{q})$, to the set of coordinates and conjugate momenta (q, p). The Hamiltonian is defined so that the equations of motion, also in Hamilton's formulation, are expressed through the partial derivatives of this function with respect to the fundamental variables. The transformation is an example of a *Legendre transformation*. Legendre transformations are also known from thermodynamics, where the changes in the choice of fundamental thermodynamical variables $p, T, V, S, \ldots$ are combined with the change in thermodynamic potentials such as Helmholtz and Gibbs free energies.

To be more specific we return to the definition of the Hamiltonian

$$H = \sum_i p_i \dot{q}_i - L , \quad p_i = \frac{\partial L}{\partial \dot{q}_i} . \tag{3.2}$$

As already discussed, we may invert the relation between the conjugate momentum and velocity in the expression for p_i, to give the velocity as a function of momentum and coordinates (and possibly time),

$$\dot{q}_i = \dot{q}_i(p, q, t) , \tag{3.3}$$

whereby the Hamiltonian may be expressed as a function of q, p and t. To see how Lagrange's equation can be reformulated in terms of partial derivatives of H, we consider first the variation in H under an infinitesimal change in the variables of the system. From the definition of H follows

$$\begin{aligned} dH &= \sum_i (dp_i \dot{q}_i + p_i d\dot{q}_i) - dL \\ &= \sum_i \left[(dp_i \dot{q}_i + p_i d\dot{q}_i) - \frac{\partial L}{\partial q_i} dq_i - \frac{\partial L}{\partial \dot{q}_i} d\dot{q}_i \right] - \frac{\partial L}{\partial t} dt \\ &= \sum_i \left[\left(p_i - \frac{\partial L}{\partial \dot{q}_i} \right) d\dot{q}_i + \dot{q}_i dp_i - \frac{\partial L}{\partial q_i} dq_i \right] - \frac{\partial L}{\partial t} dt \\ &= \sum_i \dot{q}_i dp_i - \sum_i \frac{\partial L}{\partial q_i} dq_i - \frac{\partial L}{\partial t} dt \,, \end{aligned} \tag{3.4}$$

and the important point to notice is that the differential $d\dot{q}_i$ has disappeared in the final expression, due to the definition of the canonical momentum p_i. This means that only the differentials for a set of *independent* variables (q, p) appear on the right-hand side of the equation. The coefficients in front of these can be identified as the partial derivatives of H with respect to the corresponding variables.

With H as a function of q, p and t, the general expression for the change in H due to a change in the fundamental variables is

$$dH = \sum_i \frac{\partial H}{\partial p_i} dp_i + \sum_i \frac{\partial H}{\partial q_i} dq_i + \frac{\partial H}{\partial t} dt \,, \tag{3.5}$$

and by comparing with (3.4), we find the following relations,

$$\dot{q}_i = \frac{\partial H}{\partial p_i} \,, \quad \frac{\partial H}{\partial q_i} = -\frac{\partial L}{\partial q_i} \,, \quad \frac{\partial H}{\partial t} = -\frac{\partial L}{\partial t} \,. \tag{3.6}$$

One should note that, at this point, no dynamics is involved in these equations. They are simply consequences of the definitions of the canonical momenta and the Hamiltonian. However, at the next step we make use of Lagrange's equations, which can be written as

$$\dot{p}_i = \frac{\partial L}{\partial q_i} \,. \tag{3.7}$$

By use of this, the first two equations in (3.6) can be written as

$$\dot{q}_i = \frac{\partial H}{\partial p_i} \,, \quad \dot{p}_i = -\frac{\partial H}{\partial q_i} \,. \tag{3.8}$$

These equations, which are known as *Hamilton's equations*, can be viewed as being *equivalent* to Lagrange's equations, in the sense that they constitute a complete set of equations of motion for the physical system. As is already shown, Hamilton's equations follow from Lagrange's equations, and in a similar way one can from Hamilton's equations re-derive Lagrange's equation.

Hamilton's equations (3.8) can be supplemented by a third equation,

$$\frac{dH}{dt} = \frac{\partial H}{\partial t}\,. \tag{3.9}$$

This identity follows from (3.5) by use of Hamilton's equations for $\dot{q}$ and $\dot{p}$. This shows directly that if there is no explicit time dependence, which means $\frac{\partial H}{\partial t} = -\frac{\partial L}{\partial t} = 0$, then the total time derivative of H vanishes and therefore the Hamiltonian is a constant of motion.

In the derivation of Hamilton's equations it was noticed that only the equation for $\dot{p}_i$ was dynamical, in the sense that only this equation depends on Lagrange's equation to be satisfied. However, after Hamilton's equations have been established, there is no reason for treating the equations for q and p differently. The standard way to view the equations is that both equations are parts of the full set of equations of motion for the system, with the coordinates and momenta being represented in a symmetric way.

Compared to Lagrange's formulation, it seems that we have doubled the set of equations, since now there are two equations for each degree of freedom, whereas in Lagrange's formulation there is only one. However, the two Hamilton's equations are *first order* in time derivatives, whereas Lagrange's equation is *second order*. The two first order differential equations can be replaced by a single second order differential equation, and we shall demonstrate this in a simple example.

Example

The one-dimensional harmonic oscillator

In this case there is no constraint (except for the reduction to one dimension) and we use the linear coordinate x as generalized coordinate. For kinetic and potential energy we have the expressions

$$T = \frac{1}{2}m\dot{x}^2\,, \quad V = \frac{1}{2}kx^2\,. \tag{3.10}$$

The Lagrangian is therefore

$$L = \frac{1}{2}m\dot{x}^2 - \frac{1}{2}kx^2\,, \tag{3.11}$$

and the canonical momentum conjugate to x is

$$p = \frac{\partial L}{\partial \dot{x}} = m\dot{x}\,. \tag{3.12}$$

The Hamiltonian is defined by

$$H = p\dot{x} - L = \frac{1}{2m}p^2 + \frac{1}{2}kx^2 = T + V\,, \tag{3.13}$$

and from this follows Hamilton's equations,

$$\begin{aligned} \dot{x} &= \frac{\partial H}{\partial p} = \frac{p}{m}\,, \\ \dot{p} &= -\frac{\partial H}{\partial x} = -kx\,. \end{aligned} \tag{3.14}$$

Position and momentum are therefore coupled through the two equations

$$\dot{x} = \frac{p}{m}\,, \quad \dot{p} = -kx\,. \tag{3.15}$$

From these equations p can be eliminated to give

$$\ddot{x} + \frac{k}{m}x = 0\,, \tag{3.16}$$

which is the standard harmonic oscillator equation, with $\omega = \sqrt{k/m}$ as the circular frequency of the oscillator. This is the equation we would have derived directly from Lagrange's equation,

$$\frac{d}{dt}\frac{\partial L}{\partial \dot{x}} - \frac{\partial L}{\partial x} = 0 \tag{3.17}$$

with the partial derivatives given as

$$\frac{\partial L}{\partial \dot{x}} = m\dot{x}\,, \quad \frac{\partial L}{\partial x} = -kx. \tag{3.18}$$

We note that the reduction from the two Hamilton's equations to the single Lagrange's equation is obtained by eliminating the momentum p. Although this is a very simple example, it illustrates the way in which Hamilton's equations are used, and how they relate to Lagrange's equations.

3.2 Hamiltonian description of particle in an electromagnetic field

We have in an earlier section established the form of the Lagrangian for a charged particle in an electromagnetic field,

$$L = \frac{1}{2}m\mathbf{v}^2 - e\Phi + e\mathbf{v}\cdot\mathbf{A}\,, \tag{3.19}$$

with Φ and $\mathbf{A}$ as the electromagnetic potentials, m as the mass and e as the charge of the particle. The corresponding canonical momentum is

$$\mathbf{p} = m\mathbf{v} + e\mathbf{A}\,, \tag{3.20}$$

and the Hamiltonian is

$$H = \frac{1}{2m}(\mathbf{p} - e\mathbf{A})^2 + e\Phi\,. \tag{3.21}$$

This classical Hamiltonian has the same form as its quantum counterpart, and it represents the total energy of the particle. If the potentials are time independent, the Hamiltonian H is also time independent and the energy is conserved.

We take the Cartesian coordinates of the particle as generalized coordinates, and write these as $x_i, i = 1, 2, 3$, with $x_1 = x$, $x_2 = y$ and $x_3 = z$ in the usual way. Hamilton's equations in this case give

$$\begin{aligned} \dot{x}_i &= \frac{\partial H}{\partial p_i} = \frac{1}{m}(p_i - eA_i)\,, \\ \dot{p}_i &= -\frac{\partial H}{\partial x_i} = \frac{e}{m}(\mathbf{p} - e\mathbf{A}) \cdot \frac{\partial \mathbf{A}}{\partial x_i} - e\frac{\partial \Phi}{\partial x_i}\,. \end{aligned} \tag{3.22}$$

We will check that these two equations reproduce the well-known form of Newton's second law applied to the charged particle in the electromagnetic field. We do this by eliminating $\mathbf{p}$ from the equations,

$$\begin{aligned} m\ddot{x}_i &= \dot{p}_i - e\frac{dA_i}{dt} \\ &= \dot{p}_i - e\left(\sum_j \frac{\partial A_i}{\partial x_j}\dot{x}_j + \frac{\partial A_i}{\partial t}\right) \\ &= \frac{e}{m}\sum_j (p_j - eA_j)\frac{\partial A_j}{\partial x_i} - e\frac{\partial \Phi}{\partial x_i} - e\left(\sum_j \frac{\partial A_i}{\partial x_j}\dot{x}_j + \frac{\partial A_i}{\partial t}\right) \\ &= -e\left(\frac{\partial \Phi}{\partial x_i} + \frac{\partial A_i}{\partial t}\right) + e\sum_j \left(\frac{\partial A_j}{\partial x_i} - \frac{\partial A_i}{\partial x_j}\right)\dot{x}_j\,. \end{aligned} \tag{3.23}$$

This we can write in a more familiar form by use of the expressions for the electric and magnetic fields,

$$E_i = -\left(\frac{\partial \Phi}{\partial x_i} + \frac{\partial A_i}{\partial t}\right), \quad B_k = \sum_{ij} \epsilon_{kij}\frac{\partial A_j}{\partial x_i}\,, \tag{3.24}$$

with ϵ_{ijk} as the antisymmetric Levi-Civita symbol. The last equation can be inverted to give

$$\frac{\partial A_j}{\partial x_i} - \frac{\partial A_i}{\partial x_j} = \sum_k \epsilon_{kij} B_k\,, \tag{3.25}$$

and therefore the equation of motion (3.23) can be written as

$$m\ddot{x}_i = eE_i + e\sum_{jk} \epsilon_{ijk} B_k \dot{x}_j \,. \tag{3.26}$$

In vector form it gives the standard (non-relativistic) equation of motion for a charge particle in the electromagnetic field,

$$m\mathbf{a} = e(\mathbf{E} + \mathbf{v} \times \mathbf{B}) \,. \tag{3.27}$$

This again demonstrates that Hamilton's (as well as Langrange's) equations have different forms, but are equivalent to Newton's second law when applied to the same system. We shall next see how Hamilton's equations can be used in a direct way to find the motion of a charged particle in a constant magnetic field.

Example

Charged particle in a constant magnetic field

This problem has been examined earlier in the Lagrange formulation. Here we will make a detailed discussion of the problem in the Hamilton formulation. We assume the particle to be moving in a constant magnetic field, directed along the z-axis, $\mathbf{B} = B\mathbf{k}$. The vector potential we write as

$$\mathbf{A} = -\frac{1}{2}\mathbf{r} \times \mathbf{B} \,, \tag{3.28}$$

with components

$$A_x = -\frac{1}{2}By \,, \quad A_y = \frac{1}{2}Bx \,, \quad A_z = 0 \,. \tag{3.29}$$

It is straight forward to check that the curl of this vector potential reproduces the correct magnetic field. The scalar potential vanishes, $\Phi = 0$, and the Hamiltonian (3.21) then gets the form

$$\begin{aligned} H &= \frac{1}{2m}(\mathbf{p} - e\mathbf{A})^2 \\ &= \frac{1}{2m}\left[\left(p_x + \frac{1}{2}eBy\right)^2 + \left(p_y - \frac{1}{2}eBx\right)^2 + p_z^2\right] \,. \end{aligned} \tag{3.30}$$

We note that z is a cyclic coordinate,[1] and it follows directly from Hamilton's equations that

$$\dot{p}_z = -\frac{\partial H}{\partial z} = 0 \,, \tag{3.31}$$

[1]If H does not depend on z, it is clear from the definition of the Hamiltonian that also the Lagrangian is independent of z (see (3.6)).

so that $p_z = m\dot{z}$ is a constant of motion. This means that the particle moves with constant velocity in the z-direction,

$$z = z_0 + v_{z0}t\,, \tag{3.32}$$

where the constants $v_{z0} = p_z/m$ and z_0 are determined by the initial conditions.

From this follows that the motion in the x, y-plane (the plane orthogonal to the magnetic field) is decoupled from the motion in the z-direction. We write Hamilton's equations for this motion,

$$\begin{aligned}
\dot{x} &= \frac{\partial H}{\partial p_x} = \frac{1}{m}\left(p_x + \frac{1}{2}eBy\right)\,, \\
\dot{p}_x &= -\frac{\partial H}{\partial x} = \frac{eB}{2m}\left(p_y - \frac{1}{2}eBx\right)\,, \\
\dot{y} &= \frac{\partial H}{\partial p_y} = \frac{1}{m}\left(p_y - \frac{1}{2}eBx\right)\,, \\
\dot{p}_y &= -\frac{\partial H}{\partial y} = -\frac{eB}{2m}\left(p_x + \frac{1}{2}eBy\right)\,.
\end{aligned} \tag{3.33}$$

By inspecting the right-hand side of the equations we see that they can be grouped in pairs that are essentially identical. By combining these the following equations are established,

$$\begin{aligned}
\dot{p}_x - \frac{1}{2}eB\dot{y} &= 0\,, \\
\dot{p}_y + \frac{1}{2}eB\dot{x} &= 0\,,
\end{aligned} \tag{3.34}$$

which means that there are two constants of motion,

$$\begin{aligned}
K_x &= p_x - \frac{1}{2}eBy\,, \\
K_y &= p_y + \frac{1}{2}eBx\,.
\end{aligned} \tag{3.35}$$

Combined into a vector, this vector is

$$\begin{aligned}
\mathbf{K} &= \mathbf{p} - \frac{1}{2}e\mathbf{r} \times \mathbf{B} \\
&= m\mathbf{v} + e\mathbf{A} - \frac{1}{2}e\mathbf{r} \times \mathbf{B} \\
&= m\mathbf{v} - e\mathbf{r} \times \mathbf{B}\,,
\end{aligned} \tag{3.36}$$

and it is easy to verify directly from the equation of motion (3.27) that this vector is conserved. (This is the case also when the motion in the z-direction is included, since $K_z = mv_z$ is constant.)

We consider next the linear combinations of the equations (3.33) with opposite signs of those in (3.34),

$$\begin{aligned}\dot{p}_x + \frac{1}{2}eB\dot{y} &= \frac{eB}{m}\left(p_y - \frac{1}{2}eBx\right), \\ \dot{p}_y - \frac{1}{2}eB\dot{x} &= -\frac{eB}{m}\left(p_x + \frac{1}{2}eBy\right). \end{aligned} \tag{3.37}$$

These equations can be expressed in terms of components of the *mechanical* momentum,

$$\boldsymbol{\pi} \equiv m\mathbf{v} = \mathbf{p} - e\mathbf{A} = \mathbf{p} + \frac{1}{2}e\mathbf{r}\times\mathbf{B}. \tag{3.38}$$

They get the form

$$\begin{aligned}\dot{\pi}_x &= \frac{eB}{m}\,\pi_y\,, \\ \dot{\pi}_y &= -\frac{eB}{m}\,\pi_x\,, \end{aligned} \tag{3.39}$$

which implies that each component satisfies a harmonic oscillator equation

$$\ddot{\pi}_x + \omega^2\pi_x = 0\,, \quad \ddot{\pi}_y + \omega^2\pi_y = 0\,, \tag{3.40}$$

with $\omega = -eB/m$ as the angular frequency, known as the *cyclotron (angular) frequency.*

The solution to the equations has the form

$$\pi_x = A\cos\omega t\,, \quad \pi_y = A\sin\omega t\,, \tag{3.41}$$

where A is a constant to be determined by the initial conditions, and where a convenient choice of time $t = 0$ has been chosen. These expressions may be combined with the expressions for the components of the conserved vector $\mathbf{K}$, and we focus first on the x-component,

$$p_x + \frac{1}{2}eBy = A\cos\omega t\,, \quad p_x - \frac{1}{2}eBy = K_x\,. \tag{3.42}$$

By combining these we find

$$y = \frac{1}{eB}(A\cos\omega t - K_x) \equiv y_0 + R\cos\omega t\,, \tag{3.43}$$

where, in the last expression, we have introduced the constants

$$y_0 = -\frac{K_x}{eB}\,, \quad R = \frac{A}{eB}\,. \tag{3.44}$$

Similarly we have

$$p_y - \frac{1}{2}eBx = A\sin\omega t\,, \quad p_y + \frac{1}{2}eBx = K_y \tag{3.45}$$

which gives

$$x = \frac{1}{eB}(-A\sin\omega t + K_y) \equiv x_0 - R\sin\omega t\,. \tag{3.46}$$

The solutions for the components of the position vector show that the particle moves with constant speed on a circle of radius R about a point in the x, y-plane with coordinates (x_0, y_0). These coordinates, as well as the radius R are determined by the initial conditions. The angular velocity $\omega = -eB/m$ is fixed by the strength of the magnetic field and the charge, and is independent of the energy of the particle. The direction of motion in the circular orbit is determined by the sign of eB, so that negative sign corresponds to positive orientation of the motion in the x, y-plane.

When the circular motion in the x, y-plane is combined with the linear motion along the z-axis, this gives a spiral formed orbit which winds around the magnetic field lines. The radius of the circular part of the orbit is determined by the contribution to the kinetic energy from the velocity component in the x, y-plane,

$$T_{xy} = \frac{1}{2}m(\dot{x}^2 + \dot{y}^2) = \frac{1}{2}m\omega^2 R^2\,. \tag{3.47}$$

The result is precisely the same as obtained in Sect. 2.5, by use of Lagrangian formulation, although the details of the derivation are different.

3.3 Phase space

At an earlier stage we introduced the *configuration space* of the physical system as the d-dimensional space described by the generalized coordinates $q = \{q_1, q_2, ..., q_d\}$. These d coordinates, one for each degree of freedom of the system, are all independent variables. Later, in the discussion of the Lagrangian formulation, we extended this to a larger set of $2d$ variables, by treating the velocities $\dot{q} = \{\dot{q}_1, \dot{q}_2, ..., \dot{q}_d\}$ as independent of the coordinates. The $2d$-dimensional space described by both coordinates and velocities $\{q, \dot{q}\}$ we refer to as the *phase space* of the system. Each point in phase space corresponds to a *physical state* of the system, with a well-defined time evolution.

In Hamilton's formulation coordinates and momenta are treated on equal footing. Therefore the d-dimensional configuration space seems less important than the $2d$-dimensional phase space. However, more commonly than using coordinates and velocities, one takes coordinates and conjugate momenta as the independent variables in phase space, since these are the natural variables in Hamilton's equations.

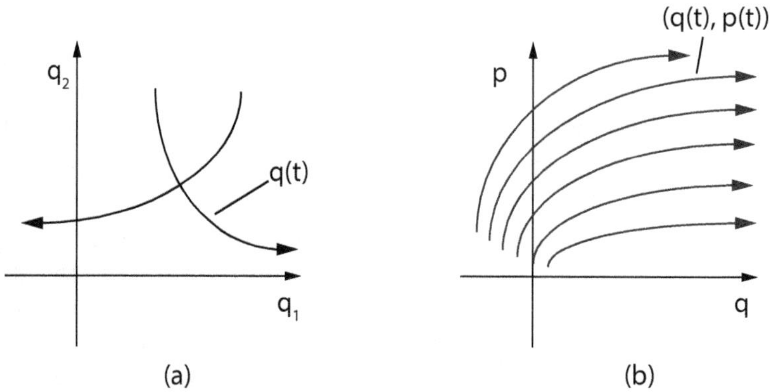

Fig. 3.1 Motion in configuration space (a) and in phase space (b). In configuration space the trajectory is determined by the time dependent coordinates $q(t)$, and many different trajectories, with different initial conditions, may pass through the same point. In phase space the trajectory is specified by the time dependent coordinates and momenta $(q(t), p(t))$. In this case only one trajectory will pass through a given point, and all dynamical trajectories will together form a flow pattern through phase space.

One of the interesting features of the phase space description becomes apparent when one considers the time evolution with given initial conditions. We know that $2d$ initial data are needed to give a unique trajectory. In the Lagrangian formulation this is because there are d *second order* differential equations to determine the motion, and in the Hamiltonian formulation since there are $2d$ *first order* equations. In configuration space this means that through a given point (determined by the d coordinates) there are many possible trajectories, as we have already discussed (see Fig. 3.1a). However, in phase space the number of coordinates needed to determine a point is $2d$, and that is also the number of data needed to determine uniquely a trajectory. This means that through a point in phase space there will pass only one dynamical trajectory (*i.e.*, a trajectory that satisfies the equations of motion).

This situation is illustrated in Fig. 3.1b for the case of a two-dimensional phase space. Through each point passes one and only one trajectory, specified by the initial conditions. If we continuously change these conditions, the trajectory will be deformed in such a way that, when we consider all possible motions of the system, the corresponding trajectories will form a flow pattern through phase space. These paths will be distinct, so that two paths will never meet (except at some singular, isolated points, which we shall discuss in an example to follow). This description of the dynamics, as a flow pattern in phase space, is particularly important in statistical mechanics, where one does not consider sharply defined initial conditions but rather a time evolution of the system with a statistical distribution over many initial data. As we shall see in some examples, the phase space description is also sometimes useful to obtain a qualitative understanding of the motion of the system, without actually solving the equations of motion. Thus, if we find the special points of the flow, corresponding to points of equilibrium, and use the general properties of the phase space flow, we can derive a good qualitative picture of the full flow pattern, and thereby the motion of the system.

Examples

Phase space for the harmonic oscillator

We write the Hamiltonian of a one-dimensional harmonic oscillator in the following form

$$H = \frac{1}{2m}(p^2 + m^2\omega^2 x^2)\,, \tag{3.48}$$

with ω as the angular frequency of the oscillator. Since the Hamiltonian has no explicit time dependence, the total time derivative of H vanishes,

$$\frac{dH}{dt} = \frac{\partial H}{\partial t} = 0\,. \tag{3.49}$$

The energy $H = E$ is therefore a constant of motion. This implies

$$p^2 + m^2\omega^2 x^2 = 2mE\,, \tag{3.50}$$

and we recognize this as the equation for an ellipse in the two-dimensional plane with x and p as coordinates, which is the phase space of the harmonic oscillator. Since x and p have different physical dimensions, the eccentricity of the ellipse has no physical significance, and we can rescale one of the coordinates, for example by redefining the x coordinate, $\tilde{x} = m\omega x$ (which gives $\tilde{x}$ it the same physical dimension as p), so that the ellipse becomes a circle,

$$p^2 + \tilde{x}^2 = 2mE\,. \tag{3.51}$$

The radius of the circle is determined by the energy and increases as $\sqrt{E}$ with energy. Since the energy is a constant of motion these circles of constant energy are the trajectories of the harmonic oscillator in phase space.

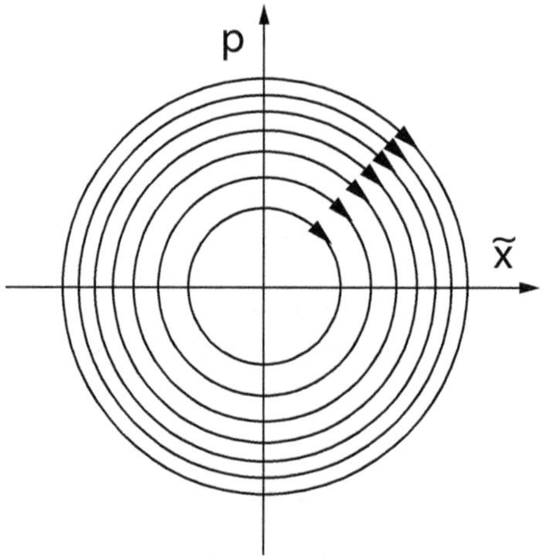

Fig. 3.2 Phase space flow for the one-dimensional harmonic oscillator. The time evolution defines circles of constant energy with motion in the clockwise direction. The curves of constant energy are here plotted with constant energy differences.

We further have Hamilton's equations,

$$\dot{\tilde{x}} = m\omega\frac{\partial H}{\partial p} = \omega p\,,$$
$$\dot{p} = -\frac{\partial H}{\partial x} = -\omega\tilde{x}\,, \tag{3.52}$$

which show that the system moves in the clockwise direction along a circle of constant energy. The initial conditions determine the energy and thereby the circle which the oscillator follows.

We may consider the Hamiltonian $H(x,p)$ as defining a *phase space potential.* Hamilton's equations show that the system moves in the direction orthogonal to the gradient of the potential, which means along one of the equipotential curves. As illustrated in Fig. 3.2 these (directed) curves of constant energy determine the phase space flow of the harmonic oscillator.

The pendulum

Let us next consider the phase space motion of a planar pendulum. With l as the length of the pendulum rod, m as the mass of the pendulum bob, and the angle of displacement θ chosen as the generalized coordinate, we find the following expression for the Lagrangian

$$L = \frac{1}{2}ml^2\dot{\theta}^2 - mgl(1-\cos\theta)\,. \tag{3.53}$$

The canonical momentum conjugate to θ is

$$p = \frac{\partial L}{\partial \dot{q}} = ml^2\dot{\theta}\,, \tag{3.54}$$

and we find the following expression for the Hamiltonian,

$$\begin{aligned} H &= p\dot{\theta} - L \\ &= \frac{1}{2}ml^2\dot{\theta}^2 + mgl(1-\cos\theta) \\ &= \frac{p^2}{2ml^2} + 2mgl\sin^2\frac{\theta}{2}\,. \end{aligned} \tag{3.55}$$

Again there is no explicit time dependence, which means that the energy $H = E$ is a constant of motion. From this follows that a trajectory of the pendulum in phase space is given by

$$p^2 + 4m^2gl^3\sin^2\frac{\theta}{2} = 2ml^2E\,. \tag{3.56}$$

For small oscillations it simplifies to

$$p^2 + m^2gl^3\theta^2 = 2ml^2E\,. \tag{3.57}$$

It has the same form as the phase space equation of the harmonic oscillator, which we have already discussed, although the coordinates are different. In the present case p has the dimension of *angular* momentum rather than *linear* momentum, and θ is a dimensionless variable. However, by a proper scaling of the variables the particle trajectory can also here be given the form of a circle, with radius determined by the energy,

$$\tilde{p}^2 + \theta^2 = \frac{2E}{mgl}\,, \quad \tilde{p} = \frac{p}{m\sqrt{gl^3}}\,. \tag{3.58}$$

When we include motion also for larger angles, we first note that the Hamiltonian $H(p,\theta)$ is a periodic function of θ, and the equipotential curves in the θ,p-plane therefore will be periodic under shifts $\theta \to \theta + 2\pi$. Therefore the point of stable equilibrium will be periodically repeated at the points $(\theta,p) = (2n\pi, 0)$ with n as an integer. As the energy, and therefore the amplitude of oscillations, is increased, the motion will be represented by circles of increasing radii around each point of stable equilibrium. Due to the periodic structure these closed curves will necessarily get deformed when the amplitude increases, and at some point a singular situation is reached, where the closed curves belonging to neighboring (stable) equilibrium points will touch. This we interpret as corresponding to the situation where the pendulum reaches the upper point, where we find an *unstable*

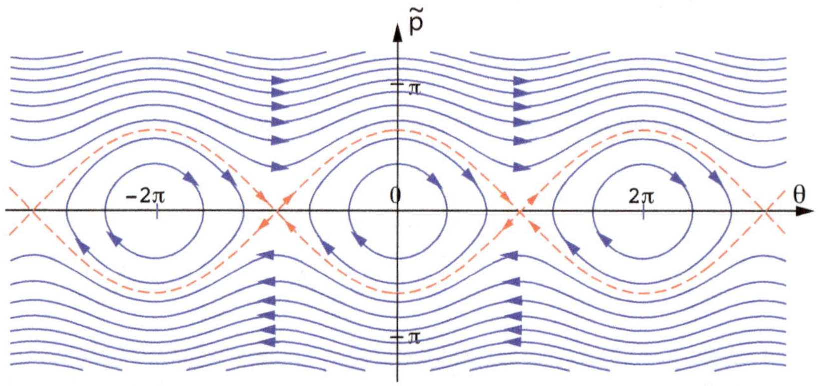

Fig. 3.3 Phase space flow of the pendulum. There are two types of motion, where the closed curves represent oscillations of the pendulum about the stable equilibrium and the open curves represent full rotations. The dashed, red curves separate the two types of motion, and are referred to as *separatrices*. The singular crossing points of these curves are the points of unstable equilibrium. They are not real crossing points of the particle trajectories, since the pendulum velocity at these points vanishes.

equilibrium point. If the energy is increased even further, the motion is no longer oscillatory but rotational, corresponding to an unbounded increase or decrease in the angular variable θ.

This qualitative picture is in full agreement with the plot of phase space trajectories shown in Fig. 3.3. There are solutions of bounded motion, corresponding to oscillations of the pendulum around the point of stable equilibrium, but there are also solutions of unbounded motion. The transition between these two different types of motion is represented by equipotential curves that intersect in singular points. These points represent the point of unstable equilibrium, with the pendulum rod at rest in an upright vertical position. We see from this discussion that we can reach a rather complete, qualitative understanding of the phase space motion by using the knowledge of what happens for small oscillations together with implications of periodicity of the motion.

A rotating pendulum

We consider now a circular hoop, which is rotating with constant angular velocity ω around a symmetry axis with vertical orientation, as shown in Fig. 3.4. Inside the hoop a planar pendulum can perform free oscillations,

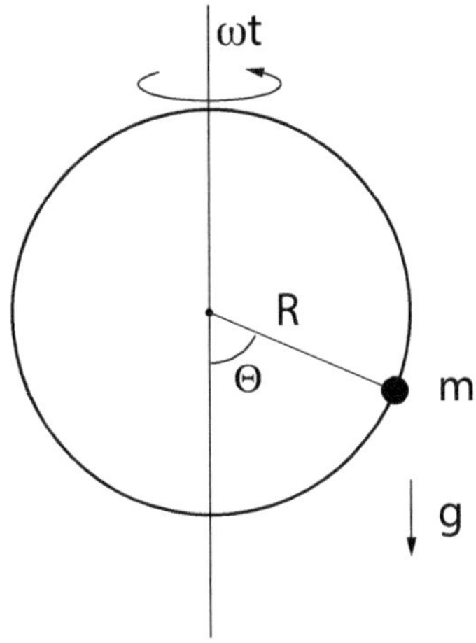

Fig. 3.4 The rotating pendulum. The pendulum bob moves without friction on a hoop which rotates with constant angular velocity ω about a vertical axis.

while the plane of the pendulum rotates with the hoop. The mass of the pendulum bob is m, the length of the pendulum rod is R and the gravitational acceleration is g. The pendulum rod is considered massless. As generalized coordinate we use the angle θ of the pendulum relative to the vertical axis. We will analyze this system in a similar way as in the previous two examples. The present system is, however, different by involving a time dependent constraint. This has the effect that the motion of the pendulum has two qualitative different phases, for small and large rotational frequencies of the hoop.

The Cartesian coordinates of the pendulum bob, when expressed in terms of θ, are

$$\begin{aligned} x &= R\sin\theta\cos\omega t\,, \\ y &= R\sin\theta\sin\omega t\,, \\ z &= -R\cos\theta\,, \end{aligned} \tag{3.59}$$

with the corresponding velocities

$$\begin{aligned} \dot{x} &= R(\dot{\theta}\cos\theta\cos\omega t - \omega\sin\theta\sin\omega t)\,, \\ \dot{y} &= R(\dot{\theta}\cos\theta\sin\omega t + \omega\sin\theta\cos\omega t)\,, \\ \dot{z} &= R\dot{\theta}\sin\theta. \end{aligned} \tag{3.60}$$

This gives the Lagrangian

$$\begin{aligned} L &= \frac{1}{2}m\dot{\mathbf{r}}^2 - mgz \\ &= \frac{1}{2}R^2(\dot{\theta}^2 + \omega^2 \sin^2\theta) + mgR\cos\theta \\ &\equiv \frac{1}{2}mR^2\dot{\theta}^2 - W(\theta)\,, \end{aligned} \tag{3.61}$$

where

$$W(\theta) = -\left(mgR\cos\theta + \frac{1}{2}mR^2\omega^2\sin^2\theta\right). \tag{3.62}$$

The Lagrangian (3.61) can be interpreted as describing the system in a rotating reference frame, which is attached to the rotating hoop. The second term in the expression for $W(\theta)$ is then the corresponding centrifugal contribution to the potential. Lagrange's equation gives the equation of motion,

$$mR^2\ddot{\theta} + \frac{dW}{d\theta} = 0 \quad\Rightarrow\quad \ddot{\theta} + \frac{g}{R}\sin\theta - \frac{1}{2}\omega^2\sin 2\theta = 0\,. \tag{3.63}$$

Clearly $\theta = 0$ is a solution to the equation of motion, and therefore an equilibrium point of the pendulum. For small displacements away from this point, the equation takes the form

$$\ddot{\theta} + \left(\frac{g}{R} - \omega^2\right)\theta = 0\,. \tag{3.64}$$

If $\omega < \omega_c \equiv \sqrt{g/R}$ the equation has oscillating solutions with angular frequency $\Omega = \sqrt{\frac{g}{R} - \omega^2}$, and $\theta = 0$ is then a *stable* equilibrium point. If $\omega > \omega_c$ there is a sign change in the equation, which implies that the equilibrium is unstable.

However, the equilibrium condition,

$$\frac{dW}{d\theta} = 0 \quad\Rightarrow\quad \sin\theta\left(\frac{g}{R} - \omega^2\cos\theta\right) = 0\,, \tag{3.65}$$

has two sets of solutions

$$\begin{aligned} &1)\quad \sin\theta = 0 \quad\Rightarrow\quad \theta = 0, \pi\,, \\ &2)\quad \cos\theta = \frac{g}{R\omega^2} \quad\Rightarrow\quad \theta = \pm\text{Arccos}\left(\frac{g}{R\omega^2}\right) \equiv \theta_\pm\,. \end{aligned} \tag{3.66}$$

The solutions 1) include, in addition to the one already discussed ($\theta = 0$), another one where the bob is at the top point of the hoop ($\theta = \pi$). This is always a point of unstable equilibrium. The solutions 2) are only valid for $\omega > \omega_c$, due to the limitations in the values of the $cos\,\theta$ function. This coincides with the values of ω where the equilibrium at $\theta = 0$ is unstable.

The two new equilibria are stable, which can be checked by evaluating the double derivative of the potential,

$$\begin{aligned}\frac{d^2W}{d\theta^2} &= mgR\cos\theta - mR^2\omega^2(2\cos^2\theta - 1)\\ \Rightarrow \quad \frac{d^2W}{d\theta^2}(\theta_\pm) &= mgR\frac{g}{R\omega^2} - mR^2\omega^2\left(2\frac{g^2}{R^2\omega^4} - 1\right)\\ &= mR^2\omega^2\left(1 - \frac{\omega_c^4}{\omega^4}\right).\end{aligned} \tag{3.67}$$

Thus, the double derivative of W is positive when $\omega > \omega_c$, which confirms that the solutions 2) are stable equilibria.

The situation is illustrated in Fig. 3.5. Panel (a) shows the form of the potential for two values of ω, one smaller and one larger than ω_c. Panel (b) shows the splitting (*bifurcation*) of the stable equilibrium at $\theta = 0$, when ω is increased through ω_c.

We next find the Hamiltonian of the system,

$$\begin{aligned}H &= p_\theta\dot\theta - L\,, \quad p_\theta = \frac{\partial L}{\partial\dot\theta} = mR^2\dot\theta \quad \Rightarrow\\ H &= \frac{p_\theta^2}{2mR^2} + W(\theta)\\ &= \frac{p_\theta^2}{2mR^2} - mgR\cos\theta - \frac{1}{2}mR^2\omega^2\sin^2\theta\,.\end{aligned} \tag{3.68}$$

In spite of the fact that the constraints of the system are time dependent, due to the rotation of the hoop, we see that the Hamiltonian has no explicit time dependence. As we have shown previously this means that the total time derivative also vanishes,

$$\frac{dH}{dt} = \frac{\partial H}{\partial t} = 0\,. \tag{3.69}$$

Thus, H is a constant of motion and therefore the equipotential curves of H define the phase space flow of the system. We illustrate this in Fig. 3.6, with a contour plot of the Hamiltonian, similar to the ones in the previous two examples. The plot is made for $\omega = 1.5\omega_c$, and the presence of two stable equilibria near $\theta = 0$ is clearly seen in the plot. It is also clear from the plot that there are in this case three different classes of motion by the pendulum. The first one corresponds to bounded motion, represented by closed curves that encircle a single equilibrium point. The second class also corresponds to bounded motion, but now encircling both stable equilibria. The third class corresponds to unbounded motion with the pendulum making full

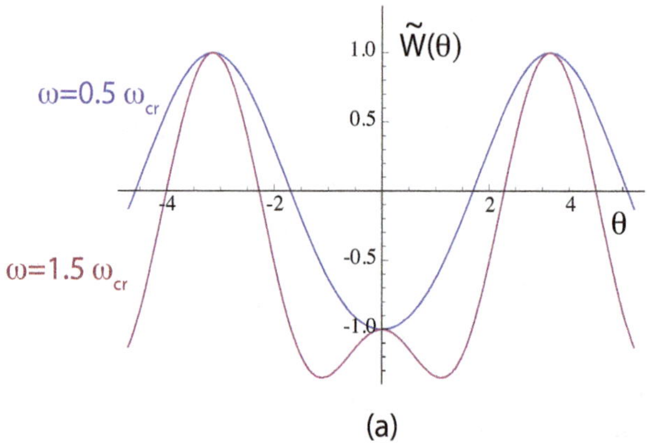

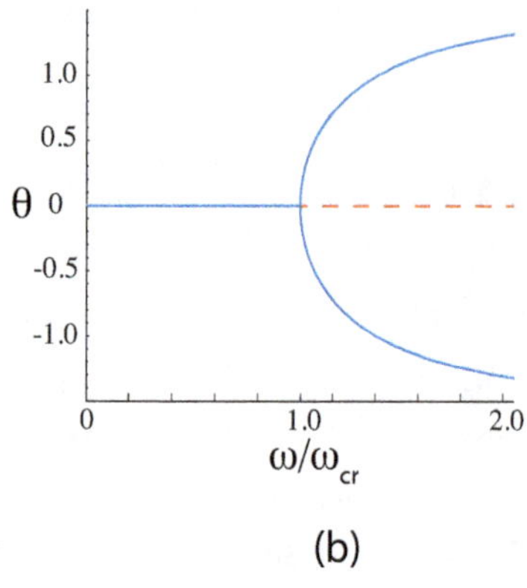

Fig. 3.5 The splitting of the stable equilibrium of the rotating pendulum. (a) Plots of the effective potential are shown for two values of the rotation frequency of the pendulum plane, $\omega = 0.5\omega_c$, which is below, and $\omega = 1.5\omega_c$, which is above the point where the stable equilibrium point splits in two. The effective potential is normalized with $\tilde{W} = W/mgR$. (b) The positions of the equilibrium points are shown as functions of the rotation frequency ω. The unbroken blue lines represent stable equilibria and the dashed red line an unstable equilibrium. The unstable equilibrium at $\theta = \pi$ is outside the interval of θ included in the plot.

rotation around the hoop. In cases where $\omega < \omega_c$ there are only two classes of motion, bounded, corresponding to oscillations about the energy minimum, and unbounded motion, corresponding to full rotations. The situation is then similar to the one shown in Fig. 3.3 for the ordinary (non-rotating) pendulum.

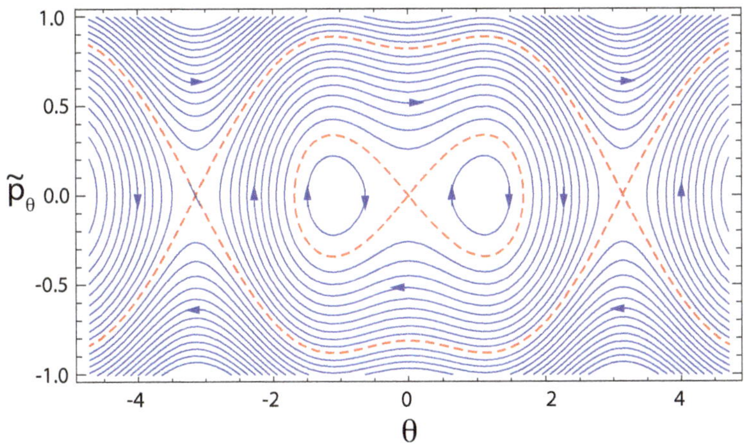

Fig. 3.6 Phase space flow for the rotating pendulum. The rotation frequency of the pendulum plane is $\omega = 1.5\omega_c$, where ω_c is the critical value, where the stable equilibrium is split in two. The separatrices, shown by the dashed red curves, separate the three different classes of motion discussed in the text. The dimensionless variable $\tilde{p}_\theta$ is proportional to the generalized momentum p_θ.

3.4 The phase space fluid

The phase space trajectories of a Hamiltonian system can be viewed as the streamlines of a (fictitious) fluid that fills phase space. Each particle of the fluid represents a specific set of initial conditions for the phase space variables, and the collection of all particles in the fluid therefore represents a (large) collection of initial conditions. These are treated simultaneously in the fluid description. If a sufficiently dense set of such initial conditions has been chosen, the time evolution of the system can be described in terms of a time dependent particle density $\rho(q,p,t)$. Hamilton's equations, which determine the trajectories of the fluid particles, give rise to a hydrodynamical equation for the fluid density. As earlier mentioned, such a picture of the phase space motion is particularly useful in statistical physics, where the system is generally described by a statistical distribution over phase space rather than by a precisely determined phase space position. The statistical distribution then corresponds to the particle density $\rho(q,p,t)$ of the fluid.

Let us introduce a unified description of the phase space coordinates as x_i, $i = 1, 2, ..., 2d$, with $2d$ as the phase space dimension. We choose odd i to correspond to the generalized coordinates and even i to correspond to the canonical momenta, so that $x_{2k-1} = q_k$ and $x_{2k} = p_k$, with (q_k, p_k) as a conjugate pair of generalized coordinates and momenta. Hamilton's equations can then be written as

$$\dot{x}_i = \sum_{j=1}^{2d} J_{ij} \frac{\partial H}{\partial x_j}, \tag{3.70}$$

with J_{ij} as the antisymmetric $2d \times 2d$ matrix

$$J = \begin{pmatrix} 0 & 1 & 0 & 0 & 0 & 0 \\ -1 & 0 & 0 & 0 & 0 & 0 \\ 0 & 0 & 0 & 1 & 0 & 0 \\ 0 & 0 & -1 & 0 & 0 & 0 \\ 0 & 0 & 0 & 0 & 0 & 1 \\ 0 & 0 & 0 & 0 & -1 & 0 \end{pmatrix}, \tag{3.71}$$

here shown with $d = 3$. We introduce next a vector notation with $\mathbf{x}$ as the $2d$-dimensional position vector with components x_i, and $\mathbf{v}$ as the corresponding velocity vector, with components $\dot{x}_i$. Hamilton's equation can be seen as defining the velocity as a vector field $\mathbf{v}(\mathbf{x})$ in phase space, which is the velocity field of the fluid particles. We note that due to the antisymmetry of the matrix J the velocity field is divergence free,

$$\nabla \cdot \mathbf{v} = \sum_i \partial_i v_i = \sum_{ij} J_{ij} \frac{\partial^2 H}{\partial x_i \partial x_j} = 0\,. \tag{3.72}$$

The number of fluid particles are conserved during the time evolution. This implies that the particle density satisfies the continuity equation

$$\frac{\partial \rho}{\partial t} + \nabla \cdot (\mathbf{v}\rho) = 0\,, \tag{3.73}$$

with the meaning that the change in density at a particular point is caused by the flow of particles to and from this point. When this equation is combined with the condition of a divergence free velocity field, we find that the time derivative of ρ, as measured in a reference frame co-moving with the fluid, vanishes,

$$\frac{d\rho}{dt} = \frac{\partial \rho}{\partial t} + \mathbf{v} \cdot \boldsymbol{\nabla} \rho = 0\,. \tag{3.74}$$

This means that the phase space fluid is incompressible. Note that this is not due to any interaction between the fluid particles, but follows from the general properties of flow patterns allowed by Hamilton's equation. There is, however, no constraint on the density of the fluid, which is determined by the chosen density function $\rho(\mathbf{x}, t)$ at some initial time t_0. The phase space fluid is rather different from normal fluids, where forces between the fluid particles are important.

The property of the phase space motion, that the fluid density ρ is constant when measured in a co-moving frame, is referred to as *Liouville's theorem.*[2]

3.5 Phase space description of non-Hamiltonian systems

So far we have linked the phase space description to the use of Hamilton's equations. However, the description of the dynamics as time evolution in phase space is not restricted to Hamiltonian systems. We may consider more generally systems, where the phase space variables satisfy a first order differential equation of the form

$$\dot{\mathbf{x}} = \mathbf{v}(\mathbf{x}, t) \,. \tag{3.75}$$

Here $\mathbf{v}(\mathbf{x}, t)$ is a fixed vector function, which determines the motion of the system, but which in general cannot be derived from a Hamiltonian. Such a system is quite generally referred to as a *dynamical system*. In this case the velocity field is not necessarily divergence free, and the continuity equation (3.73) will therefore not imply the fluid density to be conserved along the particle trajectory.

A particular class of such systems are mechanical systems with friction. Due to dissipation of energy such a system is not conservative, and therefore in general not Hamiltonian. With $\mathbf{v}(\mathbf{x})$ as a time independent function, the system will typically end up, during the time evolution, at a stable equilibrium point. Such a point is referred to as an *attractive fixed point* of the phase space flow. More generally a fixed point $\mathbf{x}_0$ is defined as a point where the velocity vanishes,

$$\mathbf{v}(\mathbf{x}_0) = 0 \,. \tag{3.76}$$

In Fig. 3.7 the flow diagram of a damped pendulum is shown. The equation of motion is

$$\ddot{\theta} + \lambda\sqrt{\frac{g}{l}}\,\dot{\theta} + \frac{g}{l}\sin\theta = 0 \,, \tag{3.77}$$

[2]Joseph Liouville (1809–1882) was a French mathematician, known for contributions in a number of fields in mathematics, but also in mathematical physics and astronomy.

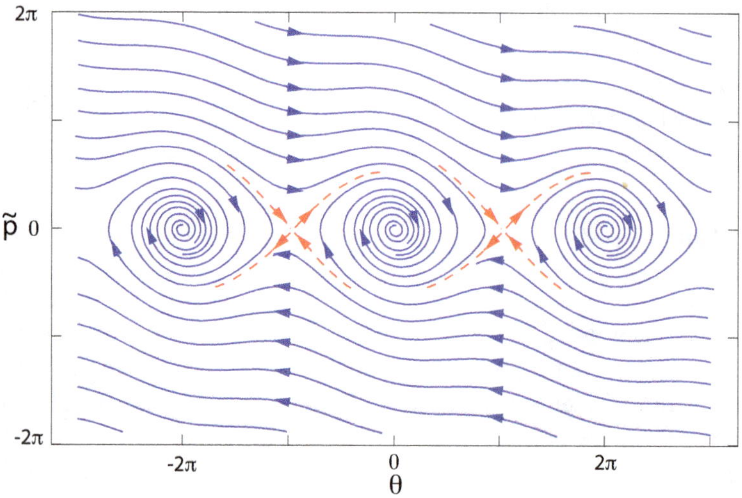

Fig. 3.7 Phase space flow of a damped pendulum. With sufficiently large initial energy the pendulum will complete a series of full rotations, corresponding to the left- and right-moving, oscillating flow lines. Eventually the damping makes the system end up at the equilibrium point. This is represented by the spiralling curves which approach one of the points with coordinates $\theta = 2\pi n$. The unstable equilibrium points are indicated by the crosses of dashed lines. They separate trajectories with different number of pendulum rotations.

with λ as a dimensionless damping parameter. Expressed in terms of the same phase space variables as used for the undamped pendulum (see Eq. (3.58)), the corresponding phase space equations are

$$\dot{\theta} = \sqrt{\frac{g}{l}}\tilde{p}\,, \quad \dot{\tilde{p}} = -\sqrt{\frac{g}{l}}(\lambda\tilde{p} + \sin\theta)\,. \tag{3.78}$$

The right-hand side of the equations determine the velocity field in phase space, which is shown in the form of stream lines in Fig. 3.7, for the parameter value $\lambda = 0.2$.

The stable equilibrium in the undamped case now is an attractive fixed point of the flow, where all trajectories will eventually end. The unstable equilibrium point (or unstable fixed point), in a similar way as in the undamped case, separates the part of the trajectories where the pendulum performs full rotations from the part where the pendulum performs oscillations. (Remember that due to the periodicity of the θ variable there is actually just one fixed point of each kind, even if in the diagram they are

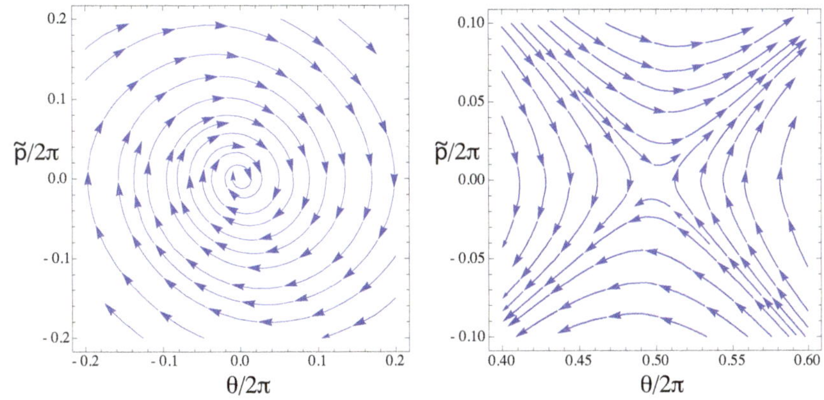

Fig. 3.8 The flow patterns around two different types of fixed points. To the left is an attractive fixed point. For the damped pendulum this corresponds to a situation with oscillations, with decreasing amplitude, about the stable equilibrium point. The diagram to the right shows a different type of fixed points, which for the damped pendulum corresponds to motion close to the unstable equilibrium point. In more general two-dimensional phase space flows other types of fixed points are also possible.

repeated with distance 2π.) The velocity fields close to the two types of fixed points are shown in Fig. 3.8.

3.6 Calculus of variations and Hamilton's principle

The motion in the configuration space of a physical system is described by the time dependent generalized coordinates $q(t)$. A specific time evolution may be determined by solving the equations of motion with initial conditions specified by the coordinates $q(t_0)$ and velocities $\dot{q}(t_0)$ for a given initial time $t = t_0$. For a d-dimensional configuration space, these $2d$ initial data uniquely specify the evolution of the system. However, the solution may be specified also in other ways, in particular by fixing the coordinates at two different times, $q(t_1)$ and $q(t_2)$. These coordinates will also give a set of $2d$ data, which specifies a unique solution, or in some cases a discrete set of solutions.

We formulate the following problem motivated by choosing the latter type of boundary conditions:

When comparing all possible paths $q = q(t)$, which satisfy the boundary conditions $q(t_1) = q_1$ and $q(t_2) = q_2$, with q_1 and q_2 as two given sets of coordinates, what characterizes the dynamical path (the one that satisfies *the equations of motion), in comparison to other continuous paths between the given endpoints?*

Hamilton gave an answer to this question in the form of a variational problem, called *Hamilton's principle.* The principle is formulated by use of the *action integral* of paths between the endpoints. The definition of the action is

$$S[q(t)] = \int_{t_1}^{t_2} L(q(t), \dot{q}(t), t)dt\,. \tag{3.79}$$

It is well defined for any continuous, differentiable path $q(t)$ between the endpoints, not only the one that satisfies the equation of motion. The action is a *functional* of the path, which means that it is a function of the function $q(t)$. Hamilton's principle refers to variations in the value of the action $S[q(t)]$ under small variations in in the path $q(t)$:

The path $q(t)$ between the fixed endpoints $q(t_1) = q_1$ and $q(t_2) = q_2$, which describes the dynamical evolution of the physical system, is characterized by the action being stationary under small variations in the path, $q(t) \to q(t) + \delta q(t)$, with $\delta q(t_1) = \delta q(t_2) = 0$. We write the condition as

$$\delta S = 0\,, \tag{3.80}$$

where the meaning of this equation is that the change in S vanishes *to first order* in the variation $\delta q(t)$, for the dynamical path $q(t)$ between the specified initial and final points.

We may say that Hamilton's principle expresses a *global* view on the evolution of the system in configuration space, with the correct, dynamical path being specified as the solution of a variational problem. Lagrange's equations, on the other hand, give a *local* condition for the dynamical evolution, in the form of a differential equation which should be satisfied at all times t during the evolution. These two ways of describing the motion of the system, although they may look rather different, are in fact equivalent, as we shall demonstrate.

In order to show this, we examine how the change in the action S for small variations of the coordinates around a given path can be expressed in terms of the Lagrangian. To first order in the variations we have

$$\begin{aligned} \delta S &= \int_{t_1}^{t_2} \delta L(q(t), \dot{q}(t), t)dt \\ &= \int_{t_1}^{t_2} \sum_k \left(\frac{\partial L}{\partial q_k}\delta q_k + \frac{\partial L}{\partial \dot{q}_k}\delta \dot{q}_k \right) dt\,. \end{aligned} \tag{3.81}$$

The integral can be manipulated in the following way,

$$\begin{aligned}
\delta S &= \int_{t_1}^{t_2} \sum_k \left[\frac{\partial L}{\partial q_k} \delta q_k + \frac{d}{dt}\left(\frac{\partial L}{\partial \dot{q}_k}\delta q_k\right) - \frac{d}{dt}\left(\frac{\partial L}{\partial \dot{q}_k}\right)\delta q_k \right] dt \\
&= \int_{t_1}^{t_2} \sum_k \left[\frac{\partial L}{\partial q_k} \delta q_k - \frac{d}{dt}\left(\frac{\partial L}{\partial \dot{q}_k}\right)\delta q_k \right] dt + \left[\sum_k \frac{\partial L}{\partial \dot{q}_k}\delta q_k \right]_{t_1}^{t_2} \\
&= \int_{t_1}^{t_2} \sum_k \left[\frac{\partial L}{\partial q_k} - \frac{d}{dt}\left(\frac{\partial L}{\partial \dot{q}_k}\right)\right] \delta q_k \, dt \,, \qquad (3.82)
\end{aligned}$$

where in the last step we have used the condition that the endpoints should be fixed during the variations, so that $\delta q(t_1) = \delta q(t_2) = 0$.

The expression we have derived for the change in the action shows that δS indeed vanishes to first order in variations of the coordinates for a path which satisfies Lagrange's equations. We note that the implication also works the other way, in the sense that if δS vanishes for *arbitrary* variations in the generalized coordinates, this implies that Lagrange's equation has to be satisfied for the path $q(t)$.

As pointed out, Hamilton's principle gives an interesting, different view on the evolution of the system. It gives a global view on the dynamical path in configuration space, and this view may add something interesting to the understanding of the evolution of the system. However, in most cases, the equations of motion, expressed in Lagrange's or Hamilton's form, will give the most convenient way to actually determine the time evolution of the system.

Variational problems are met in many fields of physics. It is interesting to note that the relation we have discussed, between the integral form of Hamilton's principle and the differential form of Lagrange's equations, may be useful for such problems more generally. Consider a problem where an integral of a physical variable (similar to $S = \int L dt$) should be stationary under variation in the physical variable (typically a minimization or maximization problem). In such a case there should be a set of differential equations (similar to Lagrange's equations) corresponding to the variational problem. This reformulation, in terms of differential equations, may be useful for solving the problem, and we shall next illustrate this by an example.

Example

Rotational surface with a minimal area

It is well-known that the surface tension of a soap film will tend to minimize the surface area of the film. We consider here a case where a soap film is formed between two circular hoops, which are placed so that the film forms a rotationally symmetric surface. The problem, to determine the form of the surface, can be regarded as a variational problem, similar to Hamilton's principle. We will here determine the shape of the surface as a solution of the corresponding Langrange's equation.

The centers of the hoops we assume to be placed on the x-axis in a symmetric way, with positions $x = \pm a$, and with the planes of the hoops both being orthogonal to the x-axis. The hoops we assume to have the same radius, which we denote by r. This set up is rotationally symmetric about the x-axis, and as a consequence the soap film, with the two hoops as boundaries, will at equilibrium form a surface with the same symmetry. Due to the rotational invariance it is sufficient to determine the function $y(x)$ which minimizes the surface area, where y is the coordinate of the surface in a direction orthogonal to the x-axis. The influence of gravity is small and will be disregarded in this problem.

The area to be minimized can be written as

$$A[y(x)] = \int_{x_1}^{x_2} 2\pi y\sqrt{1+y'^2}dx\,, \tag{3.83}$$

where we have used the notation $y' = dy/dx$. This expression is found by considering the contribution to the surface area from an infinitesimal section of the surface, with width dx in the x direction,

$$dA = 2\pi y\sqrt{dx^2+dy^2} = 2\pi y\sqrt{1+y'^2}dx\,, \tag{3.84}$$

which gives (3.83) when integrated along the x axis. The boundary conditions are

$$y(\pm a) = r\,, \tag{3.85}$$

and the variational problem can be written as

$$\delta A = 0\,. \tag{3.86}$$

This should be valid for any variation $\delta y(x)$ which vanishes at the endpoints, $\delta y(x_1) = \delta y(x_2) = 0$. The problem is of the same form as in Hamilton's principle, although with different variables and with different interpretation of the problem. To exploit the formal correspondence we write the *area functional* as

$$A[y(x)] = 2\pi \int_{x_1}^{x_2} L(y, y')dx\,, \tag{3.87}$$

with $L(y, y') = y\sqrt{1 + y'^2}$ as the function corresponding to the Lagrangian.

We note that x has taken the place of t in Hamilton's principle, and y' has taken the place of $\dot{q}$ with y as the equivalent of a generalized coordinate. (For convenience we have pulled out the constant factor 2π.) The correspondence makes it easy to write the differential equation that is equivalent to the variational problem. It has the form of Lagrange's equation,

$$\frac{d}{dx}\left(\frac{\partial L}{\partial y'}\right) - \frac{\partial L}{\partial y} = 0\,. \tag{3.88}$$

We calculate the partial derivatives,

$$\frac{\partial L}{\partial y} = \sqrt{1 + y'^2}\,, \qquad \frac{\partial L}{\partial y'} = \frac{yy'}{\sqrt{1 + y'^2}}\,, \tag{3.89}$$

and get the differential equation

$$\frac{d}{dx}\left(\frac{yy'}{\sqrt{1 + y'^2}}\right) - \sqrt{1 + y'^2} = 0\,. \tag{3.90}$$

By performing the differentiation with respect to x and simplifying the equation we get

$$yy'' - y'^2 = 1\,. \tag{3.91}$$

The equation above is a second order *non-linear* differential equation. Usually a non-linear differential equation cannot be solved by analytical methods, but in the present case it can. We will not go into details about the derivation, but simply note that the equation has the following rather simple solution,

$$y(x) = \frac{1}{k}\cosh(kx)\,, \tag{3.92}$$

with k as an unspecified constant. With this expression for $y(x)$ one readily finds that Eq. (3.91) is satisfied due to the identity $\cosh^2(kx) - \sinh^2(kx) = 1$. The expression is further symmetric under inversion $x \to -x$, consistent with the same symmetry of the boundary condition (3.85). This condition can now be written as

$$\cosh(ka) = kr\,, \tag{3.93}$$

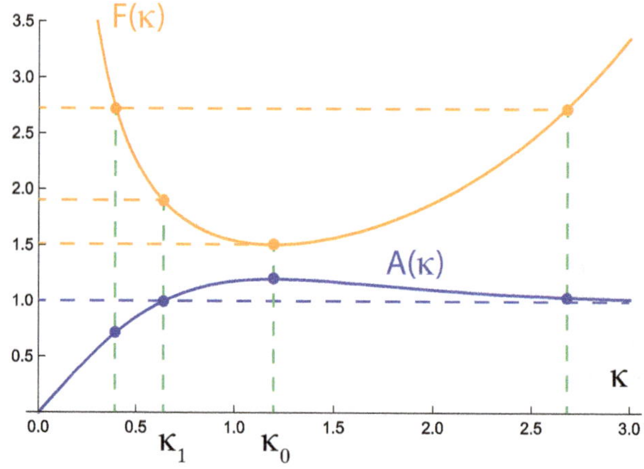

Fig. 3.9 The function $F(\kappa)$, related to the boundary condition (upper plot), and the surface area $A(\kappa)$ of the film (lower plot), both shown as functions of the dimensionless variable $\kappa = ka$. The surface area is measured in dimensionless units, by dividing with $2\pi r^2$, which is the sum of areas of the two hoops.

which can be read as an equation to determine k from the values of a and r.

There is, however, an interesting complication with regards to the boundary condition (3.93). In cases where the distance a between the two hoops is sufficiently large compared to the radius r, there is in fact no solution which satisfies the condition. To demonstrate this we introduce the dimensionless variable $\kappa = ka$ and the ratio $\alpha = r/a$. Expressed in terms of these, the boundary condition can be written as

$$F(\kappa) \equiv \frac{\cosh \kappa}{\kappa} = \alpha\,. \tag{3.94}$$

The function $F(\kappa)$ is illustrated by the upper plot in Fig. 3.9. It has a lower bound, for $F(\kappa_0) = \alpha_0$, with $\kappa_0 \approx 1.2$ and $\alpha_0 \approx 1.5$. This means that only for $a \leq r/\alpha_0 \approx 0.67\,r$ the boundary condition can be satisfied. For $\alpha = \alpha_0$ there is a single solution to Lagrange's equation (3.91), while for larger values, $\alpha > \alpha_0$, there are, as shown by the plot, two solutions for each value of α. The solutions corresponding to $\kappa < \kappa_0$ can be shown to be minimal-area solutions, while for $\kappa > \kappa_0$ the solutions are maximal-area solutions. The solutions for $\kappa > \kappa_0$ therefore describe unstable surfaces, which will spontaneously contract to a surfaces with smaller area.

The area of the surfaces can be determined from the integral (3.83), by use of the solution (3.92). This gives the following result:

$$
\begin{aligned}
A &= 4\pi \int_0^a y\sqrt{1+y'^2}dx \\
&= \frac{4\pi}{k}\int_0^a \cosh^2(kx)dx \\
&= \frac{2\pi}{k^2}(\sinh\kappa\cosh\kappa+\kappa) \\
&= 2\pi r^2\left(\tanh\kappa+\frac{\kappa}{\cosh^2\kappa}\right),
\end{aligned}
\tag{3.95}
$$

where we have exploited, at the last step, the identity $kr = \alpha\kappa = \cosh\kappa$. The surface area, regarded as a function of κ, is shown as the lower plot in Fig. 3.9. The plot shows that for all values $\kappa > \kappa_0$ where the surface is unstable, the area A is larger than $2\pi r^2$, which is the sum of the areas of the hoops. For $\kappa < \kappa_0$, the solutions are stable, but we note that in an interval $\kappa_1 < \kappa < \kappa_0$ with $\kappa_1 \approx 0.62$, the area of the surface is larger than that of the two hoops. In this interval the surface area defines a local minimum, but not the global one, which corresponds to the surface which covers separately the two hoops. (Note that this surface is not found as a solution to Eq. (3.91), since there is not any differentiable curve $y(x)$ corresponding to this divided surface.) However, for $\kappa < \kappa_1$ the solution of the variational problem defines the surface with the global minimum area.

In Fig. 3.10 the form of the minimal-area surface is indicated by the blue curves, for the case $\alpha - r/a = 1.6$. This value is slightly above the lower bound α_0. The form of the maximal-area surface, for the same value of α, is indicated by the dashed green curves. In the figure a collapsing curve, with $y \approx 0$ for all x except very close to the hoops, is also shown. The area of this surface is smaller than the area of the two other surfaces, and is very close to the minimum value $2\pi r^2$.

Based on the above discussion we have the following picture of what happens with the surface of the soap film when the distance $2a$ between the two hoops is slowly increased (with r fixed). As long as $2a \lesssim 1.3r$ it has the form similar to the minimal-area solution in Fig. 3.10. When the distance between the hoops increases beyond $2a \approx 1.3r$ the surface collapses by contracting along the symmetry axis between the two hoops, to split into surfaces which cover separately the two hoops — unless the soap film is fully destroyed in the process.

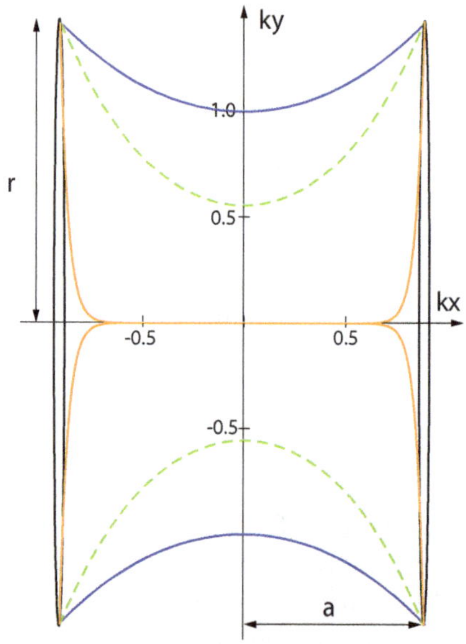

Fig. 3.10 Rotationally invariant surface with minimal area. The surface, which defines a stable equilibrium of the soap film, is represented by the blue curves. The dashed green curves represent a surface with maximal area (locally), which therefore does not correspond to a stable equilibrium. The yellow curves represent a surface which collapses into two disconnected surfaces, covering each of the two hoops.

3.7 Exercises

Problem 3.1

In Problem 2.12 the following Lagrangian was introduced

$$L = \frac{1}{2}m(\dot{r}^2 + r^2(\dot{\phi}^2 - \omega_B\dot{\phi} - \omega_0^2)) \,. \tag{3.96}$$

It describes the motion of a charged particle in a combination of a harmonic oscillator potential and a constant magnetic field, with ω_0 as the harmonic oscillator angular frequency and ω_B as the cyclotron angular frequency. The motion is restricted to a two-dimensional plane, which is perpendicular to the magnetic field, with (r, θ) as the polar coordinates of the particle. In this problem the Hamiltonian formulation of the motion of the particle is studied.

a) Find expressions for the canonical momenta p_r and p_θ corresponding to the polar coordinates, and derive the Hamiltonian of the system expressed in the chosen variables.

b) Derive the set of Hamilton's equations of the system. Show that there are two constants of motion, where the first one can be interpreted as the conserved angular momentum ℓ, and the other as the energy E. Find both of these, expressed as functions of the coordinates and conjugate momenta.

c) With ℓ fixed the energy can be expressed as a function of the radial variables (r, p_r). Make a phase-space plot of the energy in the radial variables, and give a brief description of what the diagram shows about the particle motion. (The expression to be plotted can be simplified by rescaling the physical variables to dimensionless form.)

Problem 3.2

A particle with mass m moves in a one-dimensional potential

$$V(x) = \frac{1}{4}ax^4 - \frac{1}{2}kx^2\,, \tag{3.97}$$

with a and k as positive constants, and x as the position coordinate of the particle.

a) Find the expression for the Lagrangian of the particle, and determine from this the particle's equation of motion.

b) Find the positions of the equilibrium points and determine the angular frequencies for small oscillations about the points of stable equilibrium

c) Find the expression for the Hamiltonian $H(x, p)$, and derive the corresponding set of Hamilton's equations.

d) Make a two-dimensional phase-space plot, with x and p as orthogonal coordinates, which shows equipotential curves for the function $H(x, p)$. (Simplify the expression by setting ($m = a = k = 1$.) Specify the positions of the equilibrium points, and give a qualitative description of how the different types of motion of the particle are presented in the diagram.

Problem 3.3

A particle of mass m moves in a one-dimensional periodic potential

$$V(x) = V_0(\sin x + a\sin^2 x)\,, \tag{3.98}$$

with x as the coordinate in the direction of motion, a as an external parameter that can be varied, and V_0 as a constant that measures the strength of the potential. We assume both V_0 and a to be positive.

a) Determine the equilibrium points of the potential for different values of a, and indicate which of the equilibrium points are stable and which ones are unstable. Discuss separately the cases $a < 1/2$ and $a > 1/2$.

b) Illustrate the situation by plotting the potential for the three values $a = 0,\ 1/2$ and 1. Discuss in what sense the situation changes when a increases through the value $1/2$ and relate this to the results of point a).

c) Give the expression for the Lagrangian of the particle and use Lagrange's equation to find the equation of motion of the particle.

d) Assume the particle performs small oscillations about one of the stable equilibrium points, with coordinate denoted by x_0. We write the position coordinate as $x = x_0 + \xi$ with $|\xi| << 1$. Show that this condition allows us to simplify the equation of motion so it takes the form of an harmonic oscillator equation for ξ. Determine the oscillation frequency as a function of a for $a < 1/2$ and $a > 1/2$.

e) Find the Hamiltonian $H(x, p)$ of the system, with p as the conjugate momentum of the coordinate x. Explain how the motion of the particle is represented in a two-dimensional plot of the function $H(x, p)$.

f) Make a contour plot of $H(x, p)$ for the three different situations $a = 0$, 0.5 and 1. Indicate in the diagrams the direction of motion of the particle. Discuss, based on the plots, what are the different types of motion of the particle, and indicate in the diagrams the location of the limiting curves (*separatrices*), which separate the different types of motion.

Problem 3.4

According to Fermat's Principle,[3] a light ray will follow the path between two points that makes the *optical path length* stationary. For simplicity we consider here paths constrained to a two-dimensional plane (the x, y-plane), in an optical medium with a position dependent index of refraction $n(x, y)$. The optical path length between two points (x_1, y_1) and (x_2, y_2) along $y(x)$ can be written as the integral

$$S[y(x)] = \int_{x_1}^{x_2} n(x, y)\sqrt{1 + y'^2}dx\ , \qquad y' = \frac{dy}{dx}\ . \tag{3.99}$$

[3] Pierre de Fermat (1601–1660) was a French lawyer and mathematician. He gave important contributions to the developments of analytical geometry, number theory and probability theory.

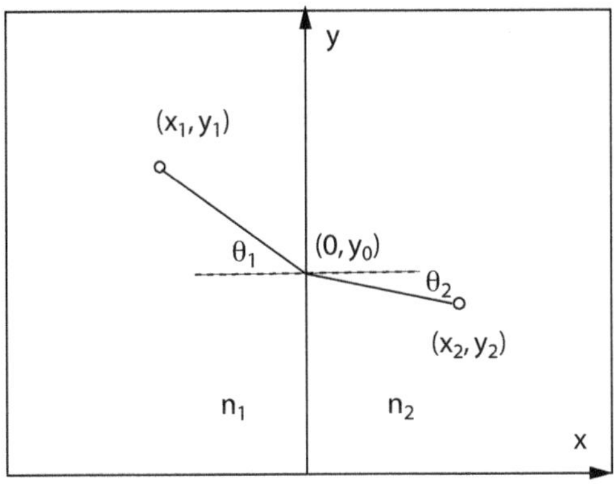

Fig. 3.11 Refraction of a light ray at the boundary between two media with different indices of refraction.

a) Find Lagrange's equation for the variational problem $\delta S = 0$, and show that if the index of refraction is constant the equation has the straight line between the two points as solution.

b) Assume the medium to have two different, constant indices of refraction, $n = n_1$ for $x < 0$ og $n = n_2$ for $x > 0$ (see Fig. 3.11). Explain why the variational problem can now be simplified to the problem of finding the coordinate $y = y_0$ for the point where the light ray crosses the boundary between the two media at $x = 0$. Find the equation for y_0 that gives the shortest optical path length. (Solving the equation is not needed.)

c) Show that the equation for y_0 implies that the path of the light ray satisfies Snell's law of refraction,[4]

$$n_1 \sin\theta_1 = n_2 \sin\theta_2 \,, \tag{3.100}$$

with θ_1 and θ_2 as the angle of the light ray relative to the normal on the two sides of the boundary.

Problem 3.5

We consider also here the use of Fermat's principle to determine the path of a light ray in an optical medium. The path is restricted to a vertical plane, with y as the vertical and x as the horizontal coordinate. The index

[4]Willebrord Snell, also called Snellius (1580–1626), was a Dutch astronomer and mathematician. He has been credited with the discovery of the law of refraction, although there is also the claim that this was earlier done by Ibn Sahl of Bagdad in the Middle Ages.

of refraction, $n(y)$, is assumed to depend only on the vertical coordinate, and the action integral therefore now takes the form

$$S[y(x)] = \int_{x_1}^{x_2} n(y)\sqrt{1+y'^2}dx\,, \qquad y' = \frac{dy}{dx}\,. \tag{3.101}$$

a) Find Lagrange's equation, which corresponds to the minimization problem for the action (3.101), and express it as a second order differential equation in the variable $y(x)$.

b) Show that the function $y(x)$, which satisfies Lagrange's equation, will also satisfy the following first order differential equation

$$\left(\frac{n(y)}{n_0}\right)^2 = 1 + \left(\frac{dy}{dx}\right)^2\,, \tag{3.102}$$

with n_0 as a constant.

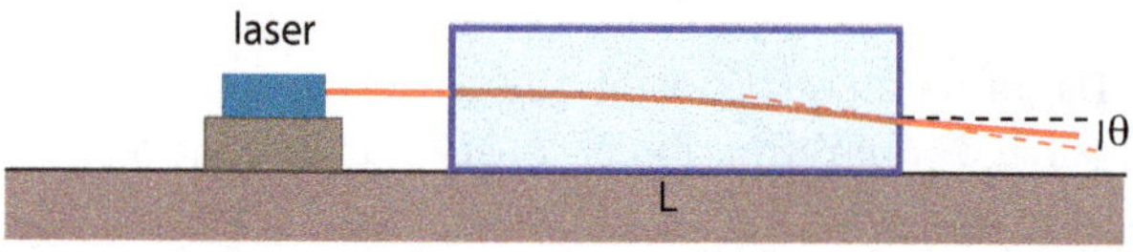

Fig. 3.12 Bending of light in a container with sugar solution. Due to variations in the strength of the solution the index of refraction decreases with height. This gives rise to a bending of the beam with θ as deflection angle.

We consider in the following the physical situation where a light ray is sent through a container with a strong sugar solution. The container has length L in the x direction. Due to the effect of gravity the strength of the solution decreases with height, and this gives rise to a variable index of refraction of the form

$$n(y) = n_0\, e^{-\alpha y}\,, \tag{3.103}$$

with n_0 and α as a constants, with $\alpha > 0$.

c) Assume a light beam is sent in the horizontal direction (x direction) into the container. To the point of entrance we give coordinates $x = y = 0$. Explain why Eq. (3.102) shows that the beam is deflected in the direction of *increasing* strength of the solution, which here means downwards.

d) Show that Eq. (3.102) is satisfied if we assume the following relation between y and x along the light path

$$e^{-\alpha y} = \frac{1}{\cos \alpha x}\,. \tag{3.104}$$

e) At the end of the container (inside the container at $x = L$) the light beam is deflected by an angle θ relative to the incoming beam. Find an expression for the deflection angle in terms of L and α.

Problem 3.6

The brachistochrone problem is a classic problem in analytical mechanics. In 1696 the problem was formulated, by Johann Bernoulli,[5] as a challenge to the mathematicians at the time.

The problem was the following:

Given two points A and B in a vertical plane, what is the curve traced out by a point acted on only by gravity, which starts at A and reaches B in the shortest time?

Five solutions were obtained from people who are well-known from the history of mathematics and physics: Newton, Jacob Bernoulli (the older brother of Johann), Leibniz and de L'Hôpital, in addition to Johann himself. Johann Bernoulli gave a formulation of the problem where he could use an analogy to Snell's law of refraction in optics to solve the problem.

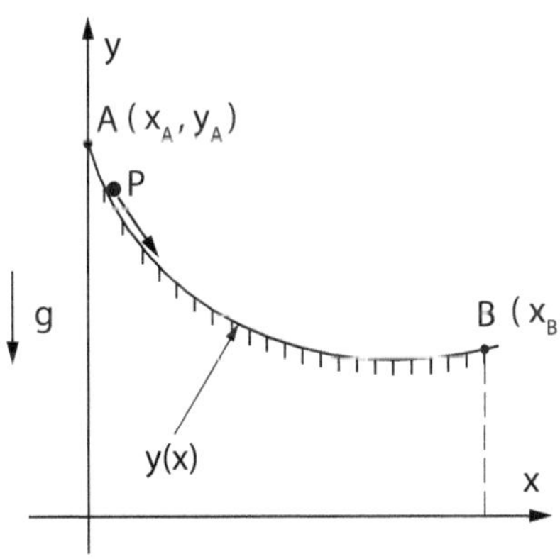

Fig. 3.13 The Brachistochrone problem.

The problem is a typical variational problem which we here reformulate in the following way: Assume a small body (P in the figure) moves in a

[5] Johann Bernoulli (1667–1748) was a Swiss mathematician. He investigated, as did his brother Jacob, the newly developed mathematical calculus, which he applied to the measurement of curves, to differential equations, and to mechanical problems.

vertical plane under the influence of gravity. It leaves a point A with zero velocity and follows (without friction) a given path in the plane which passes through a second point B, as shown in the figure. Assume the path between the two points A and B can be changed, while the points themselves stay fixed. For which path between the two points does the body spend the least time on the transit from point A to point B?

The challenge in this exercise is thus the following: Find the solution to the brachistochrone problem by using the correspondence between the variational problem (finding the "path of shortest time") and Lagrange's equation. The body P is to be treated as a point particle of mass m and the path is represented by a function $y(x)$ with x as the horizontal and y as the vertical coordinate. The boundary conditions, which fix the positions of points A and B, are specified as $y(x_A) = y_A$, $y(x_B) = y_B$. To simplify the equations, we assume that in the following $x_A = y_A = 0$. Solve the problem in a stepwise way by following the points below.

a) Show that the time T spent by the body on the way between A and B can be expressed as an integral of the form

$$T[y(x)] = \int_{x_A}^{x_B} L(y, y')dx\,, \tag{3.105}$$

with $y' = \frac{dy}{dx}$ and with

$$L(y, y') = \sqrt{\frac{1 + y'^2}{-2gy}}\,. \tag{3.106}$$

For the derivation it is convenient to set the conserved energy of the particle to zero.

b) With $L(y, y')$ interpreted as a Lagrangian (x then plays the role of t in the usual formulation) the canonically conjugate momentum is $p = \frac{\partial L}{\partial y'}$ and the Hamiltonian is $H = py' - L$. Explain why H is a constant of motion and use this fact to show that $y(x)$ satisfies a differential equation of the form

$$(1 + y'^2)y = -k^2\,, \tag{3.107}$$

with k as a constant.

c) The equation has a solution which can be written in *parametric form* as

$$\begin{aligned} x &= \frac{1}{2}k^2(\theta - \sin\theta)\,, \\ y &= \frac{1}{2}k^2(\cos\theta - 1)\,, \end{aligned} \tag{3.108}$$

where θ has been introduced as a curve parameter. Show that (3.108) is a solution of the differential equation (3.107) by changing from x to θ as a variable in the equation and by using the above expression for $y(\theta)$. In what way are the boundary conditions taken care of by this solution?

d) The curve $y(x)$ defined by the solution of the brachistochrone problem is a section of a *cycloid*, known for example as the curve traced out by a point on the periphery of a rolling wheel. Make a plot which shows the form of the curve.

e) Assume that point B lies at the lowest point of the cycloid. Show that in this case the following relation has to be satisfied, $y_B = -\frac{2}{\pi}x_B$. Calculate for this situation the time used by the body to reach point B from A, and compare this with the time used when the body instead follows a straight line between the two points.

Summary

We have in this part of the book discussed some of the basic elements of analytical mechanics. The focus has been on how to define a set of *independent*, generalized coordinates q, which describe the physical degrees of freedom of the system, and to use these in a reformulation of the equations of motion. A main motivation for introducing the generalized coordinates is to eliminate from the description the explicit reference to constraints, and thereby to the corresponding (unknown) constraint forces. Application of Newton's second law, in combination with virtual displacements of the system, then makes it possible to reformulate the dynamics in a form which refers only to time evolution of the generalized coordinates.

Two equivalent forms of dynamics are defined by Lagrange's and Hamilton's equations. Lagrange's equations form a set of differential equations, which primarily determine the motion in configuration space, $q(t)$. Hamilton's equations, on the other hand, treat the generalized coordinates q and their conjugate momenta p on equal footing, and therefore determine primarily the motion in phase space, $(q(t), p(t))$. The phase space description has the interesting property that the set of all solutions to Hamilton's equations can be viewed as forming an incompressible flow in phase space. In several simple examples, it has been shown how such flow diagrams can give a visual description of the system's dynamics.

One of the advantages of Lagrange's and Hamilton's formulations is that they specify the dynamics in a compact form through a scalar function, either the Lagrangian or the Hamiltonian. They further give explicit schemes to follow when analyzing the physical system, where only the physical degrees of freedom participate. The corresponding generalized coordinates can be chosen in many different ways, and by making a good choice, one may be able to simplify the equations and thereby the solving of the equations of motion. In particular, by exploiting the symmetries of the Lagrangian (and Hamiltonian) constants of motion of the system can be derived and used to reduce the number of independent variables of the system.

An alternative characterization of the dynamics of a Hamiltonian system is given by Hamilton's principle. This is a variational principle, which in a sense is complementary to the formulations given by Lagrange's and Hamilton's equations. Hamilton's principle selects the dynamical, time dependent path of the physical system by the condition of the action integral to be stationary. This formulation gives a global view on the time evolution of the system, which is however equivalent to the local view given by the differential equations of Lagrange and Hamilton. Hamilton's principle may be useful to study certain properties of the physical system, described by Lagrange's equations. However, for physical problems that are primarily formulated as variational problems, the correspondence to Lagrange's equations may be more important. It gives the possibility to reformulate the variational problem in terms of a set of differential equations, and in this way give a method to solve the original variational problem. Examples of this type have been given in the text and in the exercises.

In our description of the Lagrange and Hamilton formalisms we have restricted the discussion to mechanical systems described by a set of *discrete* generalized coordinates. However, these formulations can readily be generalized to classical field theory, described by *continuous* field variables, where they also play an important role. It is also interesting to note that classical physics in the formulations we have discussed are, at the formal level, closely related to the standard formulations of quantum mechanics. That is seen clearly in the fact that many of the central objects of the classical theory, like the Hamiltonian and the conjugate coordinates and momenta, are also central objects in the quantum description, although with a reinterpretation of these as Hilbert space operators.

PART 2
Relativity

Introduction

At the beginning of last century, Maxwell's equations, which give a unified description of the electromagnetic phenomena, seemed to pose a challenge to the old symmetry principle of physics, referred to as *Galilean relativity*. This symmetry principle was first formulated by Galilei, and was based on the observation that the laws of nature seemed to be the same in all *inertial reference frames*. The problem was that Maxwell's equations contain a constant with physical dimension of velocity, and the only way to make this compatible with the Galilean principle, seemed to be to assume that Maxwell's equations are valid, not generally, but only in a special inertial frame. This was thought to be the rest frame of the *luminiferous aether*, which was the name of the (imagined) physical medium in which the electromagnetic waves could propagate.

However, problems remained concerning the somewhat mysterious aether. It should fill the whole universe and it should have rather peculiar mechanical properties, but the most important problem was that there should be measurable corrections to Maxwell's equation in reference frames that moved relative to the aether. In 1887 Albert Michelson (1852–1931) and Edward Morley (1838–1923) unsuccessfully tried to find such effects experimentally. The idea was that the earth could not at all times be at rest with respect to the aether, because of its orbital motion about the sun. When the earth was moving relative to the aether, one would expect the measured speed of light to depend on the direction of the light beam. Also, daily variations as well as yearly variations were expected. However, interferometer measurements of the speed of light did not show even tiny variations in the results.

In 1905 Albert Einstein (1879–1955) offered a solution to this problem, which made the discussion about the aether completely irrelevant. He insisted on the fundamental character of Maxwell's equations and at the same time he upheld the idea of all inertial systems to be equivalent with respect to the fundamental laws of nature. His way of making this possible was to change the relations between coordinates and velocities as measured in different inertial frames. In what we know as the *special theory of relativity* he introduced a new description of space and time by assuming the *Lorentz transformations* to give the correct transformations between inertial frames. These transformations were not new, at the mathematical level they had been identified and discussed as symmetries of Maxwell's equations. George FitzGerald (1851–1901), Joseph Larmor (1857–1942), Hendrik Lorentz (1853–1928), and Henri Poincaré (1854–1912) were all involved in the study of these symmetries. Poincaré was the first to realize that the symmetries formed a mathematical group, which he named after Lorentz, but the fundamental character of the transformations was not realized until Einstein formulated his theory.

Einstein's idea was indeed revolutionary. It changed the perspective on space and time, since the transformation formula showed that space and time were not independent concepts. The idea about the larger *space-time* emerged, where a distinction between space and time is not universal, but will change from one inertial frame to another. This idea had important implications, as Einstein showed. The length contraction and time dilation of moving bodies are well-known consequences, and also the relativistic relation between mass and energy. But the impact was deeper, since the *principle of relativity* should apply to all physical laws, and all physical laws should therefore reflect, in some way, the new relation between space and time. Later, in 1915, Einstein extended his ideas further in the *general theory of relativity*, where gravitation was included in the fundamental description of space and time. In this theory the geometrical properties of space-time itself have dynamical significance, and give rise to the gravitational effects.

In this part of the book we study some of the basic elements of Einstein's special theory of relativity. Our starting point is the Lorentz transformations, which define the fundamental relations between coordinates and velocities in different inertial frames. We derive from these important kinematical relations such as length contraction and time dilation and also the relation between relativistic mass and energy. We further discuss relativistic dynamics, where the principle of relativity is used to guide us in how to bring Newton's equations into relativistic form. Our approach will be to introduce and to make use of the natural formalism for theories where space and time are treated on the same footing. This is the *four-vector formalism* where vectors in three-dimensional space are replaced with vectors in four-dimensional space-time. With the use of four-vectors (and their relatives — the *relativistic tensors*) the physical laws can be expressed in *covariant* form, a form which is explicitly invariant under transitions between inertial frames. This formalism may initially appear somewhat cumbersome, but applications show that it is useful, and if one goes deeper into relativistic theory than we do in this course it becomes indispensable. In addition to working with equations we will make extensive use of Minkowski diagrams to illustrate the space-time physics.

Chapter 4

The four-dimensional space-time

Space and time set the scene for the physical phenomena. To describe the phenomena we apply space and time coordinates, and these coordinates depend on our choice of *reference frame*. Such a reference frame we may view as a physical object, which positions and velocities are measured relative to, but in theoretical considerations we usually replace this object by an imagined frame with coordinate axes that define the origin and orientation of our reference system. A specific set of reference frames are the *inertial frames*, which we may characterize as being non-accelerated. We begin the description of the relativistic view of four-dimensional space-time by considering the coordinate transformation formulas between inertial frames, both in Galilean[1] physics and in the special theory of relativity.

4.1 Lorentz transformations

Let us for simplicity assume that all motion is restricted to one direction, which we take as the direction of the x-axis in a Cartesian coordinate system. The pre-relativistic *Galilean transformation* between two inertial frames with relative velocity v is then given by

$$x' = x - vt, \quad y' = y, \quad z' = z \quad (t' = t)\,, \tag{4.1}$$

with (x, y, z, t) as the position and time coordinates in the first inertial reference frame (S) and (x', y', z', t) as the corresponding coordinates in the second frame (S'). These are the coordinate transformations used in elementary physics (and implicitly also in every day life), and to specify that the time coordinate is the same in the two reference frames seems

[1] Galileo Galilei (1564–1642) was an Italian philosopher, scientist and mathematician. Galileo was a central figure in the scientific revolution which introduced the transition from natural philosophy to modern science.

almost unnecessary. Assume now a small body moves with velocity $u = \frac{dx}{dt}$ relative to reference frame S, and velocity $u' = \frac{dx'}{dt}$ relative to reference frame S'. The coordinate transformation (4.1) then gives us the standard velocity transformation formula

$$u' = u - v\,. \tag{4.2}$$

The transition from one inertial system to another means simply to correct the velocities by adding or subtracting the relative velocity of the two reference systems. This situation is illustrated in Fig. 4.1 with the two sets of orthogonal coordinate axes representing the inertial frames. The transformation formula clearly shows that a theory which contains a velocity as constant parameter cannot be invariant under Galilean transformations.

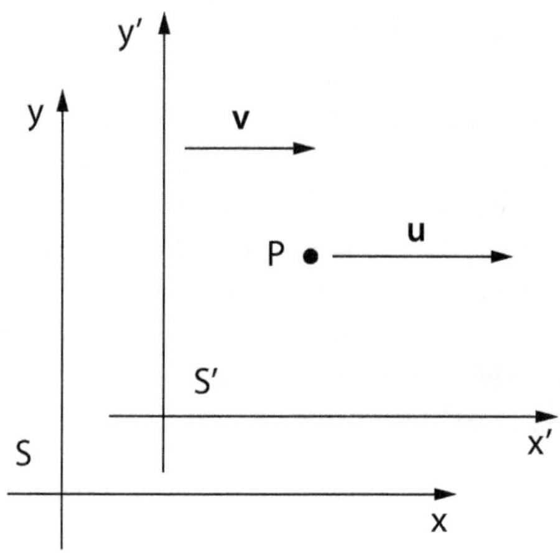

Fig. 4.1 Transition from one inertial frame S to another S', illustrated by two coordinate systems in relative motion along the x-axis. The velocity u of a particle P and and the velocity v of the reference frame S' are given relative to reference frame S. The Galilean transformation between S and S' determines the velocity u' of the particle P in S', by subtraction of the relative velocity v of the two reference frames, $u' = u - v$. In special relativity this rule for transforming velocities is no longer valid.

The Lorentz transformations, which give the correct *relativistic* formula for the transition between two inertial frames S and S' is

$$x' = \gamma(x - vt), \quad y' = y, \quad z' = z, \quad t' = \gamma\left(t - \frac{v}{c^2}x\right), \tag{4.3}$$

where γ is defined as

$$\gamma = \frac{1}{\sqrt{1 - \frac{v^2}{c^2}}}, \tag{4.4}$$

with v as the relative velocity of the two inertial frames. This transformation is not dramatically different in form from the Galilean transformation, but it is dramatically different in interpretation and in consequences.

The most prominent change in the transformation formula is that the time coordinate is no longer universal, but depends on the chosen inertial frame. It is *observer dependent.* Another important change is that the formula contains a constant c with the dimension of velocity. It has the physical interpretation as the speed of light. However, it is clear that when the relative velocity v is small compared to the speed of light c, there will be no essential difference between the Galilean and the relativistic formulas. This is seen by making an expansion in v/c

$$\gamma = 1 - \frac{v^2}{2c^2} + \dots \tag{4.5}$$

When only the leading terms in the v/c expansion are kept, the transformation equations (4.3) reduce to the Galilean equations, as one can readily check.

Let us check that an object which moves with the velocity of light will have the same velocity in all reference frames related by the Lorentz transformations (4.3). The important point is that the transformation formula for velocity now is changed. The definition of velocity is the same as before, but the Lorentz transformation between the coordinates of the two inertial frames will change the relation between u and u'. For an infinitesimal change in the position coordinates of the moving body we have

$$\begin{aligned} dx' &= \gamma(dx - vdt) = \gamma(u - v)dt\,, \\ dt' &= \gamma\left(dt - \frac{v}{c^2}dx\right) = \gamma\left(1 - \frac{uv}{c^2}\right)dt\,, \end{aligned} \tag{4.6}$$

and from this follows

$$u' = \frac{dx'}{dt'} = \frac{u - v}{1 - \frac{uv}{c^2}}\,. \tag{4.7}$$

This is the new transformation formula, which is valid when the velocity u of the object is collinear with the relative velocity v of the two inertial frames. If we now set $u = c$ in the formula it follows directly that $u' = c$. So there is no addition of the relative velocity of the two frames in this case, and the speed of light is indeed the same in all reference frames.[2]

[2] The special form of the Lorentz transformations (4.3) implies the speed of light is invariant. However, it is of interest to see that this works also the other way. Thus, if we only assume the space-time transformations to be *linear*, the assumption that the speed of light is unchanged, will imply the special form (4.3) of the transformations.

The Lorentz transformations thus imply that the speed of light is universal. It takes the same value in all inertial reference frames. This explains the classic observational results of Michelson and Morley, where the expected change of the speed of light, due to the motion of the earth, could not be detected. However, the way this problem was solved is radical, since the time coordinate now gets mixed with the space coordinates in transformations between inertial reference frames.

The relativistic velocity formula implies that if a particle moves with subluminal velocity in one inertial frame, it will move with subluminal velocity in any other inertial frame. The transformation formula does not, in a strict sense, exclude the possibility of superluminal velocity, which then, again as a consequence of the velocity formula, will have to be superluminal in all inertial frames. However, as we shall later see, this will create a problem for a causal understanding of processes where energy and momentum are exchanged over distance. The usual understanding of relativity is therefore that any signal that transports energy, or any form of information, has to propagate with a speed slower than, or equal to, the speed of light.

4.2 Rotations, boosts and the invariant distance

The Lorentz transformations (4.3) are often referred to as *boosts* or *special* Lorentz transformations. Such a transformation can be viewed as taking the first reference frame S and changing its velocity in some direction (here the x-direction) without rotating its coordinate axes, and thereby creating the new reference frame S'. The *general* Lorentz transformations are considered as transformations that include both boosts and rotations.

There is in fact a formal resemblance between rotations and the boosts. To see this we first consider a rotation in the x, y-plane, which in Cartesian coordinates takes the form,

$$\begin{aligned} x' &= \cos\phi\, x - \sin\phi\, y\,, \\ y' &= \sin\phi\, x + \cos\phi\, y\,, \end{aligned} \tag{4.8}$$

where ϕ is the rotation angle. The typical feature of the rotations is that the *distance* between two points is left invariant by the transformations. For the transformation (4.8) this invariance is expressed by

$$\Delta s'^2 \equiv \Delta x'^2 + \Delta y'^2 = \Delta x^2 + \Delta y^2 \equiv \Delta s^2\,, \tag{4.9}$$

with Δx and Δy representing the coordinate difference between two points and Δs the relative distance between the points.

For the Lorentz transformations (4.3) we introduce a new parameter χ in the following way[3]

$$\cosh\chi = \gamma\,, \quad \sinh\chi = \gamma\beta\,, \tag{4.10}$$

with β as the standard abbreviation for the dimensionless velocity $\beta = v/c$. This is a consistent parametrization, since the two expressions satisfy the requirement of hyperbolic functions,

$$\cosh^2\chi - \sinh^2\chi = \gamma^2(1-\beta^2) = 1\,. \tag{4.11}$$

The parameter χ, which is related to the relative velocity v of the two reference frames by the equation

$$v = c\tanh\chi\,, \tag{4.12}$$

is referred to as *rapidity* and is sometimes a more convenient parameter to use than the velocity. It is here introduced in order to give the Lorentz transformations a form similar to that of rotations. For the special transformation (4.3) it takes the form

$$\begin{aligned} x' &= \cosh\chi\, x - \sinh\chi\, ct\,, \\ ct' &= -\sinh\chi\, x + \cosh\chi\, ct\,. \end{aligned} \tag{4.13}$$

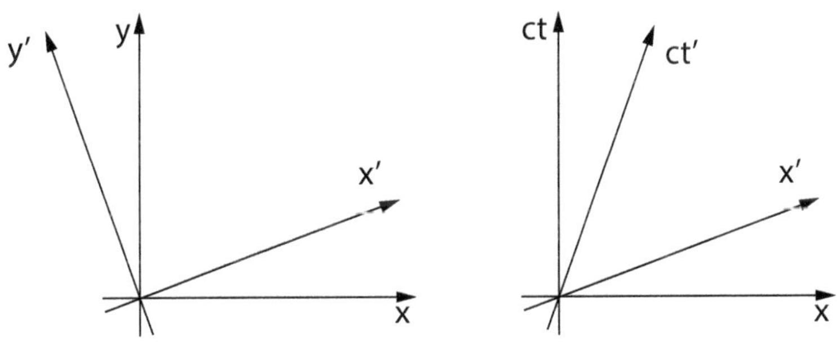

Fig. 4.2 Comparison between a rotation in the (x, y) plane and a boost in the (ct, x) plane. In the first case the rotation will transform an orthogonal coordinate frame into a rotated orthogonal frame. In the second case the orthogonal frame will not appear as orthogonal after the transformation. However, the meaning of *orthogonality* is in fact changed when time is introduced as a new *space-time* coordinate.

We note the formal similarity with the rotations (4.8), where the time coordinate ct has taken the place of the space coordinate y and the rapidity

[3]As a reminder the hyperbolic functions are defined by $\cosh\chi = \frac{1}{2}(e^{\chi} + e^{-\chi})$ and $\sinh\chi = \frac{1}{2}(e^{\chi} - e^{-\chi})$.

χ has taken the place of the angle ϕ. But χ is no angle, which is shown by the fact that the trigonometric functions are replaced by hyperbolic functions. The geometric difference between the two types of transformations are demonstrated in Fig. 4.2.

For the Lorentz transformations the distance (in three-dimensional space) between two points is no longer invariant, but another quantity, which includes also the difference in time coordinate, takes its place. Thus, with the following new definition of Δs^2, it will be invariant under the transformation (4.13) between the two inertial frames,

$$\Delta s'^2 \equiv \Delta x'^2 - c^2 \Delta t'^2 = \Delta x^2 - c^2 \Delta t^2 \equiv \Delta s^2 \,. \tag{4.14}$$

This identity follows from the properties of the hyperbolic functions. We note the important change in relative sign of the two terms, compared to that of the distance in three-dimensional space.

Distance in three-dimensional space has an immediate physical meaning as a measurable quantity that is independent of our choice of coordinate system. From a mathematical point of view it is natural to consider distance Δs as a property of space itself. It defines the *geometry* of three-dimensional space, which we then consider as equipped with a property referred to as a *metric*. The metric of three-dimensional physical space is *Euclidean*, which means that it is geometrically a flat space, with a positive measure of distance. The rotations we may regard as *symmetry transformations* of the space, which are transformations which leave the metric invariant. The Galilean transformations are time dependent extensions of these, which also leave all distances between points in three-dimensional space unchanged.

To change the fundamental transformations between inertial frames from the Galilean to the Lorentz transformations implies a change in our view of space itself. The invariant metric is no longer defined by the (Euclidean) distance between points in three-dimensional space, but rather by a generalized distance which involves also the time coordinate. The expression for the generalized distance between two points in space and time (often referred to as two *space-time events*) is given by

$$\Delta s^2 = \Delta \mathbf{r}^2 - c^2 \Delta t^2 \,. \tag{4.15}$$

This new metric, unlike the metric in three-dimensional space, does not have an immediate, physical interpretation. It can be expressed in terms of the three-dimensional distance $|\Delta \mathbf{r}|$ and the time difference $|\Delta t|$, and under certain conditions a special reference frame can be chosen where Δt vanishes and the four-dimensional distance is identical to the three-dimensional one.

But it is important to note that the metric of four-dimensional space-time, defined by the invariant (4.15), is *not* a Euclidean metric. We refer to this as a *Minkowski metric.*

The important difference between the Euclidean and Minkowski metrics is that in three-dimensional space the invariant Δs^2 is always positive, while in the four-dimensional case that is not always the case. Even so it is conventional to write the invariant as a square, Δs^2. Depending on the relative position of the two space-time points the generalized invariant (4.15) may be positive, zero or negative. If it is positive we refer to the separation of the two space-time points as being *spacelike*, if it is zero the separation is called *lightlike*, and if it is negative the separation is *timelike*. Since distance in three-dimensional space is the square root of Δs^2, this lack of positivity in four dimensions shows that the change in metric is not simply a change in the definition of distance.

The Lorentz invariance of the *line element*, $\Delta s^2 = \Delta \mathbf{r}^2 - c^2\Delta t^2$ is directly related to the fact that the speed of light is the same in all inertial frames. To see this we note that if c denotes the speed of light in a given reference frame, two space-time points on the path of a light signal through space and time will have lightlike separation,

$$\Delta s^2 = \Delta \mathbf{r}^2 - c^2\Delta t^2 = 0\,. \tag{4.16}$$

Furthermore, since Δs^2 is invariant under Lorentz transformations, if this equation is satisfied in *one* inertial frame, it will be satisfied in *all* inertial frames. This means that a signal which connects the two space-time points will travel with the same speed c in all inertial reference frames.

4.3 Relativistic four-vectors

A point in three-dimensional space can be specified by a position vector, often written as

$$\mathbf{r} = x\mathbf{i} + y\mathbf{j} + z\mathbf{k}\,, \tag{4.17}$$

with x, y and z as the Cartesian coordinates of the vector in a particularly chosen coordinate system, and with $\mathbf{i}$, $\mathbf{j}$ and $\mathbf{k}$ as the unit vectors along the orthogonal coordinate axes. These vectors define the physical three-dimensional space as a *vector space*. The vector space is defined with respect to an arbitrarily chosen reference point, corresponding to the origin $\mathbf{r} = 0$, but the position vectors $\mathbf{r}$ may otherwise be considered as being independent of any choice of coordinate system in this space. The coordinates x, y

and z, on the other hand, do depend on such a choice. This is consistent with our picture of a vector $\mathbf{r}$ in physical, three-dimensional space; it has a well-defined length and direction and can be viewed as a *geometrical object*, which exists independently of any choice of coordinate system. The coordinates are however a convenient way to characterize the vector by a set of numbers (with a physical unit), and these will then vary from one reference frame to another.

Let us write the coordinate expansion in the following way,

$$\mathbf{r} = \sum_{k=1}^{3} x_k \mathbf{e}_k \,, \tag{4.18}$$

with $\{\mathbf{e}_k, k = 1, 2, 3\}$ as a set of three orthogonal unit vectors,

$$\mathbf{e}_k \cdot \mathbf{e}_l = \delta_{kl} \,. \tag{4.19}$$

A change from one set of orthogonal vectors to another, we write as a transformation

$$\mathbf{e}_k \to \mathbf{e}'_k = \sum_{l=1}^{3} R_{kl} \mathbf{e}_l \,, \tag{4.20}$$

where orthogonality of the two sets of vectors implies that the coefficients R_{kl} satisfy the condition

$$\sum_{i=1}^{3} R_{ki} R_{li} = \delta_{kl} \,. \tag{4.21}$$

This equation gives the condition for the transformation (4.20) to be a rotation. With the vector $\mathbf{r}$ being independent of the transformation, the change of the unit vectors $\mathbf{e}_k$ has to be compensated for by a rotation of the coordinates x_k,

$$x_k \to x'_k = \sum_{l=1}^{3} R_{kl} x_l \,. \tag{4.22}$$

Due to the property (4.21) of the coefficients R_{kl} it is straight forward to check that the combined transformation of the coordinates and unit vectors leaves the vector $\mathbf{r}$ unchanged.

In a similar way as three-dimensional space is viewed as a three-dimensional vector space, space-time may be described as a four-dimensional vector space. The extension from three-dimensional space to four-dimensional space-time then leads to the extension of vectors $\mathbf{r}$ with Cartesian coordinates (x, y, z) to four-dimensional vectors with coordinates

(x, y, z, t), where t is the time coordinate. In order to have the same physical dimension for all four directions in space-time, we introduce, in the standard way, a time coordinate with dimension of length, $x^0 = ct$, where c is the speed of light.[4] Note the convention that the coordinates of space-time are written with lifted indices, so that,

$$x^0 = ct\,, \quad x^1 = x\,, \quad x^2 = y\,, \quad x^3 = z\,. \tag{4.23}$$

We shall later explain the reason for this convention.

To distinguish the 4-vectors of space-time from the 3-vectors of space, we shall in the following underline the 4-vectors. In particular, the position vector of a space-time point, when decomposed in Cartesian components, we write as

$$\underline{\mathbf{x}} = ct\underline{\boldsymbol{\tau}} + x\underline{\mathbf{i}} + y\underline{\mathbf{j}} + z\underline{\mathbf{k}}\,, \tag{4.24}$$

where we have expanded the set of three unit vectors $\mathbf{i}$, $\mathbf{j}$ and $\mathbf{k}$ with a fourth vector $\boldsymbol{\tau}$, which points in the direction of the time axis, and by underlining the unit vectors we have indicated that they are now vectors in the extended four-dimensional space-time. More often we will write the expansion in the general form

$$\underline{\mathbf{x}} = \sum_{\mu=0}^{3} x^\mu \underline{\mathbf{e}}_\mu\,, \tag{4.25}$$

with $\{\underline{\mathbf{e}}_\mu\}$ as a orthogonal set of unit vectors in four-dimensional space-time. Note that these basis vectors are written with the indices as subscript, as opposed to the coordinates where the indices are written as superscript. This is a standard convention, which means that the coordinate independent sum (4.25) appears as a sum over pairs of equal indices, where one is an upper index and the other a lower index. We shall later discuss this convention in some detail.

One important point to note is that such a set of four-dimensional unit vectors will identify uniquely an inertial reference frame. This is different from the situation with three-dimensional vectors, where a set of three orthogonal unit vectors will define the orientation of a reference frame, but not its velocity.

[4] For historical reasons the time component, in the form discussed here, is taken to be the 0th component rather than the 4th component. Originally a 4th component that was *imaginary* was introduced for time, so that $x_4 = ict$. The reason for that was to formally give boost transformations the same form as rotations. However, this convention is not so often used anymore.

A four-dimensional vector can be decomposed in its time component and its three-vector part. We often write it simply as

$$\underline{\mathbf{x}} = (x^0, \mathbf{r})\,. \tag{4.26}$$

Note, however, that this formulation is somewhat sloppy, since the four-vector $\underline{\mathbf{x}}$ is considered as being independent of the choice of a coordinate system in the four-dimensional vector space, while the decomposition (4.25) into the vector's time and space components depends on such a choice. In any case, such a decomposition, with respect to an unspecified reference frame, is often useful when making a physical interpretation of the four-vector expressions.

The space-time vector $\underline{\mathbf{x}}$, which is not linked to a specific coordinate system, we often refer to as an *abstract* vector. A concrete representation of the vector is given by its matrix representation, which is composed by its coordinates as

$$x = \begin{pmatrix} x^0 \\ x^1 \\ x^2 \\ x^3 \end{pmatrix}. \tag{4.27}$$

As opposed to $\underline{\mathbf{x}}$, this matrix does depend on the choice of reference frame, and the Lorentz transformations specify how the matrix elements change under a change of the inertial frame. In the following we shall refer to this matrix, or more generally the collection of coordinates $\{x^\mu, \mu = 0, ..., 3\}$, simply by the symbol x. It represents the set of *coordinates* of a space-time point in a particular inertial frame.

A transition between two inertial reference frames can now be viewed as a linear transformation of unit vectors and of coordinates in much the same way as transformation of three-dimensional unit vectors and coordinates given by (4.20) and (4.22). We write the relativistic transformations as

$$\begin{aligned} \underline{\mathbf{e}}_\mu \to \underline{\mathbf{e}}'_\mu &= \sum_{\nu=0}^{3} L_\mu{}^\nu \, \underline{\mathbf{e}}_\nu \\ x^\mu \to x'^\mu &= \sum_{\nu=0}^{3} L^\mu{}_\nu \, x^\nu \,. \end{aligned} \tag{4.28}$$

Again we notice the different positions of space-time indices, with the coefficient of the basis vector transformations written as $L_\mu{}^\nu$ and the coefficients of the coordinate transformation written as $L^\mu{}_\nu$. These are not identical, but closely related, as we shall later see. Here we notice that, since the space-time vector $\underline{\mathbf{x}}$ is coordinate independent, the expansion of this vector in the given basis, implies that the transformation coefficients have to satisfy the equation

$$\sum_{\mu=0}^{3} L^{\mu}{}_{\rho}\, L_{\mu}{}^{\sigma} = \delta^{\sigma}_{\rho}\,, \tag{4.29}$$

where δ^{σ}_{ρ} is the Kronecker delta written in four-vector notation. This equation is a direct generalization of the condition (4.21) satisfied by the transformation coefficients of rotations in three dimensions.

Since the transitions between inertial frames is described by Lorentz transformations, such a transformation is now identified by the set of coefficients $L^{\mu}{}_{\nu}$. It is straight forward to check that the coordinate transformation (4.3) is a special case, with coefficients given by, $L^{0}{}_{0} = L^{1}{}_{1} = \gamma, L^{0}{}_{1} = L^{1}{}_{0} = \gamma\beta, L^{2}{}_{2} = L^{3}{}_{3} = 1$, while other coefficients vanish.

4.4 Minkowski diagrams

The vector space of four-dimensional space-time, with the relativistic metric (4.15), is referred to as *Minkowski space.*[5] When discussing motion in this space, it is often useful to make a graphical representation of the space, but since we cannot make a good representation of *all* four dimensions, we usually make a restriction to the two-dimensional subspace spanned by the coordinates (x^0, x^1) or the three-dimensional subspace spanned by (x^0, x^1, x^2). Such a restricted representation may be sufficient when we consider motion in one or two (space) dimensions. The graphical representations of the subspaces are referred to as *Minkowski diagrams.* Such diagrams are especially useful in order to show the causal relations between space-time points.

In Fig. 4.3a a two-dimensional Minkowski diagram is shown, which is similar to the space-time diagram already used in Fig. 4.2, with ct and x as coordinate axes of a chosen inertial system. The coordinate axes of another inertial frame, which moves in the x-direction relative to the first one, are also shown, together with the basis vectors of the two coordinate systems. In the diagram also the lines $x = \pm ct$ are shown, which indicate space-time paths for light signals that pass through the reference point O.

[5]Hermann Minkowski (1864–1909) was a German mathematician. He worked on problems in number theory and mathematical physics and is especially known for his work on the geometry of four-dimensional space-time.

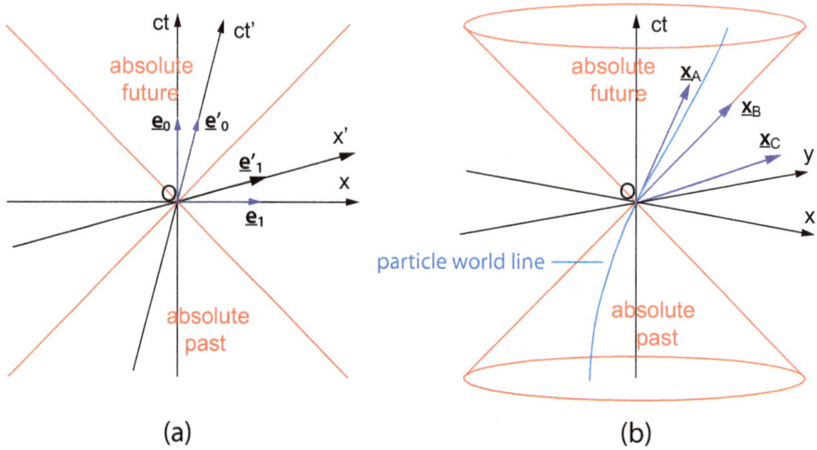

Fig. 4.3 Two-dimensional and three-dimensional Minkowski diagrams. In both diagrams the locations of the light cones relative to the point O are shown. The space-time points inside the cones are causally connected with O, while those outside are causally disconnected from O. In figure (a) the coordinate axes of two inertial frames are drawn, as well as the corresponding basis vectors. In figure (b) three different types of four-vectors are shown, with $\underline{\mathbf{x}}_A$ as a timelike vector, $\underline{\mathbf{x}}_B$ as a light like vector, and $\underline{\mathbf{x}}_C$ as a spacelike vector. In figure (b) the world line of a massive particle is also shown. It moves with subluminal velocity, which means that the four-vector velocity is timelike.

Let us first consider the information given by the direction of the coordinate axes in the diagram. The *ct* coordinate we may view as the space-time trajectory, often called the *world line*, of an (imagined) observer at rest at the origin of inertial frame S, and in the same way the ct'-axis describes the world line of an observer at rest with respect to the (moving) reference frame S'. The tilted direction of the ct'-axis simply means that the observer at rest in S' moves relative to reference frame S. However, the x'-axis is also tilted relative to the x-axis, and that is an effect which one does not see in a similar Galilean diagram. Since the x-axis describes points that are simultaneous in reference frame S, this means that the two reference frames disagree on what are simultaneous space-time events. This is one of the important predictions of relativity, that *simultaneity is not universally defined.* It is reference-frame dependent.

Let us next consider the implications of the fact that the locations of the (red) light paths in the diagram are fixed, and independent of the choice of inertial frames. The points on the lines have *lightlike* separation from the origin O. All space-time point that lie between any of these lines and the time axis, either in the upward or the downward direction, have *timelike* separation from O. Space-time points that have timelike separation from O and appear *later*, we refer to as lying in the ***absolute future*** of the point O, while points with timelike separation that appear *earlier* than O we refer to as lying in the ***absolute past***. "Absolute" here means that this ordering of events is independent of the choice of inertial reference frame.

However, for events that lie outside the light paths, either to the right or to the left, the situation is different. These are points at *spacelike* separation from the origin O. For a specific reference frame like S also these points can be characterized as being either in the past ($t < 0$) or in the future ($t > 0$), but such a characterization is now reference frame dependent. In fact for any point at spacelike separation from O there exists some inertial frames that will place this point in the past relative to the origin O and other inertial frames that will place the point in the future.

This relativity in the characterization of space-time points as being in the past or in the future may seem somewhat confusing, but is in reality not in conflict with causality, which orders events with respect to cause and effect. This is so since two points with spacelike separation are *causally disconnected* in the sense that no physical influence can propagate from one of the space-time points to the other. The speed of light sets in relativity theory an upper limit to the propagation speed of any physical signal and such a signal therefore cannot propagate between points with spacelike separation. This is a point to stress. In relativity theory there are space-time points (events) that are causally disconnected, *i.e.*, no physical signal can even in principle connect such a pair of events. In non relativistic theory this feature of space and time is absent, since there is no definite upper limit to the speed with which a signal can propagate.

In Fig. 4.3b we show a three-dimensional representation of Minkowski space. The light paths to and from the origin O now form a double cone, consisting of a *future light cone* and a *past light cone*. Space-time points inside the light cones are causally connected to O, in the sense that points inside the future light cone can be reached by a physical signal sent from O and a point inside the past light cone can reach O with a physical signal. In the diagram three four-vectors are drawn, where $\underline{\mathbf{x}}_A$ is a *timelike* vector, $\underline{\mathbf{x}}_B$ is a *lightlike* vector and $\underline{\mathbf{x}}_C$ is a *spacelike* vector. In the diagram the world line of a (massive) particle that passes through the origin is also drawn. Since its velocity at all times is lower than the speed of light this space-time curve is restricted to lie within the light cone.

In these diagrams the light cones associated with the origin O have been drawn. In reality *any* space-time point E can be associated with a past and a future light cone. These cones order the points of space-time in those that are causally connected to E and those that are causally disconnected.

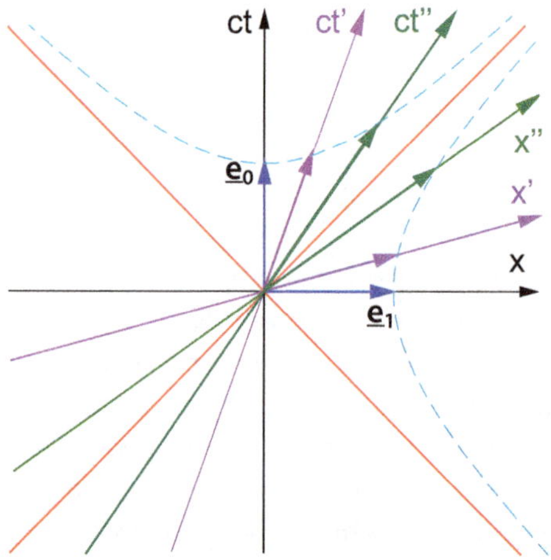

Fig. 4.4 The geometry of the two-dimensional Minkowski diagram. The coordinate axes are shown (with different colors) for three different inertial frames. The length of the unit vectors of the three frames appear as different, and also the angle between the vectors, in spite of the equivalence between the frames. This is due to the difference between the Euclidean geometry of the plane and the Minkowski geometry of space-time. Note that the set of timelike unit vectors and the set of spacelike unit vectors fall on hyperbolas in the diagram.

As mentioned above, the Minkowski diagrams are particularly well suited for showing the causal relations between space-time points. However, one should be aware of the fact that there are in other respects certain shortcomings. This has to do with the point that the Minkowski geometry of space-time is not well represented in diagrams with Euclidean geometry. This is seen quite clearly in Fig. 4.4, where the coordinate axes of three different inertial reference frames are shown. One of these seems to have a special status, since the time and space axes of this frame have orthogonal directions. That is not the case for the coordinate axes of the two other reference frames, even if we know that all the three inertial frames in reality are equivalent. The length scale along two different sets of coordinate axes are also not represented as equal in the diagram, even if they have the same length when measured in the corresponding reference frames. So one has to be aware of this, that angles and lengths are not correctly represented in the Minkowski diagrams.

Example

Minkowski diagram with spacecraft and radio signals

A spaceship passes the earth, at time $t = 0$, with the velocity $v = 0.6c$, and continues its journey with the same velocity towards a distant planet. We regard the earth and the spaceship both to define inertial reference frames, denoted S and S', respectively. The passage of the spaceship is assumed to have space and time coordinates $(0, 0)$ in both reference frames. For simplicity we disregard the (small) distance to the spaceship, and assign the coordinates $(0, 0)$ also to the earth at this point.

At time $t_A = 1$ hour a radio signal is sent from earth towards the spaceship. We denote this event as A, while the event when the signal is received is denoted B, with t_B as the time coordinate at earth for this event. A radio signal is at the same instant sent from the spaceship back to the earth, and is there received (as event C) at local time t_C.

Figure 4.5 shows the situation in a two-dimensional Minkowski diagram. The coordinates are those of the reference frame S of the earth. The space-time trajectories of the earth and the spaceship are shown, as well as the worldline of the radio signals. The light cone associated with the event where the spacecraft passes the earth is also included.

We are interested in finding the time coordinates of the three events A, B, and C, both in reference frame of the earth and of the spacecraft. In the first case the coordinates of A are given as $x_A = 0$ and $t_A = 1$ hour. The coordinates of B are related by $x_B = vt_B$, and since A and B are connected by a lightlike signal, we also have $x_B = c(t_B - t_A)$. The two equations can be used to remove x_B, and determine t_B,

$$vt_B = c(t_B - t_A) \quad \Rightarrow \quad t_B = \frac{1}{1 - \frac{v}{c}} t_A = 2.5 \text{ hours} . \tag{4.30}$$

There is clearly a symmetry in the time used by the signal sent from earth to the space ship and the signal sent back, $t_C - t_B = t_B - t_A$. This gives

$$t_C = 2t_B - t_A = \frac{1 + \frac{v}{c}}{1 - \frac{v}{c}} t_A = 4 \text{ hours} . \tag{4.31}$$

The gamma factor of space ship, as viewed from earth is

$$\gamma = \frac{1}{\sqrt{1 - \frac{v^2}{c^2}}} = \frac{1}{\sqrt{1 - 0.6^2}} = 1.25 . \tag{4.32}$$

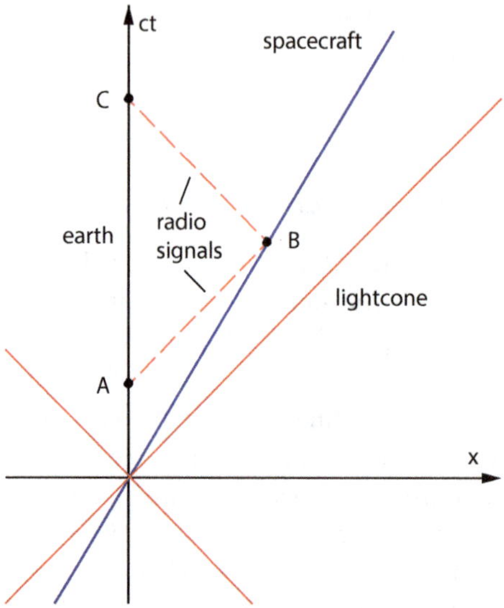

Fig. 4.5 Minkowski diagram with the worldlines of the earth and a spacecraft. Both are pictured as pointlike, with the world line of earth coinciding with the time axis of the diagram. A radio signal is shown, which is emitted from earth at the event A and received at the spacecraft at event B. Another radio signal is emitted from the spacecraft at B is received on earth at event C.

We use this next in the Lorentz transformation of the time coordinates between the reference frames. This gives for the time coordinates of the three events in the reference frame of the spacecraft

$$\begin{aligned} t'_A &= \gamma\left(t_A - \frac{v}{c^2}x_A\right) = \gamma t_A = 1.25 \text{ hours}\,, \\ t'_B &= \gamma\left(t_B - \frac{v}{c^2}x_B\right) = \gamma\left(1 - \frac{v^2}{c^2}\right) t_B = \frac{1}{\gamma} t_B = 2 \text{ hours}\,, \\ t'_C &= \gamma\left(t_C - \frac{v}{c^2}x_C\right) = \gamma t_C = 6 \text{ hours}\,. \end{aligned} \tag{4.33}$$

4.5 General Lorentz transformations

So far we have focussed on the *special* Lorentz transformations. These are the transformations that change the velocity of the inertial frame without rotating its axes. A special case of these is the boosts in the x-direction, but the velocity of a general boost can have an arbitrary direction. The special Lorentz transformations (or boosts) are therefore characterized by *three* parameters, namely the three components of the velocity vector $\mathbf{v}$

that relates the two inertial frames of the transformation. Let us denote a general transformation of this type by B (with reference to this as a *boost*).

A *general* Lorentz transformation is a transformation between inertial frames that may also include a rotation of the axes of the second reference frame with respect to the first one. Such a transformation can therefore be seen as a composite operation, first a boost and then a rotation,[6]

$$L = RB\,. \tag{4.34}$$

The Lorentz transformation L defines a linear map of the vector coordinates of the first reference frame (S) into the vector coordinates of the second reference frame (S'). We may write it as

$$x' = Lx\,, \tag{4.35}$$

where x, x' and L are matrices. Written out explicitly the matrix equation is

$$\begin{pmatrix} x'^0 \\ x'^1 \\ x'^2 \\ x'^3 \end{pmatrix} = \begin{pmatrix} L^0{}_0 & L^0{}_1 & L^0{}_2 & L^0{}_3 \\ L^1{}_0 & L^1{}_1 & L^1{}_2 & L^1{}_3 \\ L^2{}_0 & L^2{}_1 & L^2{}_2 & L^2{}_3 \\ L^3{}_0 & L^3{}_1 & L^3{}_2 & L^3{}_3 \end{pmatrix} \begin{pmatrix} x^0 \\ x^1 \\ x^2 \\ x^3 \end{pmatrix}. \tag{4.36}$$

The decomposition of the Lorentz transformation L in (4.34) can similarly be read as a matrix product of the boost matrix B and the rotation matrix R. Both these are 4x4 matrices, but the rotation matrix only mixes the space coordinates x^1, x^2 and x^3, and leaves the time coordinate x^0 unchanged.

The general Lorentz transformations, as defined above, are *homogeneous* linear transformations, which imply that the origin of the two coordinate systems are mapped into each other by the transformation. However, a transformation between inertial frames can also involve a shift of the origin. This leads to the *inhomogeneous* Lorentz transformations, which we write as

$$x' = Lx + a\,, \tag{4.37}$$

where a represents the displacement of the origin. In matrix form a is

$$a = \begin{pmatrix} a^0 \\ a^1 \\ a^2 \\ a^3 \end{pmatrix}, \tag{4.38}$$

[6]It can also be defined with the operations in opposite order, $L = B'R'$. In general B will then be different from B' and R will be different from R' since these operations do not commute.

where the four parameters define the shift of the origin in four-dimensional space-time.

The inhomogeneous Lorentz transformations depend all together on 10 parameters, 3 of these are rotation parameters, another 3 are boost parameters and finally 4 are translation parameters. In mathematical terms this set define a 10 parameter *transformation group* referred to as the *inhomogeneous Lorentz group* or the *Poincaré group*. The group property of the set implies that the successive application of two transformations will create a new transformation from the same set.[7] The homogeneous transformation define a *subgroup*, which is the 6 parameter *homogeneous Lorentz group* or simply the *Lorentz group*. The rotations form an even smaller, 3 parameter subgroup of the Lorentz group.

However, one should note that the set of boosts do not form a group, since the composition of two boosts with different directions will not be a pure boost, but will also include a rotation. This is purely relativistic effect with interesting physical consequences. A particular consequence is the *Thomas precession* effect, where a spinning particle which follows a bended path will show precession of the spin even if no force acts on the spin.

The full set of inhomogeneous Lorentz transformations define the fundamental *symmetry group* of special relativity. These symmetry transformations can in fact be interpreted in two different ways. They can be interpreted as *passive* transformations, which is the picture we use here. This means that the transformation of coordinates follows from a change of reference frame while the physical systems that are described are not changed in position or motion. When a symmetry transformation is instead interpreted as an *active* transformation this means that the change of coordinates corresponds to a physical change in the location of the processes described by the coordinates, while the reference frame is left unchanged. Such an active transformation could be to change the motion of a physical body by shifting its position, by rotating it and by changing its velocity. It is of interest to note that when working with coordinates, the formalism makes no distinction between these two types of transformation. This is a consequence of the fact that the transformations describe symmetries of the theory.

[7]The group property of the Lorentz transformations means that the composition of any two Lorentz transformations will define a new Lorentz transformation, and the inverse of a Lorentz transformation is also a Lorentz transformation. These group properties are rather obvious, with the Lorentz transformations being defined as mappings between inertial reference frames.

A common property of all the space-time transformations discussed above is that they leave invariant the line element between space-time points,

$$\Delta s^2 = \Delta \mathbf{r}^2 - c^2 \Delta t^2 \,, \tag{4.39}$$

and this was in fact, for a long time, regarded as the basic condition that defines the relativistic symmetry transformations. However, there exist some *discrete* space-time transformations, which leave the line element (4.39) unchanged without being fundamental symmetries. These are the space inversion and time reversal transformations defined by the transformation matrices,

$$P = \begin{pmatrix} 1 & 0 & 0 & 0 \\ 0 & -1 & 0 & 0 \\ 0 & 0 & -1 & 0 \\ 0 & 0 & 0 & -1 \end{pmatrix}, \quad T = \begin{pmatrix} -1 & 0 & 0 & 0 \\ 0 & 1 & 0 & 0 \\ 0 & 0 & 1 & 0 \\ 0 & 0 & 0 & 1 \end{pmatrix}. \tag{4.40}$$

Since they only change the sign of either $\Delta \mathbf{r}$ or Δt obviously Δs^2 is left unchanged. Most physical processes are in fact invariant under these transformations, but small physical effects have been experimentally detected in the physics of elementary particles, where P and T symmetries are broken. These are processes where the *weak nuclear forces* are active.

4.6 Exercises

Problem 4.1

Two inertial reference frames S and S' are moving with a relative velocity $\mathbf{v}$. The directions of the coordinate axes are chosen so that the coordinate transformation between the reference frames takes the standard form

$$x' = \gamma(x - vt)\,, \quad t' = \gamma\left(t - \frac{v}{c^2}x\right)\,, \quad y' = y\,, \quad z' = z\,, \tag{4.41}$$

with $\gamma = (1 - (v/c)^2)^{-1/2}$.

a) Equivalence between the reference frames S and S' implies that the transformation formula above should be correct if we interchange the primed and unprimed variables, and only change the sign of the relative velocity v. Show that this is indeed the case, by inverting the transformation to express x and t in terms of x' and t'.

b) A moving object is registered with velocity $\mathbf{u}$ relative to reference frame S and velocity $\mathbf{u}'$ relative to S'. With the velocity components defined in the usual way as $u_i = dx_i/dt, u'_i = dx'_i/dt'$, $i = x, y, z$, use the transformation formula (4.41) to find the corresponding transformation formula for the velocity components.

c) Assume now that the relative velocity of the two reference frames is $v = 0.5c$, and that the object moves, as measured in S', with velocity $u' = 0.8c$ in the direction with angle $\theta' = 45°$ to the x' and y' axes. What is the absolute value of the velocity u, and the angle θ to the x-axis, as measured in S? If, instead of using the Lorentz transformations, we had used Galilean transformations, what had the results been?

Problem 4.2

A boost in the x-direction will mix only the ct- and x-coordinates, and leave y and z unchanged. By leaving out y and z the boost can therefore be represented as a 2×2 matrix. We write it as

$$L = \begin{pmatrix} \gamma & -\beta\gamma \\ -\beta\gamma & \gamma \end{pmatrix} = \begin{pmatrix} \cosh\chi & -\sinh\chi \\ -\sinh\chi & \cosh\chi \end{pmatrix}, \tag{4.42}$$

where we in the last matrix have introduced the rapidity χ as velocity parameter for the boost (see Sect. 4.2).

Assume a boost L is composed of two boosts, both in the x-direction,

$$L = L_2 L_1 \,. \tag{4.43}$$

Make use of the definitions of the hyperbolic functions $\cosh\chi$ and $\sinh\chi$, to show that the rapidities, in this case, are additive,

$$\chi = \chi_1 + \chi_2 \,, \tag{4.44}$$

Compare the result with the corresponding formula for the velocities v, v_1 and v_2 (see Problem 4.1).

Problem 4.3

A thin rigid rod has rest length L_0 (length measured in its rest frame). It moves relative to an inertial reference frame S', so that the midpoint A of the rod has the time dependent coordinates $x'_A = 0, y'_A = ut', z'_A = 0$, with u as the velocity of the rod. In this reference frame the rod is at all times parallel to the x'-axis.

a) Let B be one of the endpoints of the rod. What are the time dependent coordinates of this point measured in S'?

b) The inertial frame S' moves with velocity v along the x-axis relative to another inertial frame S. (The axes of the two frames are parallel.) Find the space coordinates (x, y, z) of the points A and B as functions of the time coordinate t in the reference frame S. (Remember that if the time coordinate t in S is fixed, the time coordinate t' in S' will be different for the different space-time points A and B.)

c) Show that the rod is not oriented along the x-axis in S, by calculating ratio $\tan\phi = (y_B - y_A)/(x_B - x_A)$.

d) What is the velocity of the rod, as measured in reference frame S?

Problem 4.4

A railway carriage is moving in a straight line with constant velocity v relative to the earth. The earth is considered as an inertial reference frame S, and in this reference frame the moving carriage has the length L. The situation is shown in Fig. 4.6, where A and B indicate points on the rear wall and front wall of the carriage, respectively. C is a point in the middle of the carriage.

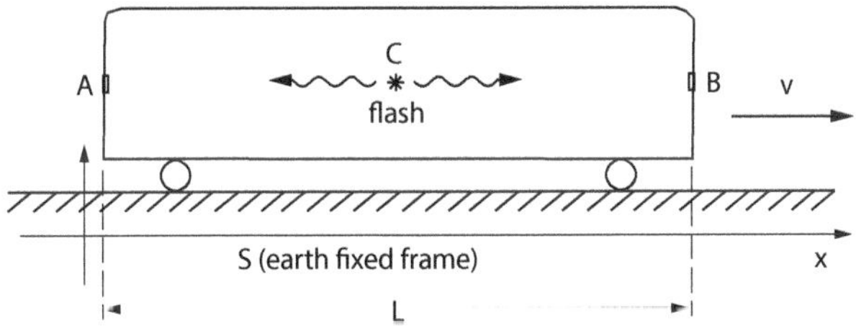

Fig. 4.6 Light paths in a moving train carriage.

a) Draw the world lines for the points A, B, and C of the carriage in a Minkowski diagram, with the orthogonal coordinate axes referring to the space and time coordinates of reference frame S. Show that the angle α between these lines and the time axis is given by $\tan\alpha = v/c$. (Choose the origin of the coordinate system in S so that A has coordinate $x = 0$ at time $t = 0$.)

At a given time t_0 a flash tube is discharged at point C. We will call this event (space-time point) E_0. Some of the light will propagate backwards in the compartment and some will propagate forwards. Let E_1 and E_2 be the events where the light signals hit the rear wall and front wall, respectively. Let us assume that the light is reflected from A and B, and that the two reflected light signals meet at a space-time point E_3.

b) Draw the world lines of the light signals as well as the four events E_0, E_1, E_2 and E_3 in the Minkowski diagram of reference frame S.

c) We introduce the co-moving reference frame S' of the carriage. Explain why E_1 and E_2 are simultaneous in this reference frame and why E_0 and E_3 fall on the same worldline C. Is this consistent with the drawing of point b)?

d) Draw the straight line from from E_1 to E_2 in the Minkowski diagram of S and show that the angle between the x-axis and this line is α.

e) Show that if a signal should connect the two space-time points E_1 and E_2 it must have the velocity c^2/v (which is greater than c).

Chapter 5

Consequences of the Lorentz transformations

The relativistic form of the fundamental space-time symmetries, expressed by the Lorentz transformations, has consequences for all physical theories at the fundamental level. Some of these we may refer to as kinematical consequences, since they are directly linked to the relativistic transformations of space and time. One of these is the *length contraction* effect, which is the effect that a body in motion appears shorter in a reference frame where the body is moving than in a reference frame where it is at rest. Another kinematical effect is the *time dilation* effect, which is the effect that time seems to run slower for a body in motion than for a body at rest. We shall discuss these effects and some further consequences of them, in particular the famous *twin paradox*, which has to do with the effect that two persons that follow different space-time paths between a common space-time point where they depart and another point where they meet again, will perceive a difference in the time spent on the journey.

5.1 Length contraction

We consider a situation where the length of a moving body is measured. For simplicity, let the body be a rod, which moves with constant velocity v relative to an inertial reference frame S. The direction of motion is along the x-axis in S, and the orientation of the rod is in the same direction. The length of the rod is L_0, when measured in a comoving rest frame S' of the rod. The situation is illustrated in Fig. 5.1.

We shall refer to the front end of the rod as A and the rear end as B. The space-time coordinates of these points in the two reference frames are

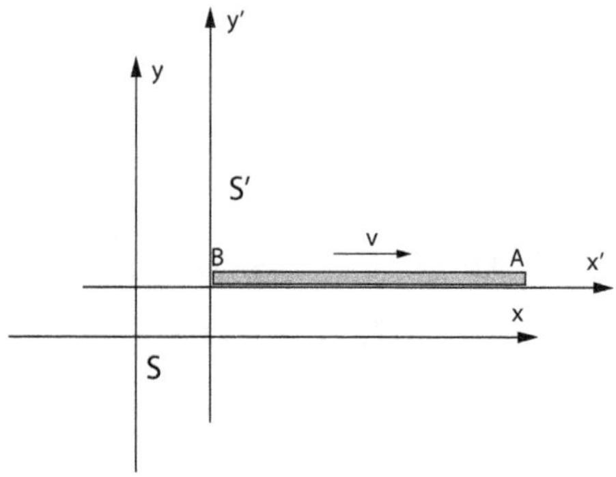

Fig. 5.1 Measurement of the length of a moving body. S' is the rest frame of the body, which has velocity v relative to the laboratory frame S. In the rest frame the measured length has its maximum value L_0, while in the lab frame it seems contracted to a shorter length $L < L_0$.

related by the Lorentz transformation

$$x'_A = \gamma(x_A - vt_A)\,, \quad t'_A = \gamma\left(t_A - \frac{v}{c^2}x_A\right)\,,$$
$$x'_B = \gamma(x_B - vt_B)\,, \quad t'_B = \gamma\left(t_B - \frac{v}{c^2}x_B\right)\,, \tag{5.1}$$

where the time coordinates of the two endpoints are independently chosen.

We note that for the measurement of length in the rest frame S' the time coordinates of the endpoints are unimportant, since the space coordinates do not change with time. The length of the rod is simply the difference between the (time independent) x-coordinates of the ends of the rod,

$$L_0 = x'_A - x'_B\,. \tag{5.2}$$

However, in S the positions of the endpoints change with time, and therefore it is meaningless to define the length as the difference in x-coordinates unless we specify for what time the positions should be determined. The natural definition is that length should be defined as the distance measured between simultaneous events on the space-time paths of the two endpoints. Note that this is how length is measured also in non-relativistic physics. If distance is measured between the positions at different times, any value could be found for the length. The important point is that in non-relativistic physics simultaneity is universally defined, whereas in relativity it is reference frame dependent. Therefore we state that

The length of a moving body measured in an inertial frame S is the space distance between the endpoints of the body measured at equal times in the same reference frame S.

This means that in order to find the correct expression for the length L in reference frame S, we should fix the time coordinates of the endpoints so that $t_A = t_B$ (rather than $t'_A = t'_B$). From the Lorentz transformation formula we then derive

$$\begin{aligned} L_0 &= x'_A - x'_B \\ &= \gamma\left[(x_A - x_B) - v(t_A - t_B)\right] \\ &= \gamma\,(x_A - x_B) \\ &= \gamma\, L\,. \end{aligned} \tag{5.3}$$

This is the length contraction formula, which we may also write as

$$L = \frac{1}{\gamma} L_0 \leq L_0\,, \tag{5.4}$$

where the last inequality follows from the fact that $\gamma = 1/\sqrt{1 - \frac{v^2}{c^2}} \geq 1$. The formula tells us that the length has its maximum when measured in the rest frame of the body. When measured in an inertial frame where the body is moving it seems length contracted in the direction of motion. In

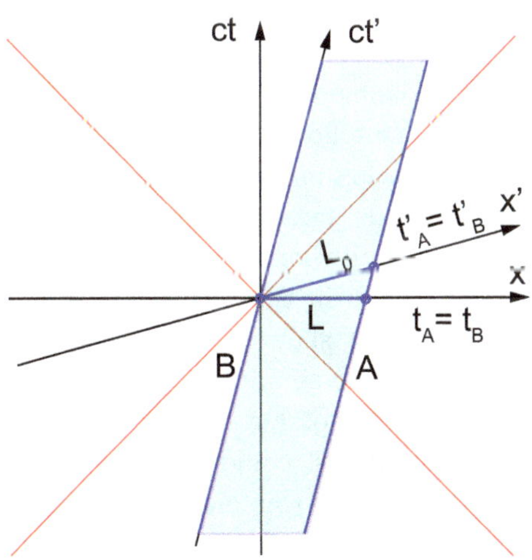

Fig. 5.2 Minkowski diagram for the length measurement. The shaded area shows the space-time trajectory of the moving rod, with A and B as the trajectories of the endpoints. Length measurement in reference frame S (with unmarked coordinates) should be performed for space-time points with $t_A = t_B$ as indicated in the figure. This is different from measurements for points with $t'_A = t'_B$, which is the natural choice in the rest frame S'.

Fig. 5.2 measurement of the length between the endpoints of the body in the two reference frames S and S' is illustrated in a Minkowski diagram.

Example

A length-contraction paradox

We consider in this example a situation where a rod of length L_0, measured in its rest frame, moves with constant, relativistic speed v relative to a box. The box has the same length L_0 as the rod, measured in its rest frame. It has two shutters, at opposite ends in the direction of the moving rod, and these are initially open so that the rod can pass through the box without colliding with it. The situation is illustrated in Fig. 5.3.

Viewed in the rest frame S of the box the rod appears length contracted, with length $L < L_0$. Due to the contraction, the rod will at a given time t be fully located inside the box. If the two shutters are rapidly closed at this instant, the rod will be trapped inside the box (see Fig. 5.3a). However, when viewed in the rest frame S' of the rod the situation looks different. Here the box is length contracted and there is no instant t' in this reference frame where the rod is fully within the box. So from this point of view it looks impossible to trap the rod inside the box (Fig. 5.3b) of the figure). We will in the following show how this apparent paradox is solved by studying the situation in the Minkowski diagrams represented in Fig. 5.4. The resolution of the paradox will also illustrate the fact that in special relativity there are limitations to the rigidity of physical bodies.

We first view the situation in the rest frame of the box, represented in the diagram to the left. It shows how the rod is trapped inside the box, when the two shutters are simultaneously closed (points a and b in the diagram). The diagram also shows what happens after the rod has been trapped. The front end first collides with the box (point c), then a deformation of the rod begins in this end, and only after a (short) time interval the deformation will affect the motion at the rear end. The light signal in the diagram indicates the earliest point at which the change in the motion of the rear end can happen. (The box is assumed to be much heavier and robust than the rod, so that it continues to be trapped inside the box after the shutters have been closed.)

The same story takes a somewhat different form when viewed from the rest frame of the rod. As illustrated in the Minkowski diagram to the right,

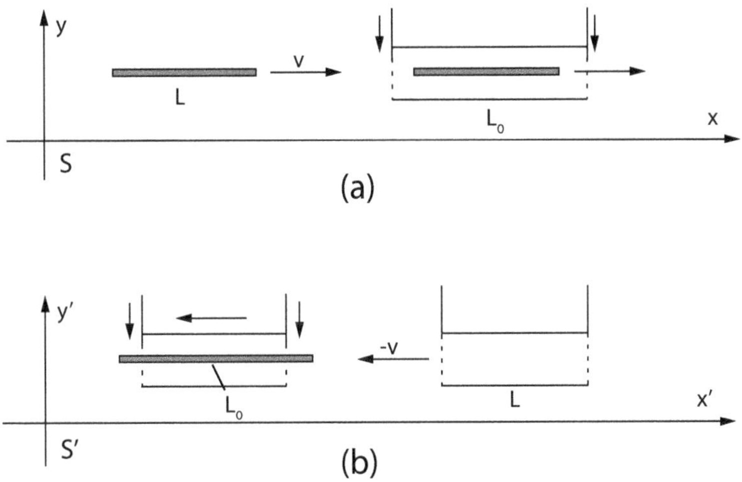

Fig. 5.3 The paradox of trapping the moving rod in the box. Panel (a) shows the system in the rest frame of the box, and Panel (b) shows the system in the rest frame of the rod.

the two shutters are then closed at different times. The shutter at the front end (point b) is now closed earlier than the shutter at the rear end (point a). The diagram also shows that the collision at the front (point c), which is simultaneous with point d in the diagram, now happens before the second shutter is closed. The deformation of the rod is therefore important for understanding what then happens. Since no immediate change in the motion, caused by the collision, can happen at the rear end of the rod, the motion at this end will continue, and bring the rod inside the box before the shutter is closed at point a. As already discussed, the relative motion of the box and this part of the rod will continue unaffected, until after the (imagined) light signal emitted from point c arrives at the end of the rod.

This shows that there is no conflict between the conclusions drawn in the two reference frames S and S'. However, in S the natural argument for the trapping of the rod inside the box is the length contraction of the rod, which brings it inside the box at the given instant, while the compression of the rod, due to the collision in front with the closed box, is a natural argument in S'. The important element from the theory of relativity, which links the two arguments, is the lack of absolute simultaneity combined with the limitation of the propagation speed of the deformation through the rod. This can be summed up as a general constraint of rigidity of physical bodies

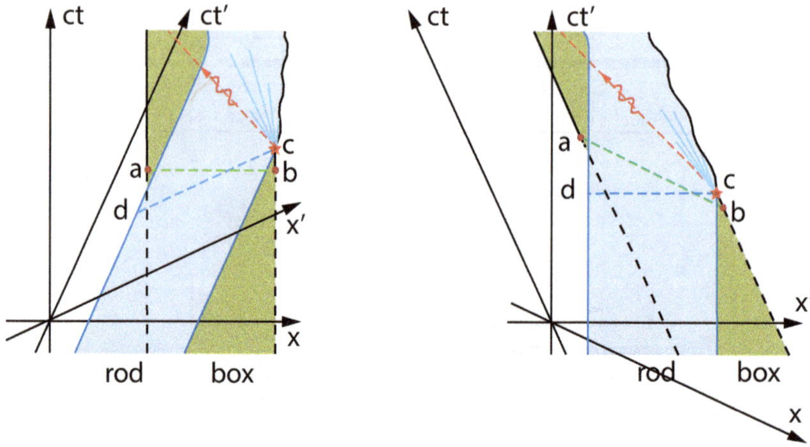

Fig. 5.4 Minkowski diagram for the trapping of the rod in the box. The figure to the left shows the process in the rest frame of the box, and the figure to the right shows the same in the rest frame of the rod. a and b label the points where the two shutters of the box are closed, c is the point of collision between the rod and the box, and d is the point at the other end of the rod which is simultaneous with c in the rest frame of the rod. The dotted red line is the world line of an imagined light signal which is sent from the collision point c at the front end towards the rear end of the rod.

in special relativity. Absolute rigidity is impossible, as a consequence of the fact that the speed of light sets an upper limit to the propagation speed of any physical influence which is transmitted from one part of the body to another.

5.2 Time dilation

Next we consider the relativistic effect that a clock in motion seems to be slower than a clock at rest. We will specify precisely how the comparison between the two clocks then is done, and also here the lack of absolute simultaneity will be important. Let us consider a situation similar to that of the previous section. An inertial reference frame S' is the rest frame of a clock (A), which is localized at the space origin ($x' = y' = z' = 0$) of S'. It measures the time coordinate t' of this reference frame. The clock and the reference frame are moving with velocity v along the x-axis relative to a second inertial reference frame S (the lab frame). The time coordinate t of S we consider to be measured by a second clock (B), which is at rest at the space origin of this reference frame. The coordinate transformation between the two reference frames is given by the same Lorentz transformation formula (5.1) as in the discussion of the length contraction effect.

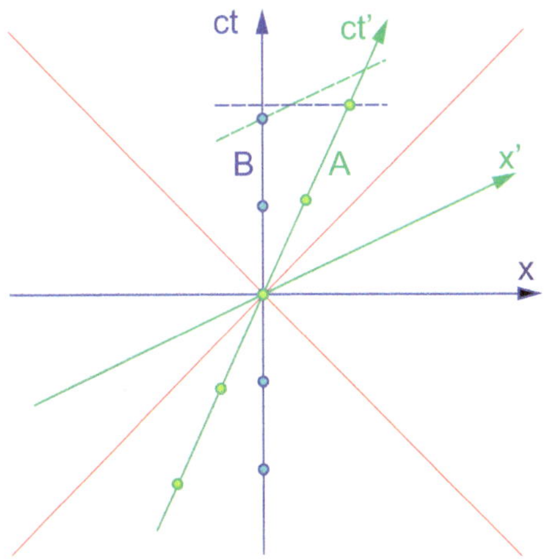

Fig. 5.5 The time dilation effect illustrated in a Minkowski diagram. The orthogonal (blue) coordinate axes define the laboratory frame S while the tilted (green) coordinate axes define a reference frame S', which moves relative to S. The green dots on the time axis denote events where a co-moving coordinate clock in S' (clock A) makes ticks at equal time intervals. Similarly the blue dots on the time axis of S denote the ticks of a clock in the lab frame (clock B). The apparent difference between the ticks in S and S' is due to the difference in how the Minkowski diagram treats the two reference frames. An observer in S finds that the ticks of the moving clock have a larger time separation than the time separation between ticks on his own clock. This is illustrated by the dashed blue line which shows that the tick of the moving clock B has larger t coordinate than the corresponding tick on clock B. Similarly an observer in S' will note that the ticks of the clock B will have larger t' coordinate than the corresponding ticks on her own clock (A). The comparison is now performed with equal times in S', as shown by the dashed green line.

The situation is illustrated in the Minkowski diagram of Fig. 5.5. The time axis of reference frame S' is the world line of the moving clock A, and the ticks of the clock at regular intervals τ are indicated in the diagram. In the same way the ticks of the clock B (also with intervals τ in its rest frame) are indicated on the time axis of S. We want to examine what the time intervals of clock A are when registered in the rest frame of clock B. Since the two clocks move relative to each other we have to make clear how this comparison is done.

Let us assume that the first tick of both clocks happen at the point of coincidence of the origins of the two reference frames: $t = t' = 0,\ \ x = x' = 0$. The second tick of the moving clock A happens at the space-time point with coordinates $(x', t') = (0, \tau)$ in S'. The corresponding coordinates of this event in S we refer to as (x, t), with $(0, t)$ as the coordinates of the simultaneous event at lab frame clock B. The time t of the second tick of the moving clock, as measured with the stationary clock, is readily found by using the inverse of the Lorentz transformation formula applied in (5.1),

$$t = \gamma\left(t' + \frac{v}{c^2}x'\right) = \gamma\,\tau \tag{5.5}$$

where $x' = 0$ at all times, at the position of the moving clock. This gives the expression

$$t = \gamma\,\tau \geq \tau\,, \tag{5.6}$$

which is the time dilation formula. Since the time t is larger than the time interval τ, the interpretation is that the moving clock (A) is slower than clock B, when viewed in the rest frame S of clock B. Note that the simultaneity of events, as viewed in reference frame S, is important in this comparison.

It is interesting to note that even if the situation, in the Minkowski diagram, may seem asymmetric between the two reference systems S and S', that is not really the case. The coordinate clock of system S' seems to be slow when viewed from reference frame S, but at the same time the coordinate clock of S seems slow when viewed from S'. The explanation for this apparent paradox is again the difference in perception of simultaneity in the two reference systems. When comparing the time difference of the two clocks in the two systems, this means in both cases making comparison between simultaneous events, but the meaning of simultaneity is different for the two, and it implies referring to different sets of points on the world lines of the two clocks. The situation is illustrated in Fig. 5.5.

The time dilation formula shows that a moving clock is running *slower* than a clock at rest, in the specific meaning discussed above. This is to be compared with the length contraction formula, which shows that the length of a body measured in the rest frame is *larger* than the length measured in any other inertial frame. These two effects are in fact closely related, and they are both related to the fact that simultaneity of events is perceived differently in reference frames that move relative to each other.

Let us now illustrate the length contraction and time dilation effects in a slightly different way. We introduce a set of coordinate clocks for each of the reference frames in the following way. With equal spacing L_0 along the x-axis of reference system S, there are clocks placed that are stationary in this system. They are *synchronized*, so they all show the coordinate time of S. This synchronization can be done by sending radio signals between the clocks. In the same way we introduce a set of (synchronized) coordinate clocks with the same spacing L_0 in reference frame S'. Since the two reference frames have a common origin for their coordinate systems, the two sets of coordinate clocks will also be synchronized. Thus, the clocks at position $x = 0$ in S and at $x' = 0$ in S' will show the same time when $t = t' = 0$.

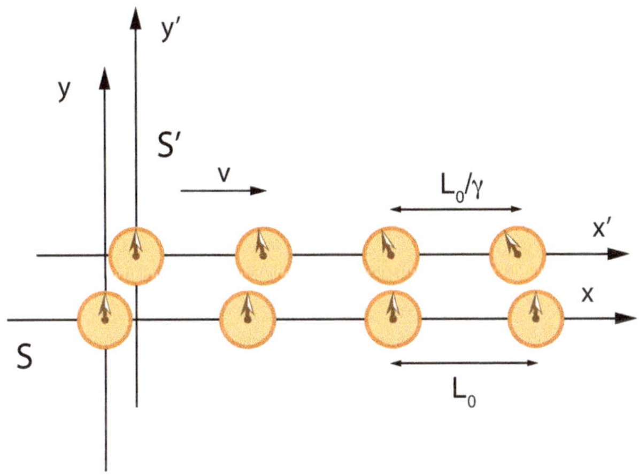

Fig. 5.6 Moving coordinate clocks. Two equivalent sets of coordinate clocks are attached to two inertial reference frames S and S' in relative motion. The situation is here registered in reference frame S, with the coordinate clocks in S all showing the same time $t = 0$. However, the coordinate clocks of S' seem, in this frame, not to be synchronized. Due to the length contraction effect they seem more densely spaced than the clocks in S, and due to the time dilation effect they seem to be running more slowly.

In Fig. 5.6 the situation is illustrated by viewing the two sets of clocks at time $t = 0$ in reference frame S. All the coordinate clocks in S show the same time $t = 0$ and are located with separation L_0. However the moving clocks (coordinate clocks of S') have a different separation L_0/γ due to the length contraction effect, and they seem to go slower due to the time dilation effect. In addition they seem not to be synchronized when viewed from S. This is demonstrated by the Lorentz transformation formula. For space-time points with $t = 0$, which are *simultaneous in S*, the coordinates in S' are

$$x' = \gamma x\,, \quad t' = -\gamma \frac{v}{c^2} x = -\frac{v}{c^2} x'\,. \tag{5.7}$$

The first equation is simply the length contraction formula. The second equation shows that the time measured on the moving clocks depends on their positions. This is again a consequence of the reference system dependence of simultaneity. The events pictured in the figure are simultaneous in S $(t = 0)$ but not in S'.

5.3 Proper time

Let us assume that a body is moving with constant velocity, and that S' is the *rest frame* of the body. By the *proper time* of the body we simply mean the coordinate time in the rest frame of the body. The time dilation effect shows that this time will be different from the coordinate time of any other inertial frame that is moving relative to the body. The definition of proper time can be generalized to moving bodies in the case where the velocity is no longer constant, as we shall now discuss.

Let us then consider a small body (a particle) with a velocity that is changing with time. As a consequence of this change in the velocity, there is no inertial reference frame which at all times is the rest frame of the particle. Even if there is no single inertial rest frame, valid for all points on the body's world line, there will be such a rest frame for any given point. This is an inertial frame that moves with the same velocity as the body *at that particular instant.* We refer to this as the *instantaneous inertial rest frame* of the body. As soon as the particle changes its velocity this inertial frame ceases to be the rest frame of the particle. The important point is that the instantaneous rest frames at different points of the world line will in general be different inertial frames.

The world line of the particle we shall consider as being divided into a sequence of small line elements. For each of these the change in velocity is negligible and the instantaneous inertial rest frame can therefore be treated as the rest frame not only at a single space-time point, but for the line element. Strictly speaking this is true only for an element of infinitesimal length, and that is what we shall consider. For such an infinitesimal element of the particle path the time dilation formula is valid, and we write it as

$$d\tau = \sqrt{1 - \frac{v^2}{c^2}}\, dt\,, \tag{5.8}$$

where $d\tau$ is the time measured in the instantaneous rest frame and dt is the time measured in an inertial frame S which we use as a fixed reference frame for the full journey of the particle.

Since the expression (5.8) is valid for any part of the particle trajectory, we can now define the proper time of this trajectory between two space-time points A and B as being identical to the integrated time

$$\tau_{AB} = \int_{t_A}^{t_B} \sqrt{1 - \frac{v(t)^2}{c^2}} dt\,. \tag{5.9}$$

The proper time is then defined as the sum (integral) of the time intervals measured in the instantaneous rest frames along the path. These do not define a single reference frame, but rather a continuous sequence of inertial frames. The variation in velocity means that the time dilation factor becomes a time dependent function.

The proper time we may consider as the time measured by an imagined clock that is fixed to the small body during its space-time journey. It should then be clear that the proper time will not depend on the choice of the reference frame S in the description of the motion. However, that is not obvious from the expression (5.9), which does seem to depend on the choice of reference frame. So, it is of interest to demonstrate more directly that proper time, as defined above, is independent of such a choice, or stated differently, that the proper time τ_{AB} is a *Lorentz invariant.*

We then focus again on an infinitesimal element of the space-time curve, and consider the corresponding Lorentz invariant line element, which we have earlier introduced. In the present case it takes the form

$$\begin{aligned} ds^2 &= d\mathbf{r}^2 - c^2 dt^2 \\ &= -c^2\left(1 - \frac{v^2}{c^2}\right) dt^2 \\ &= -c^2 d\tau^2\,. \end{aligned} \tag{5.10}$$

This shows that $d\tau^2$ is proportional to the invariant ds^2 and is therefore also a Lorentz invariant. The minus sign in the relation is explained by the fact that the world line of the particle has a timelike orientation. Since $d\tau = \sqrt{d\tau^2}$ is a Lorentz invariant quantity, the expressions for the proper time in (5.8) and (5.9) are valid in any inertial reference frame.

If we now compare the proper time for different world lines between *the same endpoints* A and B, the expression (5.9) indicates that the proper time will be *path dependent* so that the path which at average has the *largest* values of v^2 will have the shortest proper time. This is indeed a real physical effect, and it is the basis for the *twin paradox* which we shall discuss next.

Example

The twin paradox

We consider a situation where a pair of twins, who we refer to as A and B, separate for several years, with twin B leaving the earth on a spaceship, while twin A is staying behind on earth. B travels at high speed far out in the universe to visit a distant space station. After a short stay he returns to the earth, where he arrives several years after his departure. When he meets his twin sister A he realizes that she has aged more than himself. Since he is well acquainted with Einstein's theory of relativity, this does not come as a surprise. It may seem paradoxical, but he knows that twin A has been at rest with respect to the inertial rest reference frame S of the earth, while he has performed a journey with large velocities. Her proper time should therefore be longer than his own, as shown by the proper time formula (5.9). (Only in an approximate sense the earth defines an inertial frame, but since the variations in the orbital velocity of the earth are so small relative to the speed of light, he knows that it is ok to neglect this effect.)

However, there is something else that makes this situation look like a paradox. Let us assume that the velocity of the spaceship of twin B is constant and the same on the way out to the space station and on the way back, except for its direction. The time dilation factor is then the same for the two parts of the full journey. Of course, this cannot be fully correct, since there must be a period of acceleration at the beginning and at end of the journey as well as when B is close to the space station. However, we may assume these periods to be very short compared to the time spent on the rest of the journey, and therefore it seems very reasonable to assume that these short periods should only contribute with negligible corrections.

On the way out to the space station the relation between the rest frames of the two twins is symmetric, so the clock of B seems to be slow as measured with the clock of A, and *vice versa.* The time dilation factor γ is constant and it is the same whether viewed from twin A or twin B. The situation is the same on the way back from the space station, with the same value for the time dilation factor as on the way out. Based on this, twin A will find that the proper time of B is reduced by the factor γ relative to her proper time, and that is consistent with the time dilation formula (5.9). But based on the symmetry between the two twins on each of the halves of the journey and the fact that the time dilation factor is the same for the two parts, it seems that twin B could also claim that the proper time of twin A should be shorter than his. That would clearly create an inconsistency. So what is the explanation of this apparent contradiction?

In order to resolve the paradox we have to analyze the situation more carefully. First let us consider the situation from the point of view of twin A, as illustrated in Fig. 5.7a. Her own proper time is identical to the coordinate time of the inertial rest reference frame S of the earth. Let us denote her proper time for the whole journey by τ_A. The space-time path of B is assumed to be symmetric with respect to its two halves, to and from the space station, and therefore when A applies the time dilation formula to each part of the journey she obtains for the total proper time of twin B

$$\tau_B = \frac{1}{\gamma}\,\tau_A/2 + \frac{1}{\gamma}\,\tau_A/2 = \frac{1}{\gamma}\,\tau_A\,. \tag{5.11}$$

This is consistent with the proper time formula (5.9), when this is applied to the world lines of each of the two twins.

Next we consider the situation from twin B's point of view. He can also apply the time dilation formula — if he does it with some care. An important point to observe is that even if the speed of his spaceship is the same on the way out and on the way back, the inertial rest frames on the two parts of the trip *are not the same*. Let us refer to these two parts of the journey as I and II and the corresponding inertial frames as S_I and S_{II}. The main point is now to observe that when using the time dilation formula he should refer to events that are simultaneous in his own reference frame. Let us apply this to the first part of his journey, when his rest frame is S_I. The time dilation formula can be written as

$$t_A = \frac{1}{\gamma}\tau_B/2\,, \tag{5.12}$$

with t_A as the time registered on the clock on earth during the time twin B is on the way to the space station. This has the same form as the time dilation formula used by A. But note that $t_A \neq \tau_A/2$, since the event on earth that is simultaneous in S_I, with the arrival of B at the space station, does *not* correspond to the half time of the journey, as measured by A. It corresponds instead to an earlier time. A similar reasoning is applicable also for the second part for the journey. The time dilation formula (5.12) is valid also here, now with τ_B as the time measured in S_{II}. In this frame the beginning of the return journey is simultaneous with an event on earth that is *later* in time than the mid-journey event. The situation is illustrated in Fig. 5.7b.

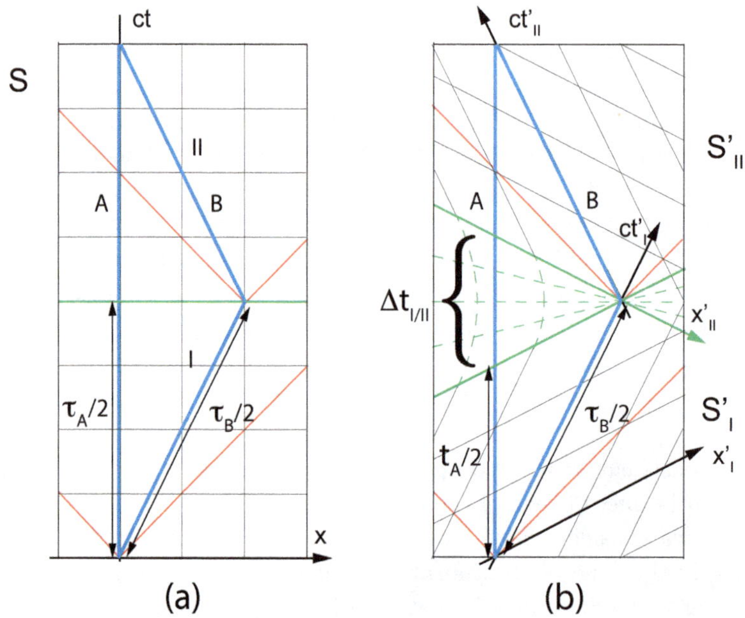

Fig. 5.7 Illustration of the Twin Paradox, with the Minkowski diagrams showing the asymmetry between the twins. The space-time journeys of the twins A and B are shown as the blue lines in two Minkowski diagrams. The world line of twin A (who remains on earth) is represented as a single straight line, while the world line of B consists of two straight lines, denoted I for the outgoing part and II for the return part of the journey. In the first diagram the coordinate lines of the rest frame S of twin A are shown, and in the second diagram the coordinate lines of the rest frames of twin B. There is a discontinuity in the coordinates of B since the rest frame of the journey out (S'_I) is different from the rest frame of the journey back (S'_{II}). The coordinate time t of twin A of the mid-journey event of B is indicated by a single green line in the first diagram, while in the second diagram the corresponding line is split in two, due to the discontinuity in the time coordinate t' of twin B. The dotted green lines indicate that in reality there is a rapid transition between the two, during the reversal of the direction of the velocity at the space station. The red lines are included in the diagrams to show the world lines of light signals emitted at the beginning of the journey and at the mid-journey event.

This means that the time dilation formula (5.12) is valid both for the travel out and the travel back, but the contributions on the left-hand side do not sum up to the total time measured on earth. The formula that relates the proper time of B, for the full journey, to the time registered on earth should therefore be written as

$$\tau_A = 2\,t_A + \Delta t_{I/II} = \frac{1}{\gamma}\tau_B + \Delta t_{I/II}\,, \tag{5.13}$$

where $\Delta t_{I/II}$ is the correction which accounts for the jump in the definition of simultaneous events when the time coordinate of B changes from reference frame S_I to S_{II}. Consistency between (5.12) and (5.13) determines this to be

$$\Delta t_{I/II} = \left(1 - \frac{1}{\gamma^2}\right)\tau_A = \frac{v^2}{c^2}\tau_A\,, \tag{5.14}$$

and a more direct calculation of the time jump based on the use of the conditions for simultaneous events in the two inertial reference frames gives the same result. The conclusion is that the situation is not symmetric with respect to describing the journey for twins A and B. Both twins may use the time dilation formula to compare the proper times of the two of them, but twin B has to be careful to add the time jump associated with the change of inertial frames. This asymmetry between the two is clearly seen when comparing the two diagrams in Fig. 5.7.

Let us finally note that if we take into account that the change between the two rest frames S_I and S_{II} of twin B in reality is not infinitely rapid, then the time jump $\Delta t_{I/II}$ will be caused by a rapid but smooth change in reference frames, beginning with S_I and ending with S_{II}. This will affect the registering performed by B of simultaneous events on earth, so that during the first and second part of the journey the clocks on earth are registered as being slower than the ones on the spaceship, but this is compensated by a very rapid speed up of the clocks on earth during the period of acceleration. The total effect is that, when correctly calculated, twin B should, like twin A, find the proper time τ_B to be shorter than the proper time τ_A between the start and endpoint of the space-time journey.

At the end, the easiest way to compare the times registered by the twins is to use the proper time formula (5.9) for the two space-time paths, since this formula does not depend on transforming between different inertial frames along the space-time trajectory.

5.4 Exercises

Problem 5.1

The length contraction and time dilation phenomena give rise to some peculiarities when applied to bodies in rotation. We consider here the situation for a rapidly rotating, circular disk of radius R, which rotates with angular velocity ω.

a) Assume a small piece of the disk has length dr in the radial direction and length $rd\theta$ in the angular direction, both lengths measured in the non-rotating lab frame, where the center of the disk is at rest. A second reference frame is introduced, which is, at a particular instant, an inertial rest frame of the small piece of the disk. Determine the lengths of the small piece, ds_r in the radial direction and ds_θ in the angular direction, both measured in the instantaneous rest frame.

b) When measured in the lab frame, the ratio between the circumference and the radius of the disk is 2π. Find the corresponding ratio when these lengths are measured on the rotating disk.

c) Clocks, which are attached to the disk, at different distances r from the center of the disk, will appear to run with different rates. Assume that at lab time $t = 0$ the clocks with angular coordinate $\theta = 0$ are all synchronized with time $t_r = 0$. At a later lab time, $t > 0$, what is the local time t_r shown by the clock at radial position r?

d) A sequence of clocks are attached to the edge of the rotating disk. We assume that there is an infinitesimal distance between neighboring clocks around the whole edge. The clocks are synchronized so that neighboring clocks show the same time for simultaneous events, with simultaneity referring to the instantaneous rest frame of the clocks.

The synchronization is performed by starting with one clock, and systematically adjusting the clocks, one by one around the disk, in the direction of the rotation. Show that there will be a jump in the time shown by the clocks when the full circle is completed, and determine this time jump? (The simplest way to find this is by first calculating the time difference in the lab frame.)

Problem 5.2

An unmanned spacecraft is sent from the earth to investigate a newly detected object which is located at a distance of $D = 1$ light month ($= 30$ light days) from the earth. By use of its very efficient engine the spacecraft is able to rapidly accelerate to a velocity v which is $3/5$ of the speed of light c. It keeps a constant velocity on the journey from the earth, and after a short stop, with investigation of the object, it returns to earth with the same speed as on the journey out from the earth. For simplicity we disregard the short periods of acceleration and consider the speed of the spacecraft to be constant, $v = 3c/5$ both on the way out and the way back.

a) What is the total time of flight of the spacecraft, measured on earth (T_e) and measured on board the spacecraft (T_s)?

b) Every hour a signal is sent from the earth station to the spacecraft, with the time interval Δ_0 between subsequent signals. Show that the time interval between the signals, when received by the spacecraft, is $\Delta_1 = 2\,\Delta_0$, on the way from the earth. What is the time interval Δ_2 between the received signals, when the spacecraft is on the way back to earth? (In both cases we refer to the time measured onboard the spacecraft.)

c) Draw a Minkowski diagram, as appearing in the rest frame of the earth, which shows the world lines of the earth and of the spacecraft, and also the sequence of signals sent from earth. (Include a reasonable number of signals by making a convenient choice for the time interval in the diagram.)

d) Signals are also regularly sent from the spacecraft to the earth, at intervals of one hour, measured on the spacecraft. Explain why, during the first part of the spacecraft's journey the time interval between the signals received on earth is identical to Δ_1, and during the second part it is identical to Δ_2. At what time T_t does the length of the time interval change, and for what reason?

e) Draw a similar Minkowski diagram as in c), but now showing the world lines of the signals sent from the spacecraft back to earth. Indicate where the time T_t is on the time axis.

Problem 5.3

A particle is circulating with constant speed in an accelerator ring of radius $R = 10\,\text{m}$. The speed of the particle corresponds to a relativistic gamma factor $\gamma = 100$.

The laboratory frame S is the rest frame of the accelerator ring, and we assume the ring to lie in the x, y-plane of a Cartesian coordinate system, with the center of the ring at the origin. Since the particle is restricted to move in this plane we simply suppress, in the following, the z-coordinate, and treat space-time as three-dimensional, with coordinates (ct, x, y).

a) Determine the velocity v/c of the particle in the lab frame S, relative to the speed of light. What is the period T of circulation and the angular frequency ω?

Explain what is meant by the *proper time* τ of the particle and relate it to the coordinate time t of the reference frame S? What is the period of circulation T_τ measured in proper time?

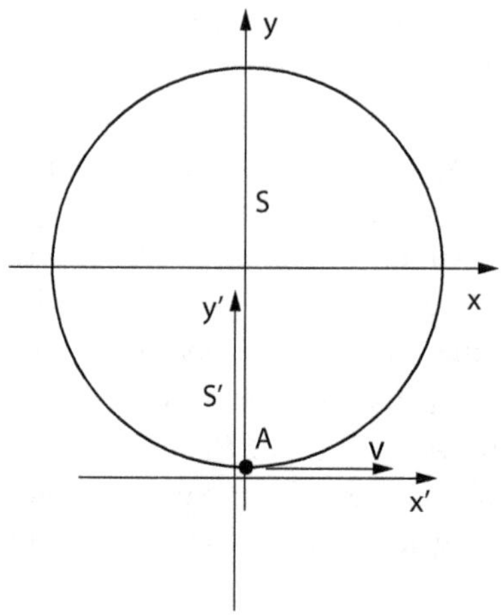

Fig. 5.8 The path of particle, which circulates in an accelerator ring. The coordinates of the lab frame S are shown, and also the coordinates of the instantaneous inertial frame S' of the particle, when it passes a given point A of the ring.

We assume that the coordinate time t and the proper time τ are defined with $t = \tau = 0$ at the instant where the particle has coordinates $(x, y) = (0, -R)$, corresponding to the point A in the figure.

A second inertial frame S' is introduced, which is the instantaneous inertial rest frame of the particle at time $t = 0$. The coordinate axes of S and S' are parallel and the particle is at the origin of Cartesian coordinate system of S' at the instant $t' = 0$ (corresponding to the point A in Fig. 5.8).

b) Explain what is meant by the instantaneous inertial rest frame, and give the transformation between the Cartesian coordinates of the two inertial frames S and S'.

c) At time $t' = 0$ the accelerator ring defines a deformed circle in S'. Based on the transformation formula between S and S', show that it is an ellipse and determine the lengths of the long and short axes. Is the result consistent with the relativistic length contraction formula?

d) Find the coordinates $x'^{\mu}(\tau)$, $\mu = 0, 1, 2$ of the particle's world line in reference system S'? Express the coordinates as functions of proper time τ, radius R and angular frequency ω.

Make a graphical representation of the trajectory in the x', y'-plane. (Use different scales in the two directions if needed.)

Problem 5.4

In a particular inertial reference frame S the coordinates of a spaceship are given as

$$t = \frac{c}{a_0} \sinh\left(\frac{a_0}{c}\tau\right), \quad x = \frac{c^2}{a_0} \cosh\left(\frac{a_0}{c}\tau\right), \quad y = z = 0\,, \qquad (5.15)$$

with a_0 as a constant and τ as a time parameter. Since the motion is linear, in the x-direction, we leave out in the following the coordinates y and z.

a) Show that the time parameter τ in Eq. (5.15) is the proper time of the spaceship.

b) A space station has coordinates $x = c^2/a_0 \equiv d,\ y = z = 0$ and is at rest in reference frame S. It sends radio messages to the spaceship at regular intervals $t_n, n = 0, 1, 2, ...$ Due to the acceleration of the spaceship the messages will be received there with increasing time difference. Show that only radio signals that are sent before a certain time, $t_n < t_{max}$, will be received in the spaceship as long as its motion is given by (5.15). What is this limit time t_{max}?

c) Draw the Minkowski diagram of reference frame S with x and ct as coordinate axes, and plot the space-time curve (*world line*) of the spaceship. Plot also the world line of the space station, with the space-time paths of radio signals sent from the space station, as well as the (limit) path of a signal sent at time t_{max}.

Chapter 6

Four-vector formalism and covariant equations

In this chapter we discuss in a more systematic way the use of four-vectors, and in particular how to give physical equations a *covariant* form. In the covariant formulation all physical variables are expressed in terms of four-vectors and related objects, called (relativistic) tensors, and this formulation secures that the equations are valid in any inertial reference frame. We discuss how tensors are defined and what are their transformation properties under Lorentz transformations.

6.1 Notations and conventions

Einstein's summation convention

When using the four-vector notation some conventions are commonly used, and we shall make use of them also here. For example when a vector index is running over all the four values taken by the *space-time* coordinates, we label the index by a greek letter, while the use of a latin letter instead will normally indicate a restriction to the three values taken by the *space* components. For example when we write x^μ, μ is allowed to take values from 0 to 3. If, however, we write x_i the index runs instead from 1 to 3.

Another convention we shall apply is Einstein's summation convention. Thus, a repeated space-time index in an expression normally means that one should sum over the index. As an example we write for the decomposition of a four-vector $\underline{\mathbf{x}}$ on an orthogonal set of basis vectors,

$$\underline{\mathbf{x}} = x^\mu \underline{\mathbf{e}}_\mu \,, \tag{6.1}$$

where the summation symbol simply is omitted. The repeated index tells us that we should sum over μ, and since it is a greek letter we know that the summation is from 0 to 3. If we at some stage should meet a case where a repeated index should *not* be taken as a summation index, we simply state that explicitly.

In the four-vector notation it is also important to place correctly the index, either up or down, while a similar distinction is not important for vectors in three-dimensional space. We shall soon have a closer look at this distinction. The consistent use of four-vectors (and tensors) we refer to as *covariant notation*, and we note as a particular rule that in the covariant notation we only sum over pairs of indices, where one is an upper index and the other a lower index. This summation is commonly referred to as a *contraction*.

Note that Einstein's summation convention is often used also in non-relativistic equations, for example in scalar and vector products, written as $\mathbf{a} \cdot \mathbf{b} = a_i b_i$ and $(\mathbf{a} \times \mathbf{b}) \cdot \mathbf{c} = \epsilon_{ijk} a_i b_j c_k$.

Metric tensor

Physical three-dimensional space is in non-relativistic physics considered to be equipped with a Euclidean metric, defined by the invariant *distance* between two neighboring points. The squared distance, which in Cartesian coordinates is

$$ds^2 = dx^2 + dy^2 + dy^2 = d\mathbf{r}^2 \,, \tag{6.2}$$

is an invariant under three-dimensional rotations. As already discussed, the four-dimensional space-time of special relativity has a different metric, called Minkowski metric. It is defined by the Lorentz invariant line element

$$ds^2 = dx^2 + dy^2 + dy^2 - c^2 dt^2 \equiv d\underline{\mathbf{x}}^2 \,. \tag{6.3}$$

When written in this way we have to remember that $d\underline{\mathbf{x}}^2$ does not have to be positive. It is positive for spacelike vectors, zero for lightlike vectors, and negative for timelike vectors.

We may write the invariant line element in the following form,

$$ds^2 = g_{\mu\nu} dx^\mu dx^\nu \,, \tag{6.4}$$

where $g_{\mu\nu}$ is referred to as the *metric tensor*. (Note that in (6.4) Einstein's summation convention has been used.) The metric tensor can be thought of as defining a 4×4 symmetric matrix, which in Cartesian coordinates is a diagonal matrix of the form

$$g = (g_{\mu\nu}) = \begin{pmatrix} -1 & 0 & 0 & 0 \\ 0 & 1 & 0 & 0 \\ 0 & 0 & 1 & 0 \\ 0 & 0 & 0 & 1 \end{pmatrix} . \tag{6.5}$$

From the decomposition of the vector $d\underline{\mathbf{x}} = dx^\mu \underline{\mathbf{e}}_\mu$ and from the writing of the invariant line element as a generalized scalar product, it follows that the basis vectors satisfy a generalized orthogonality condition,

$$\underline{\mathbf{e}}_\mu \cdot \underline{\mathbf{e}}_\nu = g_{\mu\nu} \,. \tag{6.6}$$

This means that the vectors are orthogonal and space vectors have a standard normalization $\underline{\mathbf{e}}_k^2 = 1$, $k = 1, 2, 3$, while that time vector has the normalization $\underline{\mathbf{e}}_0^2 = -1$. The last one is negative since the basis vector $\underline{\mathbf{e}}_0$ is timelike.

Upper and lower indices

We have already stressed the convention that the coordinates of a four-vector $\underline{\mathbf{x}}$ are written with upper indices, as x^μ. However also coordinates with lower indices may be defined. The precise definition is

$$x_\mu = g_{\mu\nu} x^\nu \,. \tag{6.7}$$

Thus a four-vector can be associated with two sets of coordinates, those with upper indices which are the standard ones (referred to as *contravariant* components) and those with lower indices (referred to as *covariant* components). The metric tensor acts as a lowering operator on the indices. This gives the simple relations

$$x_0 = -x^0, x_1 = x^1, x_2 = x^2, x_3 = x^3 \,. \tag{6.8}$$

Note that the only change introduced by lowering the indices is that the sign of the 0'th component is reversed.

Initially it may seem cumbersome to operate with two sets of coordinates for a four-vector, which are even so closely related. However, if one is careful to place the indices correctly, the relativistic equations can be simplified, and if the positions of the indices are consistently used in a relativistic equation, one will gain a guarantee that it keeps the form unchanged when transforming from one reference frame to another.

We note, as a special case, that the invariant line element can now be written without the metric tensor as

$$ds^2 = dx_\mu dx^\mu \,. \tag{6.9}$$

More generally, summation over a pair of four-vector indices, one lower and one upper will produce a Lorentz invariant quantity.

The metric tensor acts as a lowering operator on the vector indices. Clearly there must be an inverse to this which acts as a raising operator. We write it as

$$x^\mu = g^{\mu\nu} x_\nu \,. \tag{6.10}$$

Since it is the inverse to $g_{\mu\nu}$ we have the relation

$$g^{\mu\rho} g_{\rho\nu} = \delta^{\mu}_{\nu} \,. \tag{6.11}$$

Note that the relativistic form of the Kronecker delta is written with one upper and one lower index. This is to have the indices of the two sides of the equation consistently placed. Also note that conventionally it is written without specifying which of the indices is the left/right index. This is because the Kronecker delta is symmetric, in the sense $\delta_{\mu}{}^{\nu} = \delta^{\nu}{}_{\mu}$, as follows from the symmetry of the metric tensor.

We note from the matrix form of $g_{\mu\nu}$ that the square of the matrix is identical to the identity matrix. This means that the matrix is its own inverse and therefore $g_{\mu\nu}$ and $g^{\mu\nu}$ represent the same 4×4 matrix. Nevertheless, we insist on writing this matrix with lower indices when it is used as a lowering operator of vector indices in an equation and with upper indices when it is used as a raising operator. This is to be able to place consistently all vector indices in the relativistic equations.[1]

Metric signature

In the definition of the metric tensor given above, a particular sign convention has been used. Thus, the *metric signature*, which describes the signs of the eigenvalues of g, have been chosen as $(-,+,+,+)$. This is a choice which has often been made in textbooks on (general) relativity. It is convenient in the sense that the extension of three-dimensional physical space to four-dimensional space-time will not introduce any change in the description of the three-dimensional (sub-)space. Thus, the distinction between upper and lower indices will only affect the time coordinate, but not the space coordinates.

The metric signature $(+,-,-,-)$ is, on the other hand, commonly used in relativistic particle physics. This choice can be understood as a consequence of the fact that the particle physics description more often focuses on physical processes in momentum space than in coordinate space. The time component (energy) of the four-momentum will then typically dominate the space component of the momentum. The transition between the two sign conventions, however, quite simply means a change $g \to -g$. This will introduce a sign change for any contraction of upper and lower index

[1]The notation with covariant and contravariant components is even more important in the *general theory of relativity* where more general coordinate systems are applied. In that case the metric tensors $g_{\mu\nu}$ and $g^{\mu\nu}$ will usually not correspond to the same 4x4 matrix.

in covariant equations, and in particular change the sign of the Lorentz invariant line element ds^2.

The reason for choosing here the first mentioned of these two conventions, is that it gives the smoothest transition from the non-relativistic three-vector description to the relativistic four-vector description.

6.2 Lorentz transformations in covariant form

A Lorentz transformation, which relates the coordinates x of an inertial frame S to the coordinates x' of another inertial frame S' can be written in component form as

$$x'^{\mu} = L^{\mu}{}_{\nu}\, x^{\nu}\,, \tag{6.12}$$

where one should note the convention for placing the indices of L. In matrix form it is

$$x' = L\, x\,. \tag{6.13}$$

For a *boost* in the x-direction the Lorentz transformation matrix is

$$L = \begin{pmatrix} \gamma & -\beta\gamma & 0 & 0 \\ -\beta\gamma & \gamma & 0 & 0 \\ 0 & 0 & 1 & 0 \\ 0 & 0 & 0 & 1 \end{pmatrix}, \tag{6.14}$$

with $\beta = v/c$ and $\gamma = 1/\sqrt{1-\beta^2}$ and v as the relative velocity of the two reference frames.

If a 4×4 matrix L should represent a Lorentz transformation, it has to satisfy a certain restriction, which follows from the requirement that the velocity of light is left unchanged by the transformation. As already noted this is related to the Lorentz invariance of the line element, which implies

$$\begin{aligned} g_{\mu\nu} dx'^{\mu} dx'^{\nu} &= g_{\mu\nu} L^{\mu}{}_{\rho}\, L^{\nu}{}_{\sigma}\, dx^{\rho} dx^{\sigma} \\ &= g_{\rho\sigma} dx^{\rho} dx^{\sigma}\,. \end{aligned} \tag{6.15}$$

Since this should be valid for any displacement dx^{μ}, the L matrix has to satisfy the restriction

$$g_{\mu\nu} L^{\mu}{}_{\rho} L^{\nu}{}_{\sigma} = g_{\rho\sigma}\,. \tag{6.16}$$

In matrix form this can be written as

$$L^T g L = g\,, \tag{6.17}$$

where L^T represents the transposed matrix. This equation, which is the condition for the 4×4 matrix L to represent a Lorentz transformation, corresponds to the following condition satisfied by the 3×3 rotation matrices R in three-dimensional space,

$$R^T R = \mathbb{1}\,, \tag{6.18}$$

where $\mathbb{1}$ represents the identity matrix.

6.3 General four-vectors

So far we have considered four-vectors $\underline{\mathbf{x}}$ associated with points in four-dimensional space-time. These are in a sense the fundamental vectors of the relativistic theory. However, exactly as in three dimensions, four-vectors can represent more general objects, such as velocity, momentum, acceleration, etc. All these have the following properties, which characterize any four-vector $\underline{\mathbf{A}}$,

- it has four components A^μ, with $\mu = 0, 1, 2, 3$,
- the components transform like the coordinates x^μ under Lorentz transformations, $A^\mu \to A'^\mu = L^\mu{}_\nu A^\nu$.

The Minkowski diagram is convenient to give a geometric representation of general types of four-vectors, in the same way as with the use of a Minkowski diagram for the position vectors of space-time itself. A reference frame corresponds also here to a choice of basis vectors $\{\underline{\mathbf{e}}_\mu, \mu = 0, 1, 2, 3\}$ and a vector $\underline{\mathbf{A}}$ can be decomposed on any set of basis vectors, corresponding to different inertial reference frames,

$$\underline{\mathbf{A}} = A^\mu \underline{\mathbf{e}}_\mu = A'^\mu \underline{\mathbf{e}}'_\mu \,. \tag{6.19}$$

Thus a Lorentz transformation of the four components of $\underline{\mathbf{A}}$ simply corresponds to a change of basis in the same way as for the components of the space-time coordinates x^μ.

The Lorentz invariant *scalar product*, which is defined by

$$\underline{\mathbf{A}} \cdot \underline{\mathbf{B}} = g_{\mu\nu} A^\mu B^\nu = A^\mu B_\mu \,, \tag{6.20}$$

is indefinite, which means that the square norm is not always positive. It separates the general four-vectors, like the space-time vectors $d\underline{\mathbf{x}}$, in three classes: spacelike ($\underline{\mathbf{A}}^2 > 0$), lightlike ($\underline{\mathbf{A}}^2 = 0$) and timelike ($\underline{\mathbf{A}}^2 < 0$). In the Minkowski diagram these three classes are represented, like space-time vectors, as vectors lying outside the light cone, on the light cone or inside the light cone, respectively (see Fig. 6.1).

As already noticed, orthogonality in the sense that the scalar product of two four-vectors vanishes does not mean that they *appear* as orthogonal in the Minkowski diagram. This is illustrated in Fig. 6.1, where the two vectors $\underline{\mathbf{A}}$ and $\underline{\mathbf{B}}$ are orthogonal, $\underline{\mathbf{A}} \cdot \underline{\mathbf{B}} = 0$, and the lightlike vector $\underline{\mathbf{C}}$, with this definition of orthogonality, is orthogonal to itself, $\underline{\mathbf{C}}^2 = 0$.

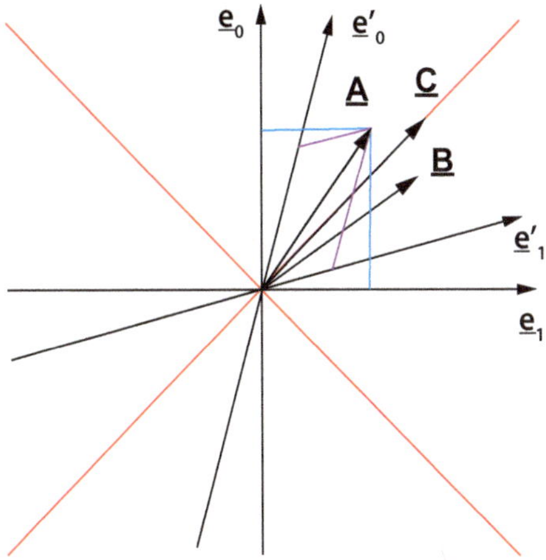

Fig. 6.1 A two-dimensional Minkowski diagram with general four-vectors. The diagram does not represent space-time itself, but the metric is the same as the space-time metric, and the vectors can be separated in the same three classes. $\underline{\mathbf{A}}$ represents a timelike vector, $\underline{\mathbf{B}}$ a spacelike vector and $\underline{\mathbf{C}}$ a lightlike vector, also referred to as a *null vector*. The null vectors define the light cone which separates the timelike and spacelike vectors. The different sets of basis vectors $\underline{\mathbf{e}}_\mu$ correspond to different inertial reference frames. The decomposition of the four-vector $\underline{\mathbf{A}}$ on two different basis sets are illustrated in the figure.

6.4 Lorentz transformation of vector components with lower index

The index of a general four-vector can be lowered by applying the metric tensor, in the same way as for the position vector x^μ,

$$A_\mu = g_{\mu\nu} A^\nu \,. \tag{6.21}$$

This relation leads to different transformation properties for vector components with upper indices (contravariant components) and lower indices (covariant components). We find the following expression for the transformed covariant components

$$\begin{aligned} A'_\mu &= g_{\mu\nu} A'^\nu \\ &= g_{\mu\nu} L^\nu{}_\rho A^\rho \\ &= g_{\mu\nu} L^\nu{}_\rho \, g^{\rho\sigma} A_\sigma \\ &\equiv L_\mu{}^\sigma A_\sigma \,. \end{aligned} \tag{6.22}$$

Note that in the last line we have introduced a modified symbol for the transformation matrix

$$L_\mu{}^\sigma = g_{\mu\nu} L^\nu{}_\rho \, g^{\rho\sigma} \,, \tag{6.23}$$

where we have followed the general rule that $g_{\mu\nu}$ acts as a lowering operator and $g^{\mu\nu}$ as a raising operator for the vector indices. With $L^\mu{}_\nu$ as the matrix elements of the 4×4 matrix L, then $L_\mu{}^\nu$ represents the matrix elements of the matrix

$$\begin{aligned} \tilde{L} &= gLg^{-1} \\ &= (L^T)^{-1} \,. \end{aligned} \tag{6.24}$$

The last expression is derived from the identity (6.17), which is satisfied by all Lorentz transformation matrices L.

Note that the covariant and contravariant components transform in inverse ways. This is in accordance with the fact that the scalar product of two vectors, which can be written as a product of the covariant components of one of the vectors and the contravariant components of the other, is invariant under Lorentz transformations. Also note that the transformation coefficients $L_\mu{}^\sigma$ of the covariant components A_μ are the same as the transformation coefficients of the basis vectors $\underline{\mathbf{e}}_\mu$, which have earlier been introduced in (4.28). This is consistent with a general property of the covariant formalism, namely that the position of the space-time index of an object, as an upper or lower index, indicates uniquely the transformation property of this object under Lorentz transformations.

6.5 Tensors

Four-vectors are in a sense the simplest geometrical objects that transform in a non-trivial way under space-time transformations. Many of the basic physical variables are represented as four-vectors, but there are also variables that cannot be represented in this way. These variables typically have more than four components that are mixed by the transformations. They are generally represented as *tensors*, which are geometrical objects related to vectors, but labeled with more than one space-time index.

Tensors are not restricted to relativistic theory. They appear also in non-relativistic theory, for example in the form of the stress tensor which measures the response of an elastic medium to external forces. However, several of the basic physical variables, which in non-relativistic theory are represented as vectors, are in relativistic theory described as tensors. For this reason one tends to meet tensors at an earlier stage in relativity theory than in non-relativistic theory.

As an example, take the angular momentum of a particle, which in non-relativistic theory is usually represented by

$$\boldsymbol{\ell} = \mathbf{r} \times \mathbf{p}\,, \tag{6.25}$$

which is a vector[2] with components ℓ_i, $i = 1, 2, 3$. We note that it can also be represented as the following two-index variable

$$\ell_{ij} = x_i p_j - x_j p_i\,, \tag{6.26}$$

which is an example of an antisymmetric *rank two* tensor, *i.e.*, a tensor with two space indices. The relation between the vector and tensor form of the angular momentum is

$$\ell_i = \frac{1}{2} \sum_{jk} \epsilon_{ijk} \ell_{jk}\,, \tag{6.27}$$

with ϵ_{ijk} as the Levi-Civita symbol. However, this mapping between vectors and antisymmetric tensors has no direct extension to four-dimensional space-time, since the Levi-Civita symbol in four dimensions has four indices rather than three. For this reason the relativistic angular momentum has a natural representation as an antisymmetric tensor rather than a vector. The form of this relativistic tensor is

$$\ell^{\mu\nu} = x^\mu p^\nu - x^\nu p^\mu\,, \tag{6.28}$$

where x^μ are the components of the position four-vector, and p^μ of the momentum four-vector.

In mathematics tensors appear in a natural way as generalizations of vectors, in the form of *tensor products.* Thus, starting with two vectors with components A^μ and B^ν, a tensor with two indices can be defined as the product

$$C^{\mu\nu} = A^\mu B^\nu\,, \tag{6.29}$$

much like what we did for the angular momentum. This is referred to as the tensor product of the two original vectors. If we further consider linear combinations of such products, they define a new vector space which we refer to as the tensor product of the two vector spaces (or here, rather of two copies of the original vector space). This vector space consists of all rank two tensors.

The transformation properties of a tensor, defined in this way, are obviously determined by the transformation properties of the original vectors.

[2]More precisely it is a *pseudovector*, which means that it transforms as a vector under rotations, but it does *not* change sign under space inversion, $\mathbf{r} \to -\mathbf{r}$.

For relativistic tensors, the rank two tensors thus transform under Lorentz transformations as

$$C^{\mu\nu} \to C'^{\mu\nu} = L^{\mu}{}_{\rho} L^{\nu}{}_{\sigma} C^{\rho\sigma} \,. \tag{6.30}$$

This property can in fact be used as a definition of tensors. An object characterized by two space-time indices, which transforms as (6.30) *is* a rank two tensor.

We have already considered the angular momentum of a particle as one example of such a tensor. Since it is antisymmetric in the two indices it has six independent components. Another example of an antisymmetric rank two tensor is the electromagnetic field tensor, conventionally written as $F^{\mu\nu}$. Also this has six independent components, which can be identified with the vector components of the electric and magnetic fields in the non-relativistic description. The energy and momentum densities of the fields define the components of yet another relativistic tensor, the symmetric energy-momentum tensor $T^{\mu\nu}$ of the electromagnetic field. A special symmetric rank two tensor that we have already encountered is the metric tensor $g_{\mu\nu}$. This has the unusual property that it is a constant, and nevertheless satisfies the transformation equation (6.30) of a rank two tensor. This peculiarity follows as a consequence of the identity (6.16) satisfied by the Lorentz transformation matrices.

Tensors may, like vectors, be written with upper indices or lower indices. These are related by the action of the metric tensor. For rank 2, we then have four related tensors

$$C^{\mu\nu}\,, \quad C^{\mu}{}_{\nu} = g_{\nu\rho} C^{\mu\rho}\,, \quad C_{\mu}{}^{\nu} = g_{\mu\rho} C^{\rho\nu}\,, \quad C_{\mu\nu} = g_{\mu\rho} g_{\nu\sigma} C^{\rho\sigma} \,. \tag{6.31}$$

In the same way as we can view the set of covariant and the set of contravariant components of a four-vector as two different representations of the same vector, we can view all the different sets of tensor components in (6.31) as being different representations of the same geometrical object.

So far we have focussed on rank two tensors, that is on variables with two space-time indices. However, there is an obvious generalization of tensors to arbitrary rank, that is to tensors with any number of space-time indices. A vector is then a special type of tensor, of rank 1, and a scalar, which carries no space-time index, is a tensor of rank 0. We have the following list of tensors of increasing rank,

A	rank 0 (scalar)	no vector index	(1 component) ,
B^{μ}	rank 1 (vector)	one vector index	(4 components) ,
$C^{\mu\nu}$	rank 2	two vector indices	(16 components) ,

$D^{\mu\nu\rho}$ rank 3 three vector indices (64 components),
etc.

What defines these as tensors is their transformation properties under Lorentz transformations. Thus, the transformed tensors are multiplied with one Lorentz matrix for each space-time index, as an obvious generalization of (6.30).

Like vectors, tensors of any rank can be considered as representing *geometrical objects*, which are well defined without specifying any particular reference frame. The components of the tensor, on the other hand refer to a specific set of basis vectors, and thus to a choice of reference frame. Equations describing physical laws in relativistic form can, in principle, be written in coordinate independent form, like vector equations in non-relativistic physics. However, usually this is not done, since it is cumbersome to introduce symbols for different types of tensors and for different types of multiplications between tensors. Relativistic equations are usually instead written in *covariant* form, which means that they are expressed in terms of tensor components, in such a way that they keep their form unchanged under Lorentz transformations.

When the equations are written in covariant form they are expressed in terms of variables with simple, standardized transformation properties. One can therefore easily check that the equations are valid in any reference system. To check that an equation has the correct covariant form we note that

- all terms that appear additively in the equation should be tensors of the same rank,
- a free index, which is an index which is not summed over, should have the same position, either up or down, in all additive terms of the equation,
- repeated indices, which are summed over, should appear in pairs, with one in the upper position and one in the lower position.

We note in particular that a contraction, *i.e.*, summation over a pair of repeated pair of indices, will reduce the rank of a tensor by 2. For example $A = A^{\mu}{}_{\mu}$ is a scalar, $B^{\mu} = B^{\mu\nu}{}_{\nu}$ is a vector etc.

As an example of a covariant equation, the equation of motion of a charged particle in an electromagnetic field can be written in the following compact form

$$m\ddot{x}^{\mu} = eF^{\mu\nu}\dot{x}_{\nu}\,, \tag{6.32}$$

where the time derivative is here with respect to the Lorentz invariant *proper time* τ of the particle. The equation is a relativistic vector equation, *i.e.*, with one free space-time index, μ, which is written in the upper position on both sides of the equation. On the right-hand side there is a contraction of index ν, with a correct repetition, where one index is in the upper and the other in the lower position. The equation clearly has a correct covariant form, which shows that it satisfies the requirement of being invariant under Lorentz transformations. The same equation can be written in the following non-covariant form

$$\dot{\mathbf{p}} = e(\mathbf{E} + \mathbf{v} \times \mathbf{B})\,, \tag{6.33}$$

where $\mathbf{E}$ is the electric field strength and $\mathbf{B}$ is the magnetic field. This equation, which is expressed in terms of three-vectors, is obviously invariant under rotations in three-dimensional space. However, the invariance under Lorentz transformations is not explicit in this equation. It depends on the transformations of $\mathbf{E}$ and $\mathbf{B}$, which generally will mix these two vector fields. A further discussion of covariance of electromagnetic equation will follow in Part 3.

6.6 Vector and tensor fields

In the same way as vectors in three-dimensional space often appear in the form of *vector fields*, vectors and tensors in four-dimensional space-time also often appear in the form of vector and tensor fields. As a particular example the electromagnetic field, in covariant relativistic form, is described by the rank two tensor field $F^{\mu\nu}(x)$. Let us list some of the tensor fields we may meet in relativistic theories:

$$\begin{aligned}
\phi &= \phi(x) && \text{scalar field}\,,\\
A^\mu &= A^\mu(x) && \text{vector field}\,,\\
F^{\mu\nu} &= F^{\mu\nu}(x) && \text{rank two tensor field field}\,,\\
&\text{etc.}
\end{aligned}$$

The fields are here written in component form and the space-time variable x here means the full set of coordinates $x = (x^0, x^1, x^2, x^3)$.

Under a change of inertial reference frames, defined by a Lorentz transformation L, the fields transform in the following way

$$\begin{array}{llcll}
\text{scalar field} & \phi(x) \;\rightarrow & \phi'(x') & = & \phi(x)\ ,\\
\text{vector field} & A^\mu(x) \;\rightarrow & A'^\mu(x') & = & L^\mu{}_\nu\, A^\nu(x)\ ,\\
\text{tensor field} & F^{\mu\nu}(x) \rightarrow & F'^{\mu\nu}(x') & = & L^\mu{}_\rho L^\nu{}_\sigma\, F^{\rho\sigma}(x)\ ,
\end{array}$$

etc.

One should note that there are two changes under the transformation. The field components transform according to the rank of the tensors, with the number of Lorentz matrices determined by their rank. But also the space-time argument changes, with $x'^\mu = L^\mu{}_\nu\, x^\nu$. This change simply means that the untransformed as well as the transformed fields refer to the same space-time point, but this point is represented by different sets of coordinates in the two inertial reference frames connected by the Lorentz transformation.

Physical fields, like the electromagnetic field, will usually satisfy a set of *field equations*, and when formulated as relativistic equations, these will often be expressed in covariant form. They are typically differential equations, and therefore we will discuss in general terms how differentiation with respect to space-time coordinates are treated in the covariant formalism.

We first examine the four-gradient of a scalar field $\phi(x)$, written as

$$A_\mu(x) = \frac{\partial \phi}{\partial x^\mu}(x) \equiv \partial_\mu \phi(x)\,. \tag{6.34}$$

The symbol ∂_μ has here been introduced to represent the derivative with respect to x^μ, and in the following we will use this as a convenient notation. We have also, by writing the partial derivative of ϕ as A_μ, indicated that the components of the derivative transform as covariant four-vector components, but that needs to be proven. In order to do so we note that the change of space-time coordinates $x \to x'$ can be viewed as a change of variables for the fields. Derivatives with respect to x' can then be related to derivatives with respect to x by the chain rule. For the differentiation operators we write this as

$$\frac{\partial}{\partial x'^\mu} = \frac{\partial x^\nu}{\partial x'^\mu}\,\frac{\partial}{\partial x^\nu}\,, \tag{6.35}$$

or simply as

$$\partial'_\mu = \frac{\partial x^\nu}{\partial x'^\mu}\,\partial_\nu\,. \tag{6.36}$$

Since the Lorentz transformation can be written as

$$x'^\mu = L^\mu{}_\nu\, x^\nu\,, \tag{6.37}$$

we find

$$\frac{\partial x'^\mu}{\partial x^\nu} = L^\mu{}_\nu\,, \tag{6.38}$$

but it is actually the derivative for the inverse transformation that we need.

To invert the transformation we make use of the property of the Lorentz transformation matrix

$$g_{\mu\nu}L^{\mu}{}_{\rho}L^{\nu}{}_{\sigma} = g_{\rho\sigma}\,, \tag{6.39}$$

and rewrite the transformation equation (6.37) as

$$g_{\sigma\rho}L^{\rho}{}_{\mu}x'^{\sigma} = g_{\sigma\rho}L^{\rho}{}_{\mu}L^{\sigma}{}_{\nu}x^{\nu} = g_{\mu\nu}x^{\nu}\,. \tag{6.40}$$

By further applying the raising operator on the μ index and changing the name of some of the indices we find the inverse transformation formula

$$x^{\nu} = g_{\mu\rho}\,L^{\rho}{}_{\sigma}\,g^{\sigma\nu}\,x'^{\mu} = L_{\mu}{}^{\nu}\,x'^{\mu}\,, \tag{6.41}$$

where we have made use of the definition $L_{\mu}{}^{\nu} = g_{\mu\rho}\,L^{\rho}{}_{\sigma}\,g^{\sigma\nu}$. As a result we find

$$\frac{\partial x^{\nu}}{\partial x'^{\mu}} = L_{\mu}{}^{\nu}\,, \tag{6.42}$$

to be compared with the transformation matrix (6.38).

The relation (6.36) between derivatives with respect to the original and the transformed space-time coordinates can then be written as

$$\partial'_{\mu} = L_{\mu}{}^{\nu}\,\partial_{\nu}\,. \tag{6.43}$$

This shows that the partial derivatives transform in the same way as the covariant components of a vector. In particular this gives for the four-gradient

$$\partial'_{\mu}\phi(x') = L_{\mu}{}^{\nu}\partial_{\nu}\phi(x)\,, \tag{6.44}$$

which is identical to the transformation equation for a covariant vector field (see (6.22)).

The rule for writing an equation in covariant form when it involves derivatives is therefore simple. The equation should be written in tensor form (including scalars and vectors) with each space-time derivative adding a covariant four-vector index to the expression. The equation (6.34) for the four-gradient therefore has a correct covariant form. In the same way the four-divergence of a vector field $A^{\mu}(x)$ can be written in the covariant form as

$$\chi(x) = \partial_{\mu}A^{\mu}(x)\,, \tag{6.45}$$

with $\chi(x)$ as a scalar field. We finally note that the transformation properties of the partial derivatives means that we can form the following Lorentz invariant quadratic differential operator,

$$\partial_{\mu}\partial^{\mu} = g^{\mu\nu}\partial_{\mu}\partial_{\nu} = \boldsymbol{\nabla}^2 - \frac{1}{c^2}\frac{\partial^2}{\partial t^2}\,. \tag{6.46}$$

It is called the *d'Alembertian* and is an extension from the *Laplacian* in three space dimensions to an operator in four space-time dimensions. As indicated by the contraction between an upper and a lower space-time index this operator transforms as a scalar under Lorentz transformation.

We finish here by writing two fundamental field equations in covariant form. The first one is the *Klein Gordon* equation,

$$(\partial_\mu \partial^\mu + \mu^2)\phi(x) = 0\,, \tag{6.47}$$

which is a relativistic wave equation for scalar particles of mass $m = \mu\hbar/c$, and the other is one of Maxwell's equations

$$\partial_\nu F^{\mu\nu}(x) = \mu_0 j^\mu(x)\,, \tag{6.48}$$

where μ_0 is the vacuum permeability and $j^\mu(x)$ is the four-vector current density. The latter we will meet again in Part 3. Here these field equations are included only to show their attractive, compact form when written in terms of relativistic tensors.

6.7 Exercises

Problem 6.1

Two four-vectors $\underline{\mathbf{A}}$ and $\underline{\mathbf{B}}$ are orthogonal in the sense $\underline{\mathbf{A}} \cdot \underline{\mathbf{B}} = A^\mu B_\mu = 0$, where $\underline{\mathbf{A}}$ is a timelike vector, $\underline{\mathbf{A}}^2 = A^\mu A_\mu < 0$. Show, by decomposing the vectors in their time and space components, that this implies that $\underline{\mathbf{B}}$ is a spacelike vector.

Problem 6.2

a) Below we have four equations, which involve tensors of different ranks. Clearly the consistency rules for covariant equations are not satisfied in all places. Show where there are errors in each equation, and show how the equations can be modified to bring them to correct covariant form.

$$C^\mu = T^\mu{}_\nu\, A^\mu\,, \quad D_\nu = T^\mu{}_\nu\, A_\mu\,, \quad E_{\mu\nu\rho} = T_{\mu\nu}\, S^\nu{}_\rho\,, \quad G = S_{\mu\nu}\, T^\nu{}_\alpha\, A^\alpha\,. \tag{6.49}$$

b) Assume A^μ and B^μ to be 4-vectors and $T^{\mu\nu}$ to be a rank 2 tensor. Show that by making products of these and by lowering and contracting indices, one can form several new 4-vectors and scalars.

c) Lorentz transformation matrices satisfy the basic relation

$$g_{\mu\nu} L^\mu{}_\rho L^\nu{}_\sigma = g_{\rho\sigma}. \tag{6.50}$$

Show, by consistently lowering and raising matrix indices, that this relation implies the identity

$$L_\mu{}^\rho L^\mu{}_\sigma = \delta^\rho_\sigma \,. \tag{6.51}$$

Problem 6.3

The following four tensor fields are functions of the space-time coordinates $x = (x^0, x^1, x^2, x^3)$,

$$f(x) = x_\mu x^\mu\,, \quad a^\mu(x) = x^\mu\,, \quad b^{\mu\nu}(x) = x^\mu x^\nu\,, \quad h^\mu(x) = \frac{x^\mu}{x_\nu x^\nu}. \tag{6.52}$$

Calculate the following derivatives,

$$\partial_\mu f(x)\,, \quad \partial_\mu a^\mu(x)\,, \quad \partial_\mu b^{\mu\nu}(x)\,, \quad \partial_\mu h^\mu(x)\,; \quad \partial_\mu \equiv \frac{\partial}{\partial x^\mu}\,. \tag{6.53}$$

Problem 6.4

Consider an inertial reference frame S in four-dimensional space-time, with coordinate axes defined by a set of unit vectors $\underline{\mathbf{e}}_\mu, \mu = 0, 1, 2, 3$. They satisfy the generalized orthonormalization condition

$$\underline{\mathbf{e}}_\mu \cdot \underline{\mathbf{e}}_\nu = g_{\mu\nu} \tag{6.54}$$

with $g_{\mu\nu}$ as the standard metric tensor. A second inertial frame S' has coordinate axes with unit vectors that mix those of S in the following way,

$$\begin{aligned} \underline{\mathbf{e}}'_0 &= \cosh\chi\, \underline{\mathbf{e}}_0 + \sinh\chi\, \underline{\mathbf{e}}_1\,, \\ \underline{\mathbf{e}}'_1 &= \sinh\chi\, \underline{\mathbf{e}}_0 + \cosh\chi\, \underline{\mathbf{e}}_1\,, \\ \underline{\mathbf{e}}'_2 &= \underline{\mathbf{e}}_2\,, \\ \underline{\mathbf{e}}'_3 &= \underline{\mathbf{e}}_3\,, \end{aligned} \tag{6.55}$$

with χ as the rapidity parameter.

a) Assume S' has velocity v relative to S. Show that

$$\cosh\chi = \gamma\,, \quad \sinh\chi = \beta\gamma \tag{6.56}$$

with $\beta = v/c$ and $\gamma = 1/\sqrt{1-\beta^2}$, by relating (6.55) to the standard Lorentz transformation formula for the coordinates of the two reference frames.

b) Assume the two sets of basis vectors $(\underline{\mathbf{e}}_0, \underline{\mathbf{e}}_1)$ and $(\underline{\mathbf{e}}'_0, \underline{\mathbf{e}}'_1)$ of S and S', are plotted in a common two-dimensional Minkowski diagram, with $\underline{\mathbf{e}}_0$ and $\underline{\mathbf{e}}_1$ defining the orthogonal, horizontal and vertical directions of the diagram. With ϕ as the angle between the two time axes $\underline{\mathbf{e}}_0$ and $\underline{\mathbf{e}'}_0$ in the diagram, show that

$$\tan\phi = \tanh\chi. \tag{6.57}$$

For the cases $\phi = 15°$ and $\phi = 30°$ determine the corresponding values of $\beta = v/c$.

c) Show that the vector $\underline{\mathbf{e}}'_0$ will trace out a hyperbola when the velocity v changes, and that $\underline{\mathbf{e}}'_1$ will trace out another hyperbola.

Chapter 7

Relativistic kinematics

We discuss in this chapter how to describe velocity and acceleration as relativistic four-vectors. The concept of *proper acceleration* is introduced and an example of motion with constant proper acceleration is investigated. We further discuss the reformulation of momentum and energy of a moving particle as a relativistic four-momentum, and derive the relativistic energy-momentum relation. Massless particles appear as a possibility in the relativistic description, and we use the four-vector description of photon momentum to derive the relativistic Doppler formula.

7.1 Four-velocity and four-acceleration

We consider the motion of a point particle through space and time. It can be described by a time dependent position vector, which we decompose in its time and space parts, defined with respect to some unspecified inertial frame,

$$\underline{\mathbf{x}}(t) = (ct, \mathbf{r}(t))\,. \tag{7.1}$$

Let us introduce the particle velocity by the time derivative of the four-vector, this also decomposed in its time and space parts,

$$\frac{d\underline{\mathbf{x}}}{dt} = \left(c, \frac{d\mathbf{r}}{dt}\right)\,. \tag{7.2}$$

However, the derivative of the four-vector $\underline{\mathbf{x}}(t)$, when differentiated with respect to time t of the chosen inertial frame, is itself not a four-vector. As a direct demonstration of this we consider the special case where a particle is moving along the x-axis with velocity u relative to a coordinate system S. Assume another inertial frame S' is moving relative to this frame with velocity v, also in the direction of the x-axis, so that the coordinates of the two frames are related by a special Lorentz transformation (boost) in the x-direction. The time derivative of the position vector, when decomposed in the coordinates of the two frames, will have the form

$$S: \ \frac{d\underline{\mathbf{x}}}{dt} = (c, u, 0, 0)\,, \qquad S': \ \frac{d\underline{\mathbf{x}}}{dt'} = (c, u', 0, 0)\,, \tag{7.3}$$

when all the four space-time components are shown. The velocities are given by $u = \frac{dx}{dt}$ and $u' = \frac{dx'}{dt'}$ and the relation between these is given by the relativistic transformation formula for velocities, (4.7),

$$u' = \frac{u - v}{1 - \frac{uv}{c^2}}. \tag{7.4}$$

Clearly the transformation of the components of $\frac{d\underline{\mathbf{x}}}{dt}$ between the two inertial frames does not have the form of a four-vector transformation.

The reason for this is easy to understand. The position vector is differentiated with respect to the time coordinate of a specific reference frame, and the resulting vector, $\frac{d\underline{\mathbf{x}}}{dt}$, will therefore not be coordinate independent. This result suggests that we need to use a Lorentz invariant time parameter in order to define velocity as a four-vector. Such a parameter is in fact available in the form of the *proper time* of the moving particle. As already discussed the proper time of a particle is directly related to the invariant line element of the particle path and is therefore also a Lorentz invariant. We therefore define the *four-velocity* of the particle as

$$\underline{\mathbf{U}} = \frac{d\underline{\mathbf{x}}}{d\tau}\,, \tag{7.5}$$

or in the component form

$$U^\mu = \frac{dx^\mu}{d\tau}. \tag{7.6}$$

With τ as a Lorentz invariant it is clear that the vector components U^μ and x^μ transform in the same way, which secures that $\underline{\mathbf{U}}$ as defined above is a four-vector.

The definition of the proper time

$$d\tau^2 = -\frac{1}{c^2} dx^\mu dx_\mu \tag{7.7}$$

furthermore shows that all the four components of $\underline{\mathbf{U}}$ cannot be independent. This is shown explicitly by evaluating the Lorentz invariant

$$\underline{\mathbf{U}}^2 = U^\mu U_\mu = -c^2\,. \tag{7.8}$$

For any motion of the particle, the four-velocity is therefore a timelike vector with a *fixed* (negative) norm squared. This can be seen in another way by expressing the four-velocity in terms of the (reference-frame dependent) velocity $\mathbf{v} = \frac{d\mathbf{r}}{dt}$. We have

$$\begin{aligned}\underline{\mathbf{U}} &= \frac{d\underline{\mathbf{x}}}{d\tau} \\ &= \frac{d}{d\tau}(ct, \mathbf{r}) \\ &= \frac{dt}{d\tau}\frac{d}{dt}(ct, \mathbf{r}) \\ &= \gamma(c, \mathbf{v}),\end{aligned} \tag{7.9}$$

where we have decomposed the four-vector $\underline{\mathbf{x}}$ into its time and space-parts (with respect to the unspecified inertial frame), and where we have made use of the time dilation formula $\frac{dt}{d\tau} = \gamma$. In this formulation the constant value for the Lorentz invariant $\underline{\mathbf{U}}^2$ follows from the identity

$$\underline{\mathbf{U}}^2 = \gamma^2(v^2 - c^2) = -c^2 . \tag{7.10}$$

We note that the presence of the factor γ in the expression (7.9) for $\underline{\mathbf{U}}$ is important in order to define it as a four-vector.

It is now fairly obvious how to define the corresponding *four-acceleration*,

$$\underline{\mathbf{A}} = \frac{d\underline{\mathbf{U}}}{d\tau} = \frac{d^2\underline{\mathbf{x}}}{d\tau^2} . \tag{7.11}$$

Again, since τ is a Lorentz invariant parameter, the transformation properties of the components of $\underline{\mathbf{U}}$ and $\underline{\mathbf{A}}$ will be the same.

We would like to relate the four-vector $\underline{\mathbf{A}}$ to the usual (reference-frame dependent) acceleration $\mathbf{a} = \frac{d\mathbf{v}}{dt}$. This we do by decomposing the four-vector into its time and space parts in a similar way as we have done for the four-velocity $\underline{\mathbf{U}}$,

$$\underline{\mathbf{A}} = \gamma\left(\frac{dU^0}{dt}, \frac{d\mathbf{U}}{dt}\right), \tag{7.12}$$

and we examine the two parts separately. For the 0-component we have

$$\frac{dU^0}{dt} = c\frac{d\gamma}{dt}, \tag{7.13}$$

and for the three-vector part

$$\frac{d\mathbf{U}}{dt} = \frac{d}{dt}(\gamma\mathbf{v}) = \gamma\frac{d\mathbf{v}}{dt} + \frac{d\gamma}{dt}\mathbf{v} = \gamma\mathbf{a} + \frac{d\gamma}{dt}\mathbf{v} . \tag{7.14}$$

The time derivative of the γ-factor is

$$\begin{aligned}\frac{d\gamma}{dt} &= \frac{d}{dt}\left(1-\frac{v^2}{c^2}\right)^{-\frac{1}{2}} \\ &= \left(1-\frac{v^2}{c^2}\right)^{-\frac{3}{2}} \frac{v}{c^2}\frac{dv}{dt} \\ &= \frac{1}{2}\gamma^3\,\frac{1}{c^2}\,\frac{d(\mathbf{v}^2)}{dt} \\ &= \gamma^3\,\frac{1}{c^2}\,\mathbf{v}\cdot\frac{d\mathbf{v}}{dt} \\ &= \gamma^3\,\frac{1}{c^2}\,\mathbf{v}\cdot\mathbf{a}\,. \end{aligned} \tag{7.15}$$

This gives for the time and space components of the four-acceleration

$$\begin{aligned} A^0 &= \gamma c\frac{d\gamma}{dt} = \gamma^4\frac{\mathbf{v}\cdot\mathbf{a}}{c}\,, \\ \mathbf{A} &= \gamma^2\mathbf{a} + \gamma\frac{d\gamma}{dt}\mathbf{v} = \gamma^2\mathbf{a} + \gamma^4\frac{\mathbf{v}\cdot\mathbf{a}}{c^2}\mathbf{v}\,. \end{aligned} \tag{7.16}$$

These expressions are valid in any inertial frame, with $\mathbf{v}$ as the (time dependent) velocity of the particle in this frame and $\mathbf{a}$ as the time derivative of the velocity in the same frame. If we focus on the space part of the four-vector $\underline{\mathbf{A}}$, we note that it has one part which is proportional to the acceleration $\mathbf{a}$ (in the chosen inertial frame), with a proportionality factor that can be interpreted as a time dilation factor of the proper time relative to the coordinate time. However there is another term, which in direction is proportional to $\mathbf{v}$ rather than $\mathbf{a}$. This comes from the time derivative of the time dilation factor.

There are now two new Lorentz invariants that we can construct with the help of $\underline{\mathbf{A}}$ (and $\underline{\mathbf{U}}$). The first one is

$$\underline{\mathbf{U}}\cdot\underline{\mathbf{A}} = \underline{\mathbf{U}}\cdot\frac{d\underline{\mathbf{U}}}{dt} = \frac{1}{2}\frac{d\underline{\mathbf{U}}^2}{dt} = 0\,. \tag{7.17}$$

This result follows from the fact that $\underline{\mathbf{U}}^2$ is a constant. The other Lorentz invariant is

$$\begin{aligned}\underline{\mathbf{A}}^2 &= \mathbf{A}^2 - {A^0}^2 \\ &= \gamma^4\mathbf{a}^2 + \gamma^8\left(\frac{\mathbf{v}\cdot\mathbf{a}}{c^2}\right)^2 v^2 + 2\gamma^6\left(\frac{\mathbf{v}\cdot\mathbf{a}}{c}\right)^2 - \gamma^8\left(\frac{\mathbf{v}\cdot\mathbf{a}}{c}\right)^2 \\ &= \gamma^4\mathbf{a}^2 + \gamma^6\frac{(\mathbf{v}\cdot\mathbf{a})^2}{c^2}\,, \end{aligned} \tag{7.18}$$

where the last expression is valid in any inertial reference system.

We have already noticed that the four-velocity $\underline{\mathbf{U}}$ is a timelike vector. Since $\underline{\mathbf{A}}$ is orthogonal (in the relativistic sense) to a timelike vector, it has itself to be a spacelike vector (see Problem 6.1). Since $\underline{\mathbf{A}}$ is spacelike, this means that one can, by properly choosing the inertial frame, transform the time component A^0 to zero. As shown by the expressions (7.16) this happens when $\mathbf{v}$ and $\mathbf{a}$ are orthogonal. In particular this is the case in the instantaneous inertial rest frame of the particle, where $\mathbf{v} = 0$. The acceleration measured in the instantaneous inertial rest frame is referred to as the *proper acceleration* of the particle.

Let us denote the proper acceleration as $\mathbf{a}_0$. We should stress that this acceleration is for any point on the particle's world line measured in the *inertial* reference frame where the particle is instantaneously at rest. This means that when we follow the motion of the particle, the proper acceleration refers to a (continuous) sequence of inertial frames, each of them associated with a particular point on the particle path. The proper acceleration will in general vary along the path, so that it can be regarded as a function of the proper time of the particle's world line, $\mathbf{a}_0 = \mathbf{a}_0(\tau)$. When decomposed in the instantaneous rest frame at proper time τ, the four-acceleration then gets a particularly simple form,

$$\underline{\mathbf{A}}(\tau) = (0, \mathbf{a}_0(\tau)) \qquad \text{(instantaneous rest frame)}\,. \tag{7.19}$$

This means that we can identify the Lorentz invariant (7.18) with $\mathbf{a}_0^2$, and therefore we have the following relation between the proper acceleration and the acceleration measured in another inertial frame

$$\mathbf{a}_0^2 = \gamma^4 \mathbf{a}^2 + \gamma^6 \frac{(\mathbf{v}\cdot\mathbf{a})^2}{c^2}\,. \tag{7.20}$$

This shows that the proper acceleration is larger than the acceleration measured in any other inertial frame, $a_0 \geq a$.

Let us consider two special cases. For motion in a circular orbit with constant speed we have $\mathbf{v}\cdot\mathbf{a} = 0$, and therefore

$$a_0 = \gamma^2 a\,. \tag{7.21}$$

In the rest frame the acceleration is enhanced by the factor γ^2, and this we may see as a time dilation effect, due to the double differentiation with respect to proper time rather than coordinate time. The other special case is linear acceleration, where $\mathbf{v}\cdot\mathbf{a} = v\,a$. In this case we find

$$\begin{aligned} a_0^2 &= \gamma^4 \left(a^2 + \gamma^2 \frac{v^2}{c^2} a^2\right) \\ &= \gamma^4 (1 + \gamma^2\beta^2) a^2 \\ &= \gamma^6 (1 - \beta^2 + \beta^2) a^2 \\ &= \gamma^6 a^2\,, \end{aligned} \tag{7.22}$$

so in this case the enhancement factor is even larger,

$$a_0 = \gamma^3 a\,. \tag{7.23}$$

Example

Hyperbolic motion through space and time

We will illustrate the discussion of the previous section by considering a space travel with constant proper acceleration. Let us therefore assume that a spaceship is leaving earth for a travel far out in the universe. The ship is maintaining a constant direction of velocity and the engines are providing a thrust so that the effective gravitational field on board is kept constant and equal in strength to the gravitational field on the surface of the earth. This means that the proper acceleration of the spaceship, which is the acceleration measured relative to an instantaneous inertial rest frame, is the same at all times of the travel, with $a_0 = g = 9.8m/s^2$. The problem to be discussed is how the travel appears in a non-rotating, earth-fixed frame, which we can assume to be (to a good approximation) an inertial reference frame.

Since the motion of the spaceship is assumed to be linear, the velocity and acceleration, as seen in the earth-fixed reference frame, are collinear, with $\mathbf{v}\cdot\mathbf{a} = va$. The relation between the (constant) proper acceleration and the acceleration measured in the earth-fixed frame is then, as shown in (7.23),

$$a = \frac{a_0}{\gamma^3} = \frac{g}{\gamma^3}\,. \tag{7.24}$$

The acceleration therefore seems to decrease with time, when measured at earth, and this we can view as a consequence of the time dilation effect. By integrating the above equation we can find the position of the spaceship as a function of its proper time τ. We choose the x-axis of the inertial frame to be in the direction of the motion.

First we rewrite the equation as a differential equation for $\beta = v/c$,

$$\frac{d\beta}{d\tau} = \frac{dt}{d\tau}\frac{d\beta}{dt} = \gamma\frac{1}{c}a = \frac{g}{c}\frac{1}{\gamma^2} = \frac{g}{c}(1-\beta^2)\,. \tag{7.25}$$

It is convenient to substitute β with the rapidity χ, which we have earlier introduced. It is related to β by $\beta = \tanh\chi$, which gives

$$\frac{d\beta}{d\tau} = \left(\frac{d}{d\chi}\tanh\chi\right)\frac{d\chi}{d\tau} = \frac{1}{\cosh^2\chi}\frac{d\chi}{d\tau} \tag{7.26}$$

and

$$1-\beta^2 = 1-\tanh^2\chi = \frac{1}{\cosh^2\chi}\,. \tag{7.27}$$

The differential equation for β therefore gets the following simple form when expressed in term of the rapidity

$$\frac{d\chi}{d\tau} = \frac{g}{c}\,. \tag{7.28}$$

with solution

$$\chi = \frac{g}{c}\tau\,. \tag{7.29}$$

We have then assumed that the time coordinates are $t=\tau=0$ at the beginning of the journey, when the velocity vanishes, and therefore $\beta = \chi = 0$. The solution for the velocity is then

$$\beta = \tanh\left(\frac{g}{c}\tau\right), \tag{7.30}$$

which for the γ factor gives

$$\gamma = \cosh\left(\frac{g}{c}\tau\right). \tag{7.31}$$

The relation between the coordinate time t and the proper time τ can be determined from the time dilation formula,

$$dt = \gamma d\tau = \cosh\left(\frac{g}{c}\tau\right)d\tau, \tag{7.32}$$

which by integration gives

$$t = \frac{c}{g}\sinh\left(\frac{g}{c}\tau\right). \tag{7.33}$$

In a similar way the x-coordinate can be found by integrating the expression for the velocity,

$$\frac{dx}{d\tau} = \gamma\frac{dx}{dt} = \gamma\beta c = c\sinh\left(\frac{g}{c}\tau\right), \tag{7.34}$$

which gives

$$x = \frac{c^2}{g}\cosh\left(\frac{g}{c}\tau\right). \tag{7.35}$$

In the last expression we have for simplicity chosen the integration constant to be zero. Note that this means that the x-coordinate is not zero at the beginning of the journey, but rather $x(0) = c^2/g$.

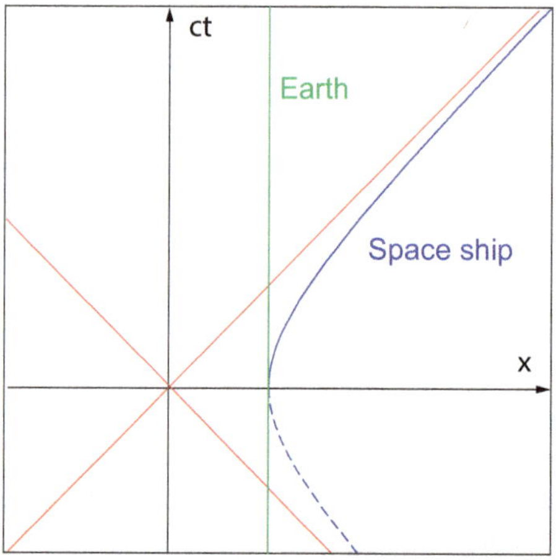

Fig. 7.1 The hyperbolic space-time path of the accelerated spaceship. In the Minkowski diagram the world line of the earth is the solid green line parallel to the time axis. The world line of the spaceship, which has constant proper acceleration, defines a part of a hyperbola, shown by the solid blue line in the diagram. The dashed blue line shows the remaining part of the hyperbola. The asymptotes of the hyperbola (red lines) correspond to motion with the speed of light.

To sum up, the coordinates of the spaceship in the inertial frame of the earth are given by

$$ct = \frac{c^2}{g}\sinh\left(\frac{g}{c}\tau\right), \quad x = \frac{c^2}{g}\cosh\left(\frac{g}{c}\tau\right), \quad y = z = 0\,, \tag{7.36}$$

when the proper time τ is used as the time parameter of the spaceship's world line. From this follows that the coordinates satisfy the equation

$$x^2 - (ct)^2 = \frac{c^4}{g^2}\,. \tag{7.37}$$

In a two-dimensional Minkowski diagram, with ct and x as coordinate axes, the world line of the spaceship will therefore define a hyperbola. This is illustrated in Fig. 7.1.

To get some feeling for what this means, let us consider how time and position of the spaceship, as registered on earth, develop as functions of the time coordinate τ registered on the spaceship. We first note that the proper acceleration $a_0 = g$ defines a time constant

$$\tau_0 = \frac{c}{g} = 0.97 \text{ year}\,, \tag{7.38}$$

when $g = 9.81\,\text{m/s}^2$. So this time constant is very close to 1 year. This also means that the start value of the spaceship's x-coordinate, which is also the x-coordinate of the earth is

$$x_0 = \frac{c^2}{g} = 0.97 \text{ light year}. \tag{7.39}$$

In the table the change in position and coordinate time is shown for a sequence of increasing proper times of the spaceship.

Table 7.1 Space-time positions of a spaceship with hyperbolic motion. The table shows a list of distances and coordinate times for increasing proper times τ. For large τ the distance and coordinate time increase exponentially with the proper time of the spaceship. The notation *yr* means *year*, and *ly* means *light year*.

τ	1 yr	2 yr	3 yr	5 y	7 yr	11 yr	15 yr
t	1.2yr	3.6 yr	10.0 yr	74 yr	548 yr	30 000 yr	$1.6 \cdot 10^6$ yr
$x - x_0$	0.5 ly	2.8 ly	9.1 ly	73 ly	547 ly	30 000 ly	$1.6 \cdot 10^6$ ly

The numbers shown in the table are quite remarkable. Even if the acceleration as felt in the spaceship is quite modest, it is no more than the acceleration of gravity experienced at the surface at the earth, the speed and the distance to the earth increases rapidly. Already after one year, as measured onboard the spaceship, the distance to the spaceship is half a light year. After a little more than 2 years the spaceship will have a distance equal to the distance to our nearest star. Then the velocity really becomes large. After 11 years onboard the spaceship it will pass the distance to the center of the galaxy and after 15 years the distance to the Andromeda galaxy. All this is due to the time dilation effect, or alternatively due to the length contraction effect, since distances between heavenly objects seem to shrink when observed from the spaceship. As shown by the Minkowski diagram the speed of the spaceship seems to approach the speed of light, so that already after 3 years it has reached a velocity $v = 0.995c$.

The numbers of the table also seem to indicate that space travels even into distant parts of the universe may be possible with a travel time of a few years and under conditions that seem quite agreeable. However, as shown by the corresponding coordinate time on earth, and by comparison with conclusions made in the discussion of the twin paradox, it is clear that if the ship returns to the earth it will experience a major jump forward in earth time as compared to the time experienced on board the spaceship.

There is another major obstacle to carrying out such a travel. If the time dilation effect should cut down the time of the journey in a substantial way, the spaceship has to reach velocities close to the speed of light. This creates a serious energy problem. How should it be possible to feed the engines with the large amount of energy needed? It seems impossible to bring all this fuel along, even with the most efficient conversion of fuel into energy. So the only possibility seems for the spaceship to be recharged with energy during the travel. But the safest conclusion may seem to be that for a spaceship to maintain a constant proper acceleration on the time scale of years, even at the modest value of $a = g$, is outside the reach of any practical setting. However, we shall include a further discussion of this energy problem in the next chapter.

7.2 Relativistic energy and momentum

The relativistic space-time symmetries introduce important changes in the description of energy and momentum as compared to that of non-relativistic physics. This leads, in particular, to a new understanding of the relation between energy and mass, as captured by the famous Einstein formula $E = mc^2$. We will here study how relativity influences the kinematical relations between energy and momentum for single particles. In the next chapter we will then examine the consequences of conservation of these physical quantities for systems of particles.

Consider a point particle of mass m. When moving, this particle will, in the non-relativistic description, carry momentum $\mathbf{p} = m\mathbf{v}$ and kinetic energy $E = \frac{1}{2}m\mathbf{v}^2$. To find the correct relativistic form of the particle's energy and momentum, we apply the formalism of four-vectors, with the idea to rewrite the non-relativistic three-vector momentum as a relativistic four-vector. The four-vector form makes the expression independent of any particular inertial frame, and if it reproduces correctly the non-relativistic three-momentum in reference frames where v/c is small, this gives a strong indication that the correct relativistic expression has been found. That this expression indeed is correct has been demonstrated experimentally in many ways, in relativistic processes where energy and momentum are conserved. A similar formal approach will later be used when non-relativistic equations are updated to their *covariant* relativistic form.

The natural assumption is to replace the three-vector velocity $\mathbf{v}$ by the four-velocity $\underline{\mathbf{U}}$ in the definition of the momentum. The expression for the *four-momentum* of a particle will then be

$$\underline{\mathbf{P}} = m\underline{\mathbf{U}}\,. \tag{7.40}$$

We consider the non-relativistic limit of this four-vector. It is convenient to separate the time component from the space component (in an arbitrarily chosen inertial frame) in the same way as we have earlier done with the four-velocity (see (7.9)),

$$\underline{\mathbf{P}} = (\gamma mc, \gamma m\mathbf{v})\,. \tag{7.41}$$

Since γ approaches the value 1 for low velocities, the three-vector part has the correct non-relativistic limit,

$$\mathbf{p} = \gamma m\mathbf{v} \quad \underset{v << c}{\longrightarrow} \quad m\mathbf{v}\,. \tag{7.42}$$

We therefore conclude that the correct three-vector part of the relativistic momentum is

$$\mathbf{p} = \gamma m\mathbf{v} = \frac{m\mathbf{v}}{\sqrt{1-\frac{v^2}{c^2}}}\,. \tag{7.43}$$

At this point we make a comment on the notations that we apply. When decomposing the four-velocity and four-acceleration, we write these with capital letters,

$$\begin{aligned} \underline{\mathbf{U}} &= (U^0, \mathbf{U}) \\ \underline{\mathbf{A}} &= (A^0, \mathbf{A})\,. \end{aligned} \tag{7.44}$$

This is because the space components of these four-vectors *are not identical* to the three-vectors $\mathbf{v}$ and $\mathbf{a}$. Even in the relativistic context the original definitions of velocity and acceleration are valid as the quantities measured in a specific inertial frame, and we therefore make a distinction between these and the three-vector parts of $\underline{\mathbf{U}}$ and $\underline{\mathbf{A}}$. As far as the momentum is concerned the situation is different. The measured three-vector part is identical to $\mathbf{p} = \gamma m\mathbf{v}$, and the expression $m\mathbf{v}$ is only to be considered as the non-relativistic approximation. For this reason we do not make any distinction between $\mathbf{P}$ and $\mathbf{p}$, and use in the following the relativistic definition for $\mathbf{p}$, with the old expression valid only for velocities $v << c$.

When we make the transition from non-relativistic to relativistic theory by replacing three-vectors with four-vectors, an additional component is introduced, which is the time component of the vector. It is of interest to understand the meaning of this additional component. For the four-momentum this is

$$P^0 = \gamma mc = \frac{mc}{\sqrt{1 - \frac{v^2}{c^2}}} \,. \tag{7.45}$$

To see the physical interpretation we consider its non-relativistic form by making an expansion to first order in v^2/c^2,

$$P^0 = mc + \frac{1}{2} m \frac{v^2}{c} + \dots \,. \tag{7.46}$$

When multiplied with c this gives

$$cP^0 = mc^2 + \frac{1}{2} mv^2 + \dots \,. \tag{7.47}$$

The second term is identical to the (non-relativistic) kinetic energy of the particle, while the first term is a constant with physical dimension of energy. It is called the *rest energy* of the particle and is here simply a constant. We refer to the full expression as the *relativistic energy* of the particle,

$$E = \gamma mc^2 = \frac{mc^2}{\sqrt{1 - \frac{v^2}{c^2}}} \,, \tag{7.48}$$

and the expression for the rest energy is

$$E_0 = mc^2 \,. \tag{7.49}$$

Since E_0 is a constant we may simply subtract it to get the correct relativistic form for the kinetic energy,

$$T = E - E_0 = (\gamma - 1) mc^2 \,. \tag{7.50}$$

When T is expanded in powers of $\frac{v^2}{c^2}$, the first terms are

$$T = \frac{1}{2} mv^2 + \frac{3}{8} m \frac{v^4}{c^2} + \dots \,. \tag{7.51}$$

So for small velocities the expression for the kinetic energy reduces to the non-relativistic expression, but there are higher order relativistic corrections.

However, even if the rest energy here only appears as an innocent looking constant, the formula indicates the presence of a relation between mass and energy, and we know that this relation has far-reaching implications. Mass can be converted to energy, and as we shall discuss, this can be seen already in a study of inelastic collisions. But the true significance is, as we all know, in the field of nuclear physics, where large amounts of free energy are created by converting small amounts of mass, either in nuclear reactors or in nuclear bombs.

To sum up, the relativistic four-momentum can be separated in a time component which is the energy of the particle divided by c, and a space component which is the relativistic momentum three-vector. The expressions are

$$\underline{\mathbf{P}} = \left(\frac{E}{c}, \mathbf{p}\right) = \left(\frac{mc}{\sqrt{1-\frac{v^2}{c^2}}}, \frac{m\mathbf{v}}{\sqrt{1-\frac{v^2}{c^2}}}\right). \tag{7.52}$$

7.3 The relativistic energy-momentum relation

From the four-moment $\underline{\mathbf{P}}$ we can form the following Lorentz invariant,

$$\underline{\mathbf{P}}^2 = P_\mu P^\mu = \mathbf{p}^2 - \frac{E^2}{c^2}. \tag{7.53}$$

A direct calculation gives

$$\begin{aligned}\underline{\mathbf{P}}^2 &= \gamma^2 m^2 v^2 - \gamma^2 m^2 c^2 \\ &= -m^2 c^2 \gamma^2 \left(1 - \frac{v^2}{c^2}\right) \\ &= -m^2 c^2 .\end{aligned} \tag{7.54}$$

From this follows the relativistic relation between energy and momentum for a freely moving particle,

$$E^2 - c^2\mathbf{p}^2 = m^2 c^4 , \tag{7.55}$$

or

$$E = \sqrt{\mathbf{p}^2 c^2 + m^2 c^4} . \tag{7.56}$$

This replaces the non-relativistic relation

$$E = \frac{1}{2m}\mathbf{p}^2 . \tag{7.57}$$

The connection between the two expressions is found by making an expansion of the relativistic energy in powers of p^2/m^2c^2,

$$E = mc^2 + \frac{1}{2m}\mathbf{p}^2 + \dots , \tag{7.58}$$

which is essentially the same as the expansion (7.47). The first term is the rest energy and the second term the non-relativistic kinetic energy.

The presence of the rest energy in the energy-momentum relation has one important consequence. This is seen by considering the limit $m \to 0$.

In this limit the expansion in powers of p^2/m^2c^2 makes no sense, and that is reflected in the difference the limit $m \to 0$ makes for the relativistic and non-relativistic energy. In the relativistic case we get in this limit

$$E = \sqrt{p^2c^2 + m^2c^4} \to cp\,, \quad p = |\mathbf{p}|\,. \tag{7.59}$$

The limit is well defined and gives an energy which is proportional to the absolute value of the momentum. In the non-relativistic case the limit gives instead

$$E = \frac{1}{2m}\mathbf{p}^2 = \frac{1}{2}m\mathbf{v}^2 \to 0\,, \tag{7.60}$$

where we have assumed that the velocity is finite also in this limit. Similarly the momentum tends to zero, $\mathbf{p} = m\mathbf{v} \to 0$. Since both momentum and energy vanish when $m \to 0$, the reasonable conclusion is that the non-relativistic formalism has no place for massless particles. The conclusion is different in the theory of relativity, where the formalism is open for the presence of particles with zero mass. That is fortunate, since nature seems to provide such particles, with the photons being the most well-known example.

Let us derive some further consequences for massless particles. We first note that the relativistic expressions for E and $\mathbf{p}$ give the following expression for the velocity,

$$\mathbf{v} = c^2\frac{\mathbf{p}}{E}\,, \tag{7.61}$$

which should be compared with the non-relativistic expression $\mathbf{v} = \mathbf{p}/m$. In the limit $m \to 0$ the relativistic expression gives

$$\mathbf{v} = \frac{\mathbf{p}}{p}c\,, \tag{7.62}$$

which means that in absolute value the speed of the particle is identical to the speed of light. Thus *a massless particle always moves with the speed of light*, and this is independent of what the energy carried by the particle is. Therefore, we cannot think of a massless particle as being accelerated to the speed of light, it has simply to be born with the speed of light. This is contrasted by the property of massive particles: A particle with mass $m \neq 0$ can never reach the speed of light. This is demonstrated by the form of the relativistic energy,

$$E = \frac{mc^2}{\sqrt{1-\frac{v^2}{c^2}}} \underset{v \to c}{\longrightarrow} \infty\,. \tag{7.63}$$

Example

Spaceship with constant proper acceleration

We return to the situation discussed in Section 7.1, where a spaceship was assumed to perform a space-time journey with constant proper acceleration far out in the universe. The acceleration would give a monotonic increase in the velocity of the spaceship, which then would asymptotically approach the speed of light. In the discussion of the space-time motion we only briefly commented on the point that such a journey cannot go on indefinitely, since the limitation of available energy will end the journey after a finite time. Let us now consider this limitation in some detail.

We assume that the total mass of the spaceship at the beginning of the journey is m_0 with m_1 as the mass of the ship without fuel. Since we do not know what kind of engine the spaceship has, we only seek an upper limit to its efficiency. Let us for simplicity assume that *all* the mass of the fuel is converted to energy according to the Einstein formula, $E = mc^2$, and that the energy is used to send the exhaust gas with *maximum momentum* in the opposite direction of the velocity of the spaceship. The energy momentum relation (7.56) tells us that this happens if massless particles are emitted from the spaceship. So we assume that photons are emitted in one direction, and the spaceship due to this emission is accelerated in the opposite direction.

Let us consider what happens in a short time interval $d\tau$ on the spaceship. In this time interval an amount of mass is converted to energy, with $dm\,(< 0)$ as the corresponding change in the fuel mass. The photons that carry the energy away also carry an amount of momentum, $dp = dm\,c$. This gives the same amount of momentum to the ship, but in the opposite direction. Measured in the instantaneous inertial rest frame of the spaceship at time τ, the change in momentum can be written as $mdv = -dm\,c$, where dv is the small increase in velocity of the spaceship in the time interval $d\tau$. Thus, the velocity increase in this time interval is $dv = -dm\,c/m$, and this gives for the proper acceleration,

$$a_0 = g = \frac{dv}{d\tau} = -\frac{c}{m}\frac{dm}{d\tau}\,, \tag{7.64}$$

where we have assumed that the proper acceleration is kept fixed at the level of the gravitational acceleration on the surface of the earth. We note that this gives a differential equation for the change with time of the mass of the spaceship

$$\frac{dm}{d\tau} = -\frac{mg}{c}, \tag{7.65}$$

with an exponential function as solution

$$m(\tau) = m_0 \exp\left(-\frac{g}{c}\tau\right). \tag{7.66}$$

We denote by T the time onboard when all fuel has been consumed, so that $m(T) = m_1$. This gives

$$m_1 = m_0 \exp\left(-\frac{g}{c}T\right). \tag{7.67}$$

If we make the assumption that 90% of the spaceship's weight at the beginning of the journey is fuel this gives the following (proper) time onboard the ship when it runs out of fuel after word

$$T = \frac{c}{g}\ln 10 \approx 2.3 \text{ years.} \tag{7.68}$$

The speed of the spaceship is then (see (7.30))

$$v = \tanh(\frac{g}{c}T)c = \frac{m_0^2 - m_1^2}{m_0^2 + m_1^2} \approx 0.98c\,, \tag{7.69}$$

and the time dilation factor is

$$\gamma = \cosh\left(\frac{g}{c}T\right)c = \frac{1}{2}\left(\frac{m_0}{m_1} + \frac{m_1}{m_0}\right) \approx 5\,. \tag{7.70}$$

This is indeed a large velocity and gamma factor. The coordinate time at earth and the distance to the ship at this point are (see (7.36))

$$t = \frac{c}{g}\sinh\left(\frac{g}{c}T\right) \approx 5\,\text{years}\,,$$

$$x - x_0 = \frac{c^2}{g}\cosh\left(\frac{g}{c}T\right) \approx 4\,\text{light years}\,. \tag{7.71}$$

Even if this does not bring the spaceship out to distant galaxies, the distance is still very impressive, comparable to the distance to the closest star. One should, however, note that the assumptions we have made are rather unrealistic. In particular this is so for the assumption that all mass of the fuel is converted to energy, which should be compared to the efficiency of mass conversion of about 1% for the nuclear fusion process where hydrogen is transformed into helium. A more realistic estimate would definitely limit the space travel much more than shown by the numbers above. However, the idea that a rocket engine based on emission of photons could give a constant acceleration over a long time, and thereby bring a spaceship in an efficient way far outside the solar system, may not be such a bad idea.

7.4 Doppler effect with photons

Even if the speed of a light signal is unchanged when changing from one inertial reference frame to another, the frequency of the light will generally be different in the two frames. This is due to the Doppler effect,[1] which is well known for wave propagation also in non-relativistic physics. The correct relativistic Doppler shift formula can be found by considering light as a propagating wave, but another way to derive it, which is in fact simpler, is to make use of the transformation formula for relativistic four-momenta. This is the approach we take here, when we consider the transformation of four-momentum for a massless photon between two inertial frames, and use the de Broglie relations to translate this to a transformation of frequencies.

Let us then consider the situation where a photon is emitted from a space-time point O, which is the origin of an inertial reference frame S. In this frame the photon momentum is directed with angle θ relative to the x-axis in the x, y plane,

$$\mathbf{p} = p(\cos\theta\mathbf{i} + \sin\theta\mathbf{j})\,. \tag{7.72}$$

Since the photon is massless the components of the four-momentum in this frame are

$$\underline{\mathbf{P}} = p\,(1, \cos\theta, \sin\theta, 0) \qquad (S \text{ frame})\,. \tag{7.73}$$

Let us assume that the photon is absorbed by a detector in another inertial reference frame S', which moves with velocity v in the x-direction relative to S. The four momentum in this frame has components

$$\underline{\mathbf{P}}' = p'(1, \cos\theta', \sin\theta', 0) \qquad (S' \text{ frame})\,. \tag{7.74}$$

The components of the two reference frames are related by the Lorentz transformation

$$\begin{aligned} p'^0 &= \gamma(p^0 - \beta p^1)\,, \\ p'^1 &= \gamma(p^1 - \beta p^0)\,, \\ p'^2 &= p^2\,, \\ p'^3 &= p^3\,, \end{aligned} \tag{7.75}$$

which gives

$$\begin{aligned} p' &= \gamma p(1 - \beta\cos\theta)\,, \\ p'\cos\theta' &= \gamma p(\cos\theta - \beta)\,, \\ p'\sin\theta' &= p\sin\theta\,. \end{aligned} \tag{7.76}$$

[1] Christian Andreas Doppler (1803–1853) was an Austrian mathematician and physicist. He is celebrated for his principle, the Doppler effect, which he used to explain the color of binary stars.

Only two of these are independent equations, as one can readily check, and these two equations can be used to solve for p' and $\cos\theta'$,

$$p' = \gamma(1 - \beta\cos\theta)p\,, \tag{7.77}$$

$$\cos\theta' = \frac{\cos\theta - \beta}{1 - \beta\cos\theta}\,. \tag{7.78}$$

The first one of these gives the Doppler shift formula. To show this we make use of the de Broglie formula which gives the link between the particle and wave nature of the photon, $p = E/c = h\nu/c$, with ν as the photon frequency. Equation (7.77) then gives the frequency transformation formula

$$\nu' = \gamma(1 - \beta\cos\theta)\nu\,. \tag{7.79}$$

This equation shows how the frequency of a light signal changes between two inertial frames in relative motion. The frame S' moves with velocity βc relative to S and θ is the angle between the photon and the relative velocity of the two frames, as measured in S. Clearly the same formula should be applicable if we interchange the two frames. This gives

$$\nu = \gamma(1 + \beta\cos\theta')\nu'\,, \tag{7.80}$$

where we have introduced a sign change for the relative velocity. The formula can be rewritten as

$$\nu' = \frac{1}{\gamma(1 + \beta\cos\theta')}\nu\,. \tag{7.81}$$

Consistency between (7.79) and (7.81) then gives a relation between the angles measured in the two frames, and this is the same as the equation (7.78).

We conclude that the Doppler shift can be expressed either as in (7.79) or in (7.81), depending on whether the angle of the light signal refers to the inertial frame S where it is emitted or the inertial frame S' where it is absorbed. We consider now some special cases.

a) $\theta = 0$: The light signal is emitted in the direction of motion of reference frame S'. Seen from S' the emitter of the signal is moving away from the receiver. The formula is

$$\nu' = \gamma(1 - \beta)\nu = \sqrt{\frac{1-\beta}{1+\beta}}\,\nu\,. \tag{7.82}$$

The light is now redshifted since the frequency in S' is lower than in S.

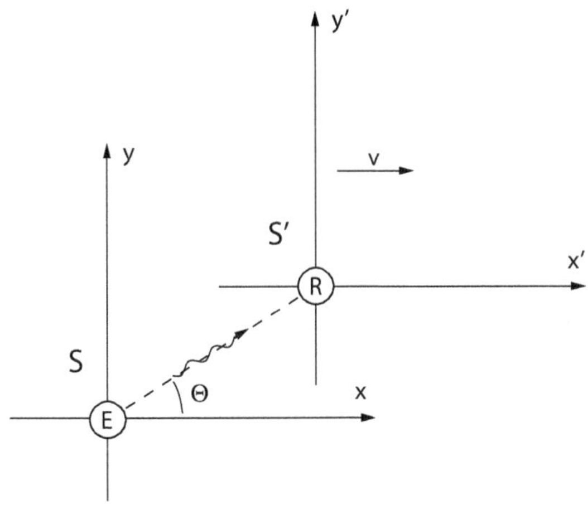

Fig. 7.2 The Doppler effect. A photon is emitted from a sender E in an inertial frame S at an angle θ with respect to the x-axis. The photon is registered by a receiver R in another inertial frame S', which moves with velocity v relative to the first reference frame. The photon frequency registered in R is different from the frequency of the photon when emitted from E. Also the direction of the photon appears different in the two frames, as discussed in the text.

b) $\theta = \pi$: The light signal is emitted against the direction of motion of reference frame S', so that the emitter is moving towards the receiver. The formula is

$$\nu' = \gamma(1+\beta)\nu = \sqrt{\frac{1+\beta}{1-\beta}}\,\nu\,, \tag{7.83}$$

and the light is blue shifted in reference frame S'.

c) $\theta' = \pi/2$: The light signal is now *received* with direction orthogonal to the velocity of reference frame S'. This gives

$$\nu' = \frac{1}{\gamma}\nu\,. \tag{7.84}$$

Even in this case there is a Doppler shift. We may view this as a time dilation effect, where time in S is seen as slow when viewed from reference frame S'. The light signal is redshifted.

d) $\theta = \pi/2$: The light signal is now *emitted* at 90° degrees in S, and formula is now

$$\nu' = \gamma\nu\,. \tag{7.85}$$

The time dilation effect works the other way, and the light signal is blue shifted. In reference frame S' the angle θ' is larger than 90°, as follows from Eq. (7.78). This means that the signal is received with a velocity component against the motion of the frame, which is consistent with the blue shift.

Examples

Reflections from moving mirrors

A monochromatic light source is at rest in the laboratory and sends light signals with frequency ν_0 towards two mirrors, which both have their reflective surfaces oriented so that the light is reflected back to the source. The frequencies of the reflected signals are measured there by detectors which are at rest in the lab frame. The situation, as viewed in the lab frame, is illustrated in Fig. 7.3.

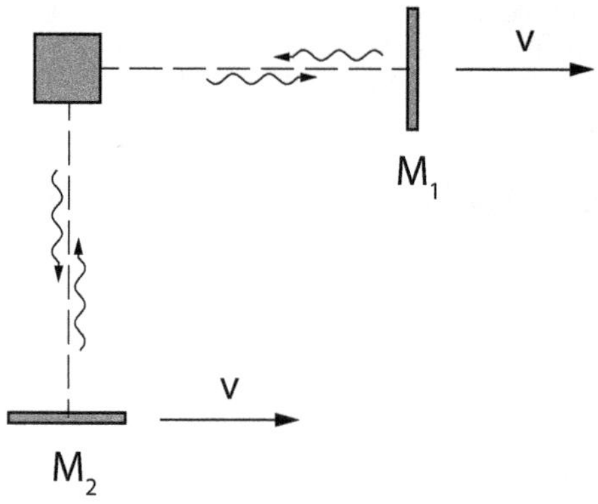

Fig. 7.3 Reflection of light on moving mirrors. The reflected light ray in mirror M_1 is emitted in the same direction as the motion of the mirror. The light ray reflected in M_2 is emitted perpendicular to the motion of the mirror.

The first mirror (M_1) moves away from the source, so that the light signal is oriented in the same direction as the velocity vector $\mathbf{v}$ of the mirror. The second mirror (M_2) has the same velocity $\mathbf{v}$, but it follows a path parallel to the first one. The light signal sent to this mirror has direction orthogonal to the velocity vector, when viewed in the lab frame. The problem we address here is what the frequencies of the reflected light rays are in the two cases, when detected at the light source.

We first consider the reflection on M_1. The emitted light signal has angle $\theta = 0$ relative to the velocity of the mirror. This corresponds to case a) in the discussion above. The frequency of the signal, when viewed from the reference frame of the mirror is then given by the Doppler formula

$$\nu_1 = \sqrt{\frac{1-\beta}{1+\beta}}\,\nu_0\,, \quad \beta = v/c\,. \tag{7.86}$$

In the reference frame of the mirror the reflected beam will have the same frequency ν_1 as the incoming signal, but with opposite direction of propagation. However, in this reference system the lab frame will also move in the opposite direction, with velocity $-\mathbf{v}$. Thus, with the reflected signal viewed as emitted by the mirror, the situation for emission and detection of this signal is also described by case a). This means that the frequency detected there is

$$\nu_1' = \sqrt{\frac{1-\beta}{1+\beta}}\,\nu_1 = \frac{1-\beta}{1+\beta}\,\nu_0\,. \tag{7.87}$$

This shows that for mirror M_1 the red-shift effect is repeated. It happens both for the outgoing and the reflected light signal.

We next consider the situation with reflection of the light signal on mirror M_2. For the outgoing signal we have the situation described by case d) with the light signal emitted at angle $\theta = \pi/2$, when viewed in the rest frame of the emitter. This gives as the frequency of this signal, when viewed in the rest frame of the mirror,

$$\nu_2 = \gamma\nu_0\,. \tag{7.88}$$

When viewed in the lab frame the reflected signal will propagate in the direction opposite of the outgoing signal. However, now the mirror acts as the emitter of the light signal, while the direction of propagation of the signal is $\pi/2$ in the rest frame of the detector. This corresponds to case c), which means that the frequency detected there for the reflected light signal is,

$$\nu_2' = \frac{1}{\gamma}\nu_2 = \nu_0\,. \tag{7.89}$$

We conclude that in this case the Doppler effect on the outgoing light signal is cancelled by the Doppler effect on the reflected signal.

Stellar aberration

The Doppler formulas show that for light sent between an emitter and a receiver, the light ray appears with different frequencies for the two, if they move relative to each other. However, the Doppler formulas also imply that the direction of the light ray appears to be different for the emitter and the receiver. This is explicitly shown in Eq. (7.78). The effect is referred to as *aberration* of light. In astronomy it causes the apparent positions of stars to make small changes during the year, due to the variation in the velocity of the earth under the motion around the sun. We will study this effect here.

For simplicity, let us consider the effect for a star which is located in the ecliptic plane, that means in the plane defined by the earth's orbit. The light ray from the star will then make an angle θ with the velocity vector of the earth, which describes a full circle, $0 \to 2\pi$, during the year. The velocity of the earth is then measured relative to the rest frame of the sun, and the Doppler formula relates the frequency of the light wave registered on earth to the frequency of the light in the reference frame of the sun.

The speed of the earth in its orbit around the sun is $v = 30 \cdot 10^3$ m/s, which corresponds to $\beta = 1.0 \cdot 10^{-4}$. This is very small, and an approximation, by expanding Eq. (7.78) to first order in β, therefore seems reasonable,

$$\cos\theta' \approx (\cos\theta - \beta)(1 + \beta\cos\theta) \approx \cos\theta - \beta\sin^2\theta\,. \tag{7.90}$$

Let us further introduce η as the difference between the angles θ' and θ. It describes the yearly variations in the angular position of the star, as it appears on earth. To first order in η we have $\cos(\theta + \eta) = \cos\theta + \eta\sin\theta$, and inserting this in (7.90) we get

$$\eta(\theta) = \beta\sin\theta\,. \tag{7.91}$$

We note that η vanishes for $\theta = 0$ and $\theta = \pi$, which means when the earth's velocity vector points directly towards or away from the star. The maximum deviation happens when the velocity vector is orthogonal to the direction of the star. Measured in radians the maximum deviation is

$$\eta(\pm\pi/2) = \pm\beta = \pm 1.0 \cdot 10^{-4}\,, \tag{7.92}$$

and measured in arcseconds it is

$$\eta(\pm\pi/2) = 20.6''\,. \tag{7.93}$$

7.5 Exercises

Problem 7.1

a) An electron is moving in a storage ring of radius $R = 10$m, with a speed that corresponds to the gamma factor $\gamma = 30$. Find the velocity of the particle. What is the time, measured in the lab frame, spent on one circulation in the ring, and what is the proper time. Find the acceleration a, measured in the lab frame, and the proper acceleration a_0 of the electron.

b) Another particle moves in the x, y-plane with coordinates

$$x = ut, \qquad y = \frac{1}{2}gt^2, \tag{7.94}$$

(u and g are constants) in an inertial frame. What is the acceleration in this frame, and what is the proper acceleration a_0. What will happen with the proper acceleration as t increases?

Problem 7.2

A spacecraft leaves the earth at local time $t = 0$ and heads for the closest star, *Proxima Centauri*, at a distance of $d = 4.2$ light years. The spacecraft starts with a velocity $v = 0$ relative to earth, and it moves during the journey along the x-axis in an earth fixed reference frame. The initial value of the position coordinate is $x = 0$, and the (proper) time coordinate measured on the space craft is set to $\tau = 0$ on departure.

a) During the first part of the journey (part I), the coordinates of the spacecraft, in the earth-fixed reference frame, define a hyperbolic space-time orbit, given by

$$x - x_I = \frac{c^2}{a}\cosh\left(\frac{a}{c}(\tau - \tau_I)\right), \quad t - t_I = \frac{c}{a}\sinh\left(\frac{a}{c}(\tau - \tau_I)\right), \tag{7.95}$$

with x_I, t_I and τ_I as constants. Use the initial conditions to determine the constants, and rewrite the equations with these values.

b) Show that the time parameter τ satisfies the criterion of being the proper time of the spacecraft. Find, for part I of the journey, the 4-velocity and 4-acceleration as a functions of τ, and show that the parameter a is identical to the proper acceleration of the spacecraft.

At time τ_0 measured in the proper time of the spacecraft, the spacecraft is halfway to Proxima Centauri. At this point the acceleration is reversed, so that $a \to -a$, and the velocity then decreases until it reaches Proxima Centauri with final velocity $v = 0$. This part of the trip, with acceleration

$-a$, we refer to as part II. The spacecraft visits Proxima Centauri only for a short time, and we neglect this time interval in our description of the journey. The return travel to earth is carried out in the same way as the travel towards Proxima Centauri, and we refer to these parts of the journey as parts III and IV. Similar expressions as (7.95) are valued for parts $II - IV$, but with other constants.

c) Based on the symmetry between the four parts of the journey, draw a two-dimensional Minkowski diagram, which shows the full journey to Proxima Centauri and back.

d) In the following assume the acceleration is identical to the gravitational acceleration on earth, $a = 9.8m/s^2$. Determine the proper time value τ_0 at half way to the star. Find the total time of the journey as measured on earth and on the spacecraft.

e) What is the maximum speed reached by the spacecraft on the journey, as measured on earth?

Problem 7.3

A straight rod is moving along the x-axis of an inertial reference frame S. The two endpoints A and B follow hyperbolic space-time trajectories, described the following time dependent x-coordinates in S,

$$x_A = c\sqrt{t^2 + c^2/a^2}\,, \quad x_B = c\sqrt{t^2 + c^2/b^2}\,, \tag{7.96}$$

where c is the speed of light, and a and b are positive constants, with $b < a$.

a) A second inertial frame S' moves along the x-axis with velocity v relative to S. The origins of the two reference frames ($x = t = 0$ and $x' = t' = 0$) are chosen to coincide.

Show that the motion of A and B, when expressed in terms of the coordinates of S', has precisely the same form as in S,

$$x'_A = c\sqrt{t'^2 + c^2/a^2}\,, \quad x'_B = c\sqrt{t'^2 + c^2/b^2}\,. \tag{7.97}$$

(To demonstrate this it may be convenient to rewrite the above relations in terms of the squared coordinates x^2 and t^2.)

b) At time $t = 0$ the frame S is an instantaneous rest frame of both A and B. Show this and find the distance between A and B measured in S at this moment. Show that the same results are valid for the reference frame S' at time $t' = 0$.

Based on this we may conclude that for any point on the space-time trajectory of A, the instantaneous inertial rest frame of A is a rest frame also for B. Furthermore, the distance between A and B, when measured in the instantaneous inertial rest frame, is constant. Explain these conclusions.

c) Use the above results to show that the proper accelerations of A and B are constants, and give the values of these.

d) At $t = 0$ a light signal with frequency ν_0 is sent from A and is subsequently received at B. What is the velocity of B (measured in S) when the signal is received, and what is the frequency of the signal, measured at B?

Problem 7.4

A spacecraft is passing the earth at great speed, $v = 0.8c$. The motion is linear, with d as the shortest distance between the spacecraft and the earth. In an earth-fixed frame S the position of the spacecraft is described by coordinates

$$x(t) = vt\,, \quad y = d\,, \quad z = 0. \tag{7.98}$$

When passing, the spacecraft is continuously submitting messages to the earth by radio, on frequency ν_0. An antenna on earth, located at the origin of the coordinate system, receives the messages and registers the frequency $\nu(t)$ and direction of the received signal during the passage. This direction is measured by the angle $\theta(t)$ between the signal and the x-axis (the direction of motion of the spacecraft).

a) A radio signal, which is received at time t on earth, is emitted a bit earlier, at time $t_e = t - \Delta t$. Determine Δt as function of t.

b) Express $\cos\theta$ in terms of t and Δt, and plot $\cos\theta$ as a function of time for an interval around $t = 0$. Use $\tilde{t} = ct/d$ as a dimensionless time coordinate.

c) Similarly give the relevant Doppler formula for the frequency ratio $\nu(t)/\nu_0$ and plot also this as a function of $\tilde{t}$. What are the asymptotic values for $\nu(t)/\nu_0$ when $t \to \pm\infty$?

Chapter 8

Relativistic dynamics

In non-relativistic physics energy and momentum are both conserved in physical processes, and so is the total mass of an interacting system. Also in special relativity energy and momentum are conserved, even if the definition of these physical variables are changed, but mass is no longer conserved in an absolute sense. This is really one of the most important consequences of Einstein's theory. We will in the first part of this chapter examine these modified conservation laws by applying them to relativistic collision processes. We will next discuss how Newton's second law can be correctly updated to relativistic form. Our approach is based on the general idea of rewriting the non-relativistic equation in a *covariant* relativistic form. This means that we express the equation in terms of four-vectors and tensors, in such a way that it has the correct non-relativistic limit for low velocities $v << c$. The covariant form secures that the equation is valid in all inertial frames, and therefore satisfies the condition of relativistic invariance. Even if this is a formal approach, it leads to correct results in concrete cases, as for example for charged particles in electromagnetic fields, which we analyze in some detail.

8.1 Conservation of relativistic energy and momentum

To examine the different implications of energy and momentum conservation in non-relativistic and relativistic physics, we consider a collision process of the form schematically shown in Fig. 8.1. In this process a set of particles are initially freely moving, but then enter a region of interaction. From this region another set of particles is emerging, and these particles are again moving freely. The outgoing particles will generally be different from the incoming, but for the sake of simplicity we assume that no massless particle (radiation) is emitted in the process.

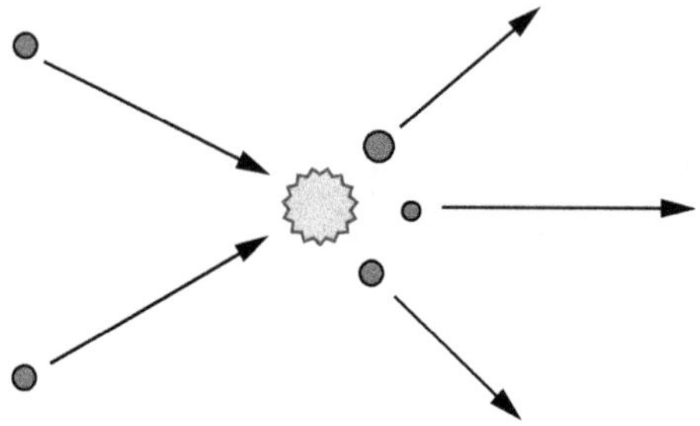

Fig. 8.1 A collision process. Two particles are moving freely until they reach a collision region. As a result of the collision a set of new particles emerges. Relativistic four-momentum is conserved in the process.

In the non-relativistic case we may formulate three conservation laws which apply to the collision process. They are

$$\sum_i \mathbf{p}_i = \sum_f \mathbf{p}_f \,, \tag{8.1}$$

$$\sum_i \frac{\mathbf{p}_i^2}{2m_i} = \sum_f \frac{\mathbf{p}_f^2}{2m_f} + Q \,, \tag{8.2}$$

$$\sum_i m_i = \sum_f m_f \,, \tag{8.3}$$

where the index i refers to the incoming particles and f to the outgoing ones. The first equation states that total momentum is conserved. The second one states that total energy is conserved. This does not mean that total *kinetic* energy needs to be conserved. In inelastic collisions that is not the case, and Q measures the amount of energy that is transformed from kinetic to other forms of energy (internal energy) in such a collision. The third equation expresses the conservation of total mass.

In the relativistic setting these conservation laws are replaced by a single four-vector equation,

$$\sum_i \underline{\mathbf{P}}_i = \sum_i \underline{\mathbf{P}}_f \,, \tag{8.4}$$

which states that the total four-momentum is preserved in the process. If the non-relativistic limit should be reached in the correct way from this equation, then Eqs. (8.1)-(8.3) should follow from (8.4) when the particle velocities are small compared to the speed of light. We shall check the conditions for this to be the case.

The space component of the four-vector equation has the same form as the non-relativistic equation for conservation of momentum. But the meaning is different because of the relativistic form of momentum. Expressed in terms of velocities it is

$$\sum_i \gamma_i \, m_i \, \mathbf{v}_i = \sum_f \gamma_f \, m_f \, \mathbf{v}_f \,, \tag{8.5}$$

where the gamma factors of the particles are missing in the non-relativistic equations. However, in the non-relativistic limit, $v/c \to 0$, we have $\gamma \to 1$ and the relativistic equation reproduces the non-relativistic equation.

We next consider the time component of the four-vector equation (8.4), which we may write as

$$\sum_i \gamma_i \, m_i \, c^2 = \sum_f \gamma_f \, m_f \, c^2 \,. \tag{8.6}$$

Obviously, if the non-relativistic limit also here is taken as $\gamma \to 1$ the equation will reproduce the non-relativistic mass conservation equation. However, this raises the question of how the non-relativistic equation for conservation of energy should be reproduced. The answer is that this is also contained in the time component of (8.4), but we have to keep the first order contributions in v^2/c^2 when we make an expansion in this small quantity. This gives the following equation

$$\sum_i m_i \, c^2 + \frac{1}{2} m_i v_i^2 = \sum_f m_f \, c^2 + \frac{1}{2} m_f \, v_f^2 \,. \tag{8.7}$$

By use of the non-relativistic form of the momentum $\mathbf{p}$ it can be rewritten as

$$\sum_i \frac{\mathbf{p}_i^2}{2m_i} = \sum_f \frac{\mathbf{p}_f^2}{2m_f} + \left(\sum_f m_f \, c^2 - \sum_i m_i \, c^2 \right) \,. \tag{8.8}$$

This is seen to have the same form as the non-relativistic energy conservation equation, but with an explicit expression for the Q term,

$$Q = \sum_i m_i \, c^2 - \sum_f m_f \, c^2 \,. \tag{8.9}$$

This result is interesting and important. Consistency between the relativistic and non-relativistic equations implies that the total mass cannot be conserved in a strict sense in special relativity. Instead, in an inelastic collision, where $Q \neq 0$, there will be a mass difference between the initial and final states, which is determined by the ratio Q/c^2. So mass is in such a process converted to kinetic energy or kinetic energy is converted to mass.[1] This gives a concrete, physical interpretation of the rest energy $E = mc^2$ of a physical body, as the total energy stored in the body.

As a simple example we consider a completely inelastic collision, where two bodies collide and create a single larger body. In the process heat is created, and we assume that the heat energy is stored in the body as internal energy. For simplicity, let us assume the two bodies before the collision to have equal masses m. We consider the collision process in the *center-of-mass system*, where the two bodies before the collision have momenta of equal size but with opposite directions. The larger body that is produced in the collision has mass M and sits at rest in this reference frame. The three-vector part of the relativistic momentum conservation equation simply states that the total momentum vanishes in this frame. The conservation equation for energy is

$$2\gamma mc^2 = Mc^2 \,, \tag{8.10}$$

with γ as the (common) gamma factor of each of the two colliding particles. The equation shows that the mass of the body that is formed in the collision is larger than the sum of the masses of the two particles before the collisions,

$$\Delta m = M - 2m = 2m(\gamma - 1) > 0 \,. \tag{8.11}$$

When expanded in powers of v^2/c^2, we have $\gamma = 1 + \frac{1}{2}\frac{v^2}{c^2} + \ldots$. This gives for $v << c$

$$2\left(\frac{1}{2}mv^2\right) = \Delta m\, c^2 \,, \tag{8.12}$$

which shows that the kinetic energy of the colliding particles is present, after the collision, as an increase in the mass of the larger body that is formed by the collision. This result is independent of how energy is stored in the body, but in the present case it seems natural to identify $Q = \Delta mc^2$ with the heat created by the collision. The mass energy formula in fact suggests quite generally that if a body is heated, the increase in internal energy will lead to an increase in its mass. However, the mass increase obtained by heating the body is under normal conditions extremely small. In the present case, if we assume the two bodies collide with the velocity $v = 100\,\text{km/h}$ we have $v/c = 9.26 \cdot 10^{-8}$, which gives for the relative increase in the mass, $\Delta m/(2m) = v^2/(2c^2) = 4.29 \cdot 10^{-15}$, a truly small number.

[1]In general the kinetic energy may in part take the form of radiation, *i.e.*, massless particles.

The inverse of the process considered above is a fission process where a body is split in parts by an explosion of some sort. In that case the total mass after the explosion is smaller than the total mass before the explosion, and the missing mass is converted to kinetic energy (and radiation) according to the mass conversion formula $\Delta E = \Delta mc^2$. Under conventional explosions, the mass converted to energy is extremely small even for strong explosions. However, as is well known, it is in the case of nuclear explosions that the mass to energy conversion is really relevant. For example, in the case of the Hiroshima bomb, with the strength of explosion corresponding to that of 20 000 tons of TNT, the converted mass was about $\Delta m = 1$ gram. This gives, according to Einstein's formula for the energy released, $E = \Delta mc^2 = 0.9 \cdot 10^{14}$ J.

To conclude, in special relativity the total four-momentum of an isolated system is always conserved. This conservation law reduces to the standard expressions for conservation of energy and momentum in the non-relativistic limit. However, a consequence of the relativistic formula is that the total mass is not strictly conserved. The change of mass in a physical process is related to the difference in total kinetic energy, including radiation energy, between the initial and final states.

8.2 The center-of-mass system

Consider a composite system which is isolated from the surroundings so that no external force acts on the system. In non-relativistic physics the center of mass of the system, **R**, is defined by

$$M\mathbf{R} = \sum_k m_k \mathbf{r}_k \,, \tag{8.13}$$

with M as the total mass of the system, and where $\mathbf{r}_k, k = 1, 2, ...$ denote the position vectors of the small parts of the system, with masses m_k. Without external forces the center of mass is non-accelerated and therefore we can find an inertial reference frame where it is at rest. This is the *center-of-mass system*, which is then characterized by

$$\mathbf{P} = M\dot{\mathbf{R}} = \sum_k m_k \dot{\mathbf{r}}_k = 0 \,. \tag{8.14}$$

In the center-of-mass system (the *CM system* for short) the total momentum **P** of the physical system therefore vanishes.

In relativistic physics the center of mass is not a well defined concept. This can be seen in the following way. In the definition of the CM-position vector **R** it is essential that the sum (8.13) is performed at *equal times* for all parts for the system. This is a definition which is independent of choice of inertial frame in non-relativistic physics, since there *equal time* is a universal concept. However in relativistic physics that is no longer true. If we therefore define the sum over a three-dimensional space with time coordinate $t = constant$ in one inertial frame, that is in general different from defining the sum over three-dimensional space with $t' = constant$ in another inertial frame. There will in general be no simple relation between the result of such different summations. Therefore we simply give up the idea of defining, in general, the center of mass of an extended system.

Even so, the center-of-mass system (also referred to as the center-of momentum system) is both a well defined and useful concept in relativistic physics. This follows from the fact that the total four-momentum $\underline{\mathbf{P}}$ of the system is well defined, and the condition that the space part vanishes, as in (8.14), specifies an inertial frame which we identify as the center-of-mass system. The condition that identifies the center-of-mass system therefore is

$$\mathbf{P} = \sum_k \mathbf{p}_k = 0\,, \tag{8.15}$$

which is now written in a form that is correct both in non-relativistic and relativistic physics, but in the latter case one has to remember that for massive particles the right definition of relativistic momentum is $\mathbf{p} = \gamma m \mathbf{v}$.

The total momentum $\underline{\mathbf{P}}$ is a reference-frame independent four-vector, in spite of the fact the sum over contributions from all parts of the physical system seems to depend on the choice of reference frame. This is in fact a consequence of conservation of the total four-momentum for an isolated physical system. For a system of *non-interacting* particles this is quite clear, since the momentum is conserved for each particle individually. This means that the sum of the particles' four-momenta will be independent of the points on the particle world lines that are chosen when performing the sum. In particular the result is the same whether they are summed at equal times in one inertial frame or another. In the general case the same conclusion can be reached by use of momentum conservation, expressed as a *local* conservation law.

We conclude that the total four-momentum of an isolated system, defined as a sum of four-momenta of all the (small) parts of the system, is a well defined, conserved four-vector, and since it is a timelike vector, an inertial frame can be found where the space part of the vector vanishes.

(See however Problem 8.1 for an exceptional case.) This is the center-of-mass system, and it is a unique reference frame, up to its orientation in three-dimensional space.

Example

Pi meson decay

Pi mesons (pions) are unstable elementary particles. We consider here a decay process of a charged pion π^+ into a muon μ^+ and a neutrino ν_μ. The masses of the particles are $m_\pi = 273m_e$ and $m_\mu = 207m_e$, with $m_e = 0.51\,\text{MeV/c}^2$ as the electron mass. (The standard energy unit in particle physics, eV = electron volt is used.) The mass of the neutrino is so small that the particle can be regarded as massless. Note that in the following the labels π, μ and ν are used to specify the types of particles rather than the space-time coordinates.

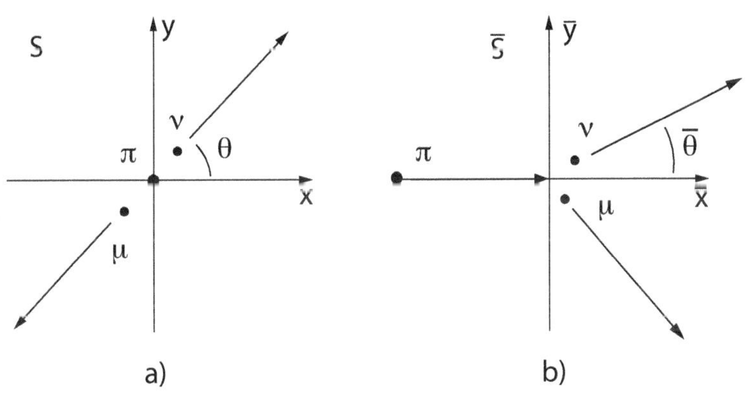

Fig. 8.2 The decay of a pion into a neutrino and a muon, as seen in the CM system S, and in the lab frame $\bar{S}$. The x-axis is chosen in the direction of motion of the pion in $\bar{S}$. All particles are viewed as moving in the x, y plane.

In the figure the decay process is shown both in the rest frame S of the pion, and in the laboratory frame $\bar{S}$. We assume that the pion, in this reference frame, moves with the velocity $v = 0.8c$ along the x-axis. To distinguish the variables of the two reference frames S and $\bar{S}$ we mark the variables of the latter with a "bar", so that for example the angle of the neutrino relative to the x-axis in S is θ and the corresponding angle in $\bar{S}$

is $\bar{\theta}$. The particles are viewed as moving in the x, y-plane, with the pion velocity in the x-direction.

We study first the process in the rest frame S, where we set up the equations for conservation of relativistic energy and momentum and use them to determine the energy and the momentum of the muon and of the neutrino in this reference system. The equations are

$$\begin{aligned} E_\mu + E_\nu &= m_\pi c^2 \,, \\ \mathbf{p}_\mu + \mathbf{p}_\nu &= 0 \,, \end{aligned} \tag{8.16}$$

with the two equations related by the relativistic energy-momentum relations

$$\begin{aligned} E_\mu^2 &= p_\mu^2 c^2 + m_\mu^2 c^4 \,, \\ E_\nu &= p_\nu c \,. \end{aligned} \tag{8.17}$$

Since the momenta of the muon and the neutrino are equal in absolute value, we get from (8.17),

$$E_\nu^2 = p_\mu^2 c^2 = E_\mu^2 - m_\mu^2 c^4 \,, \tag{8.18}$$

and from (8.16)

$$E_\nu^2 = (m_\pi c^2 - E_\mu)^2 \,. \tag{8.19}$$

Combined they give

$$E_\mu = \frac{m_\pi^2 + m_\mu^2}{2m_\pi} c^2 = 215 m_e c^2 = 109.6\,\mathrm{MeV} \,, \tag{8.20}$$

and

$$E_\nu = m_\pi c^2 - E_\mu = \frac{m_\pi^2 - m_\mu^2}{2m_\pi} c^2 = 58 m_e c^2 = 29.6\,\mathrm{MeV} \,. \tag{8.21}$$

The absolute value of the momentum of the two particles is

$$p_\mu = p_\nu = E_\nu / c = 29.6\,\mathrm{MeV/c} \,. \tag{8.22}$$

Due to rotational invariance of the decay process the direction of the momentum vector $\mathbf{p}_\nu$ is undetermined, but momentum conservation restricts $\mathbf{p}_\mu$ to be directed opposite to $\mathbf{p}_\nu$.

In the lab frame $\bar{S}$ the situation is only axially symmetric about the x-axis, which is the direction of motion of the pion. We will find the neutrino energy as a function of energy in $\bar{S}$, and also the probability distribution over the directions of the emitted neutrino. For the momentum four-vectors

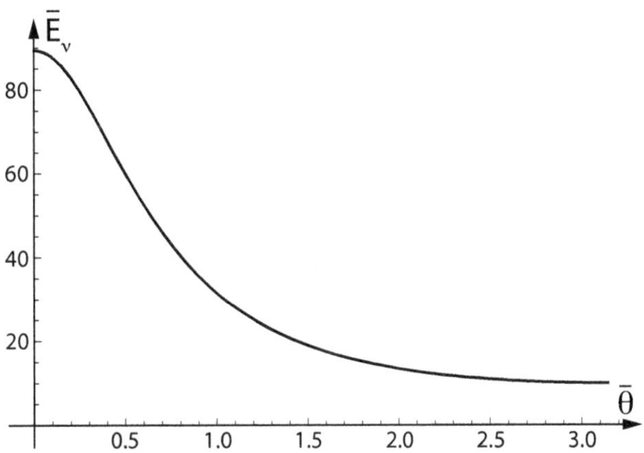

Fig. 8.3 Energy of the neutrino as a function of the angle between the neutrino momentum and the direction of motion of the decaying pion. The energy is measured in *MeV*.

we use the Lorentz transformations between the S and the $\bar{S}$ frames in the form

$$\begin{aligned} \bar{p}^0 &= \gamma(p^0 + \beta p^1)\,, \\ \bar{p}^1 &= \gamma(p^1 + \beta p^0)\,, \end{aligned} \tag{8.23}$$

with $\beta = 0.8$, which gives $\gamma = 1.67$. The component p^2 is unchanged in the transformation, while p^3 vanishes for all the particles.

For the neutrino we have the relation $p^0 = E/c$, and the transformation can be expressed as

$$\begin{aligned} \bar{E}_\nu &= \gamma(E_\nu + \beta c p_\nu^1) = \gamma(1 + \beta\cos\theta)E_\nu\,, \\ \bar{p}_\nu^1 &= \gamma(p_\nu^1 + \beta E_\nu/c) = \gamma(\cos\theta + \beta)p_\nu\,. \end{aligned} \tag{8.24}$$

The ratio gives the relation between the angles in S and $\bar{S}$,

$$\cos\bar{\theta} = \frac{\bar{p}_\nu^1 c}{\bar{E}_\nu} = \frac{\cos\theta + \beta}{1 + \beta\cos\theta}\,, \tag{8.25}$$

with inverse

$$\cos\theta = \frac{\cos\bar{\theta} - \beta}{1 - \beta\cos\bar{\theta}}\,. \tag{8.26}$$

This determines the energy in the lab frame as function of $\bar{\theta}$

$$\bar{E}_\nu = \frac{1}{\gamma(1 - \beta\cos\bar{\theta})}E_\nu = \frac{m_\pi^2 - m_\mu^2}{2m_\pi\gamma(1 - \beta\cos\bar{\theta})}c^2\,. \tag{8.27}$$

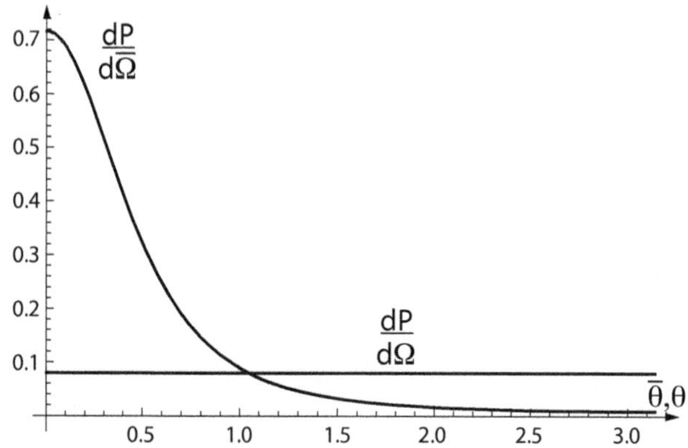

Fig. 8.4 Probability distribution for the direction of the neutrino momentum, as function of the angle θ. The flat curve shows the distribution in the rest frame S of the pion, and the peaked curve shows the distribution in the lab frame $\bar{S}$ of the decaying pion.

The result is shown graphically in Fig. 8.3. We consider next the probability distribution, described by the probability per unit solid angle for the direction of the emitted neutrino. In reference frame S where the situation is rotationally invariant, the distribution is uniform, with

$$\frac{dP}{d\Omega} = \frac{1}{4\pi}\,. \tag{8.28}$$

In the lab frame $\bar{S}$, the probability distribution depends on the angle $\bar{\theta}$. The two distributions are related by

$$\frac{dP}{d\bar{\Omega}} = \frac{dP}{d\Omega}\frac{d\Omega}{d\bar{\Omega}}\,, \tag{8.29}$$

with $d\Omega = \sin\theta d\theta d\phi$ and $d\bar{\Omega} = \sin\bar{\theta} d\bar{\theta} d\phi$. This follows since the integrated probabilities are equal in S and $\bar{S}$. This further gives for the probability distribution in $\bar{S}$

$$\begin{aligned}\frac{dP}{d\bar{\Omega}} &= \frac{1}{4\pi}\frac{d(\cos\theta)}{d(\cos\bar{\theta})}\\ &= \frac{1}{4\pi\gamma^2}\frac{1}{(1-\beta\cos\bar{\theta})^2}\,.\end{aligned} \tag{8.30}$$

The distribution, shown in Fig. 8.4, has a clear peak in the forward direction of the decaying pion.

8.3 Newton's second law in relativistic form

Our starting point is the (non-relativistic) Newton's second law, which we write as

$$\mathbf{F} = \frac{d\mathbf{p}}{dt}, \tag{8.31}$$

with $\mathbf{p} = m\mathbf{v}$. Let us assume that the equation is applied to the motion of a small body (point particle), with $\mathbf{p}$ as the momentum of the body, and $\mathbf{F}$ as the force acting on the body. The equation is a vector equation in three dimensions, which in a relativistic generalization should take the form of a four-vector equation. An obvious suggestion for the corresponding relativistic equation is

$$\underline{\mathbf{K}} = \frac{d\underline{\mathbf{P}}}{d\tau} \tag{8.32}$$

where the non-relativistic momentum is replaced by the relativistic four-momentum, and the coordinate time is replaced by the proper time of the particle. The derivative of $\underline{\mathbf{P}}$ with respect to the proper time is a four-vector, and to make the equation consistent, we have simply replaced the three-vector force $\mathbf{F}$ in the equation with an unspecified four-vector $\underline{\mathbf{K}}$. This we refer to as the four-vector force, or simply the four-force acting on the body. We shall examine what constraints physics puts on this vector, but for the moment we just note that the new equation has a correct covariant form.

The equation can also be written as

$$\underline{\mathbf{K}} = m\underline{\mathbf{A}}, \tag{8.33}$$

with m as the mass of the particle and $\underline{\mathbf{A}}$ as the four-acceleration. This follows since $\underline{\mathbf{P}} = m\underline{\mathbf{U}}$ and therefore $\frac{d\underline{\mathbf{P}}}{d\tau} = m\frac{d\underline{\mathbf{U}}}{d\tau}$, with $\underline{\mathbf{U}}$ as the four-velocity of the particle. However, note that since the three-vector part of $\underline{\mathbf{A}}$ is generally not identical to the acceleration $\mathbf{a}$ in a chosen inertial frame, the three-vector part of the right-hand side of (8.33) is *not* simply $m\mathbf{a}$.

The next step is to relate the four-force $\underline{\mathbf{K}}$ to the (non-relativistic) three-vector force $\mathbf{F}$. To this end we decompose the four-vector in its time and space components, with reference to some unspecified inertial reference frame,

$$\underline{\mathbf{K}} = (K^0, \mathbf{K}). \tag{8.34}$$

The three-vector part of the equation (8.32) is then

$$\mathbf{K} = \frac{d\mathbf{p}}{d\tau} = \gamma\frac{d\mathbf{p}}{dt}, \tag{8.35}$$

with $\mathbf{p} = \gamma m \mathbf{v}$ as the relativistic momentum. The factor γ appears in the equation as a time dilation effect.

Let us now return to the original form (8.31) of Newton's second law, and assume that the three-vector force $\mathbf{F}$ is defined so that this equation is correct also in relativistic physics. Since $\mathbf{p}$ then is the relativistic momentum, we have $\mathbf{F} \neq m\mathbf{a}$, but (8.31) will obviously have the correct non-relativistic limit, since $\gamma \to 1$ means that $\mathbf{p}$ is changed from its relativistic to its non-relativistic form. By comparing (8.31) and (8.35), we find the relation

$$\mathbf{K} = \gamma \mathbf{F} \,. \tag{8.36}$$

We now examine the time component of $\underline{\mathbf{K}}$,

$$K^0 = \frac{dP^0}{d\tau} = \gamma \frac{1}{c} \frac{dE}{dt} \,. \tag{8.37}$$

The relativistic energy-momentum relation of the particle is

$$E^2 = p^2 c^2 + m^2 c^4 \,, \tag{8.38}$$

and the time derivative of this equation gives

$$E \frac{dE}{dt} = c^2 \, \mathbf{p} \cdot \frac{d\mathbf{p}}{dt} \,. \tag{8.39}$$

This further gives

$$\frac{dE}{dt} = c^2 \frac{\mathbf{p}}{E} \cdot \frac{d\mathbf{p}}{dt} = \mathbf{v} \cdot \mathbf{F} \,, \tag{8.40}$$

where we have made use of the relativistic relation $\mathbf{v} = c^2 \mathbf{p}/E$. It is interesting to note that with the relativistic generalization introduced for the three-vector force $\mathbf{F}$, the expression for the power is $\mathbf{v} \cdot \mathbf{F}$, precisely as in non-relativistic physics. The four-force, when decomposed in time and space components, can then be written as

$$\underline{\mathbf{K}} = \gamma \left(\frac{1}{c} \mathbf{v} \cdot \mathbf{F}, \mathbf{F} \right) . \tag{8.41}$$

While the space part is proportional to the three-vector force $\mathbf{F}$, the time component is proportional to the power of the force, $\mathbf{v} \cdot \mathbf{F}$.

One should note from the above expressions that even when the three-vector force $\mathbf{F}$ is a velocity *independent* force, the four-force $\underline{\mathbf{K}}$ will quite generally depend on the velocity of the particle. This point is also demonstrated by the fact that the four-force, which is proportional to the four-acceleration, is always orthogonal, in the relativistic sense, to the four-velocity,

$$\underline{\mathbf{K}} \cdot \underline{\mathbf{U}} = m \, \underline{\mathbf{A}} \cdot \underline{\mathbf{U}} = 0 \,. \tag{8.42}$$

We further note that since $\underline{\mathbf{U}}$ is a timelike vector this implies that $\underline{\mathbf{K}}$ is a spacelike vector (see Problem 6.1). From this follows that we can always find an inertial frame where the time component of the force vanishes. The expression (8.41) shows that this happens in the instantaneous rest frame of the particle, where the four-force reduces to the form

$$\underline{\mathbf{K}} = (0, \mathbf{F}) \qquad \text{(rest frame)}\,. \tag{8.43}$$

The Lorentz force

We consider here, as a special case, the force acting on a charged particle in the electromagnetic field, and derive the relativistic equation of motion in covariant form. The non-relativistic form of the three-vector force is

$$\mathbf{F} = e(\mathbf{E} + \mathbf{v} \times \mathbf{B})\,, \tag{8.44}$$

with e as the charge and $\mathbf{v}$ as the velocity of the particle, and this is a valid expression also for relativistic velocities. The corresponding expression for the four-force (8.41), decomposed in its time and space components, is

$$\underline{\mathbf{K}} = \gamma e \left(\frac{1}{c}\mathbf{v} \cdot \mathbf{E},\, \mathbf{E} + \mathbf{v} \times \mathbf{B} \right). \tag{8.45}$$

We would now like to write this in covariant form, and that requires that we introduce the *electromagnetic field tensor*. This is an antisymmetric tensor where the electric field appears as the time components and the magnetic field as the space components in the following way,

$$\begin{aligned} F^{0k} &= \frac{1}{c} E_k \quad k = 1,2,3\,, \\ F^{kl} &= \sum_m \epsilon_{klm} B_m \quad k,l = 1,2,3\,, \end{aligned} \tag{8.46}$$

where ϵ_{klm} is the three-dimensional Levi-Civita symbol and the Cartesian components are here numbered 1,2,3. In matrix form the field tensor is

$$F = (F^{\mu\nu}) = \begin{pmatrix} 0 & \frac{1}{c}E_1 & \frac{1}{c}E_2 & \frac{1}{c}E_3 \\ -\frac{1}{c}E_1 & 0 & B_3 & -B_2 \\ -\frac{1}{c}E_2 & -B_3 & 0 & B_1 \\ -\frac{1}{c}E_3 & B_2 & -B_1 & 0 \end{pmatrix}. \tag{8.47}$$

Constructed in this way $F^{\mu\nu}$ transforms indeed as a relativistic tensor under Lorentz transformations.

We consider first how the time component of the four-force can be expressed in terms of the electromagnetic field tensor. The expression is

$$K^0 = e\gamma \frac{\mathbf{v}}{c} \cdot \mathbf{E} = eF^{0\nu} U_\nu\,, \tag{8.48}$$

with U_ν as the four-velocity of the particle. The space part we rewrite in a similar way,

$$K^k = e\gamma\left(E_k + \sum_{lm} \epsilon_{klm} v_l B_m\right)$$
$$= e\gamma\left(cF^{0k} + \sum_l F^{kl} v_l\right). \tag{8.49}$$

We next make use of the following identities,

$$U^0 = -U_0 = \gamma c\,, \quad U^l = U_l = \gamma v_l\,, \quad F^{0k} = -F^{k0}\,. \tag{8.50}$$

This gives

$$K^k = e\left(F^{k0}U_0 + \sum_l F^{kl}U_l\right) = eF^{k\nu}U_\nu\,. \tag{8.51}$$

We can now combine the expressions for K^0 and K^k into a single equation for the components of the four-force,

$$K^\mu = eF^{\mu\nu}\,U_\nu\,. \tag{8.52}$$

With the above expression for the four-force the equation of motion of the charged particle can be rewritten in covariant form. The general relativistic form of Newton's second law is

$$\underline{\mathbf{K}} = m\underline{\mathbf{A}} = m\frac{d^2\underline{\mathbf{x}}}{d\tau^2}\,, \tag{8.53}$$

and in the electromagnetic case this gives the covariant equation

$$m\ddot{x}^\mu = eF^{\mu\nu}\dot{x}_\nu\,, \tag{8.54}$$

where the time derivative marked by the *dot* here means differentiation with respect to the proper time of the particle.

We finally note that this covariant equation is equivalent to the non-covariant equation of motion,

$$\frac{d\mathbf{p}}{dt} = e(\mathbf{E} + \mathbf{v}\times\mathbf{B})\,, \tag{8.55}$$

which has the same form in the non-relativistic limit. However, in the relativistic case the left-hand side of the equation cannot simply be replaced by $m\mathbf{a}$, due to the presence of the gamma factor in the expression for the momentum, $\mathbf{p} = \gamma m\mathbf{v}$.

The two versions, (8.54) and (8.55), of the relativistic equation of motion for a charged particle in the electromagnetic field, we have earlier referred to at the end of Sect.6.5.

Example

Relativistic motion of a charged particle in a constant magnetic field

The non-relativistic motion of a charged particle in a constant magnetic field has earlier been studied in Sects. 2.5 and 3.2. Here we derive the motion in a more direct way from Newton's second law, in its relativistic form. The equation of motion is

$$\frac{d\mathbf{p}}{dt} = e\mathbf{v} \times \mathbf{B}\,, \tag{8.56}$$

with $\mathbf{p} = \gamma m\mathbf{v}$ and e as the charge of the particle. The power of the force vanishes, $\frac{dE}{dt} = \mathbf{F} \cdot \mathbf{v} = 0$, which means that γmc^2 is constant. Thus, the kinetic energy is constant in the relativistic description as well as in the non-relativistic approximation. The force has no component along the direction of the magnetic field, and the motion in this direction is therefore a constant drift. For simplicity we assume initial conditions with no velocity in this direction, and the motion is therefore restricted to a plane orthogonal to $\mathbf{B}$.

The equation of motion can be written as

$$\frac{d}{dt}(\gamma m\mathbf{v} - e\mathbf{r} \times \mathbf{B}) = 0\,, \tag{8.57}$$

where the expression inside the bracket thus is a constant of motion. We write this constant as

$$\gamma m\mathbf{v} - e\mathbf{r} \times \mathbf{B} \equiv -e\mathbf{r}_0 \times \mathbf{B}\,, \tag{8.58}$$

with $\mathbf{r}_0$ as a constant vector. The form we have chosen for the right-hand side of the equation is consistent with the fact that this vector is restricted to the plane orthogonal to $\mathbf{B}$ when $\mathbf{v}$ has no component along the magnetic field. We rewrite the equation as

$$\gamma m\mathbf{v} - e(\mathbf{r} - \mathbf{r}_0) \times \mathbf{B} = 0\,, \tag{8.59}$$

and this shows that $\mathbf{r}_0$ can be absorbed in a shift of the origin in the plane of motion. In the following we assume this to have been done, so the motion satisfies the equation

$$\gamma m\mathbf{v} = e\mathbf{r} \times \mathbf{B}\,. \tag{8.60}$$

This equation shows that the velocity is orthogonal to the position vector, $\mathbf{r} \cdot \mathbf{v} = 0$, so that $\mathbf{r}^2$ is a constant of motion.

From the arguments given above we conclude that the particle moves in a circle with constant speed. With $\mathbf{r}_0 = 0$ the center of the orbit is at the origin of the coordinate system, but without this restriction the center can be placed anywhere in the plane. The angular velocity is

$$\omega = \frac{v}{r} = \frac{eB}{\gamma m} \equiv \frac{\omega_0}{\gamma}\,, \tag{8.61}$$

where we have introduced the symbol $\omega_0 = eB/m$ for the non-relativistic cyclotron frequency.

It is interesting to note that the relativistic effect is only to replace the mass m with the velocity dependent mass γm. Thus the motion in the plane orthogonal to $\mathbf{B}$, is, in the relativistic as well as in the non-relativistic case, circular motion with constant speed. But whereas the angular velocity is energy independent in the non-relativistic description, it decreases with energy when the speed is relativistic.

Increasing energy means increasing radius of the orbit, and the frequency can be found as a function of the radius if we write the equation for ω as

$$\omega = \omega_0\sqrt{1 - \frac{\omega^2 r^2}{c^2}}\,. \tag{8.62}$$

Solving this for ω we find

$$\omega = \frac{\omega_0}{\sqrt{1 + \frac{\omega_0^2 r^2}{c^2}}}\,, \tag{8.63}$$

which gives

$$\gamma = \frac{1}{\sqrt{1 - \frac{\omega^2 r^2}{c^2}}} = \sqrt{1 + \frac{\omega_0^2 r^2}{c^2}}\,. \tag{8.64}$$

Thus the expression for the relativistic energy of the particle is

$$E = \sqrt{1 + \frac{\omega_0^2 r^2}{c^2}}\, mc^2\,. \tag{8.65}$$

As shown by this expression the energy is a quadratic function of r for non-relativistic velocities but this changes to a linear dependence in the relativistic regime.

The decrease of the angular frequency with energy of the circulating charge is important in a type of particle accelerator called *cyclotrons*. In these accelerators charged particles are circulating in a strong magnetic field and energy is fed to the particles by applying an electric field which oscillates with the angular frequency of the particles. In the early cyclotrons, where the particles moved non-relativistically, the frequency of the field was kept fixed. Later on accelerators were built to accelerate beams of particles to relativistic speeds. In these accelerators, called *synchrocyclotrons*, the frequency of the accelerating electric field was synchronized with the decreasing circular frequency of the accelerated particles. In *isochronous cyclotrons* a different approach was taken to compensate for the relativistic effect. These accelerators work with constant electric field frequency, but the strength of the magnetic field is increased with time. As shown by Eq. (8.61) the angular frequency of the circulating particles can be kept fixed by compensating for the increase of γ with a similar increase in the value of B.

8.4 Lagrangian for a relativistic particle

In the Lagrangian formulation of Newton's mechanics, the time coordinate plays a role which is different from that of the generalized coordinates of the system. Time there acts as a parameter for the paths that the system can follow through configuration space. For a particle moving through three-dimensional space this means that the space coordinates and the time coordinate appear in different ways in the formalism. This difference seems to create a problem when extending the formalism to relativistic theory, where time and space coordinates are mixed by the Lorentz transformations. We consider now the question of how to introduce a relativistic particle Lagrangian, and in order to do so we make an attempt to follow the same approach as in the previous section, where relativistic generalizations of physical formulas were introduced by re-writing the equations in covariant form. We apply this first to a freely moving particle of mass m.

Instead of considering the Lagrangian directly we start with the action, which is the integral of the Lagrangian for an arbitrarily chosen time dependent path in configuration space. For a free particle it has the form

$$S = \int T dt = \int \frac{1}{2} m \dot{\mathbf{r}}^2 dt \,, \tag{8.66}$$

with T as the kinetic energy of the particle. As a first step in a relativistic modification of the integral we note the following correspondence

$$T dt \to E dt = c P^0 dt \,, \tag{8.67}$$

where the expressions to the right have the one to the left as the non-relativistic limit, except for the presence of the rest energy $E_0 = mc^2$. However, a constant can always be added to the Lagrangian, since it does not affect Lagrange's equations. This immediately suggests that one could add the term $-\mathbf{P}\cdot d\mathbf{r}$ to make the expression Lorentz invariant

$$T dt \to -P^\mu dx_\mu \,. \tag{8.68}$$

However, with this modification we have to check again the non-relativistic limit. We have

$$\begin{aligned} -P^\mu dx_\mu &= cP^0 dt - \mathbf{P}\cdot d\mathbf{r} \\ &= (cP^0 - \mathbf{P}\cdot\mathbf{v})dt \\ &= \left(mc^2 + \frac{1}{2}mv^2 - mv^2 - ...\right) dt \\ &= \left(mc^2 - \frac{1}{2}mv^2 - ...\right) dt \,, \end{aligned} \tag{8.69}$$

where $-...$ denotes higher order terms in v^2/c^2 which are omitted in the non-relativistic approximation. We note that the term we have added has in fact changed the sign of the non-relativistic kinetic energy. To compensate for this we may simply change the sign in the correspondence (8.68).

As a result we have found the following expression for the action integral which is Lorentz invariant and has the correct non-relativistic limit,

$$\begin{aligned} S &= \int P^\mu dx_\mu \\ &= m \int U^\mu dx_\mu \\ &= m \int U^\mu U_\mu d\tau \\ &= -mc^2 \int d\tau \,. \end{aligned} \tag{8.70}$$

We have here applied the identities,

$$P^\mu = mU^\mu, \quad dx^\mu = U^\mu d\tau, \quad U^\mu U_\mu = -c^2 \,, \tag{8.71}$$

with τ as the proper time of the particle.

It may initially seem very natural that in the Lorentz invariant form the proper time rather than the coordinate time should be used as a time parameter in the Lagrangian. However, this choice of time parameter creates a problem, since the Lagrangian then is simply a constant, which cannot be used to derive Lagrange's equations. The origin of the problem is the following. When Lagrange's equation of motion is derived from Hamilton's principle, one considers changes in the action under variations in the space-time path where the endpoints are kept fixed. Also the time parameter along the paths has to take fixed values at the endpoints. However, proper time does not satisfy this requirement, since the proper time between two space-time points depends on the path between the points, as we have previously discussed.

The conclusion we therefore draw is that the expression we have found for the action may be fine, but proper time is not a good parameter to use if we want to derive the equation of motion from the action integral. To circumvent the problem we may simply introduce a new, unspecified parameter λ for the space-time paths, $x^\mu(\lambda)$, where λ take fixed, path independent, values at the endpoints ot the paths. For this new parameter we find the following expression,

$$cd\tau = \sqrt{-dx^\mu dx_\mu} = \sqrt{-\frac{dx^\mu}{d\lambda}\frac{dx_\mu}{d\lambda}}d\lambda\,, \tag{8.72}$$

where one should note that the minus sign under the square root is simply to compensate for the fact that dx^μ is a timelike vector for the path of the particle. The expression given by the equation is clearly invariant under arbitrary changes in the parameter, $\lambda \to \lambda'$.

The action can then be written as

$$S = -mc\int_{\lambda_0}^{\lambda_1} \sqrt{-g_{\mu\nu}\dot{x}^\mu\dot{x}^\nu}d\lambda\,, \tag{8.73}$$

where the parameter values at the endpoints are called λ_0 and λ_1, and where we here define $\dot{x}^\mu = \frac{dx^\mu}{d\lambda}$. The corresponding Lagrangian is then

$$L = -mc\sqrt{-g_{\mu\nu}\dot{x}^\mu\dot{x}^\nu}\,, \tag{8.74}$$

and Lagrange's equations have the standard form

$$\frac{d}{d\lambda}\left(\frac{\partial L}{\partial \dot{x}^\mu}\right) - \left(\frac{\partial L}{\partial x^\mu}\right) = 0\,, \quad \mu = 0,1,2,3\,. \tag{8.75}$$

It should be straight forward to check whether the equations found in this way are the correct equations of motion for a free particle. We first note that all the coordinates x^μ are cyclic, $\frac{\partial L}{\partial x^\mu} = 0$, so that we have the following set of constants of motion

$$\frac{\partial L}{\partial \dot{x}^\mu} = mc\frac{\dot{x}^\mu}{\sqrt{-\dot{x}^\nu\dot{x}_\nu}} \equiv k^\mu\,, \tag{8.76}$$

where k^μ satisfies the condition $k^\mu k_\mu = -m^2c^2$. This gives

$$k^\mu = mc\,n^\mu\,, \tag{8.77}$$

with n^μ as a timelike unit vector, $n^\mu n_\mu = -1$. If we now introduce the four-velocity, $U^\mu = \dot{x}^\mu \frac{d\lambda}{d\tau}$, Eq. (8.76) implies

$$U^\mu = c\, n^\mu \,, \tag{8.78}$$

which shows that the four-velocity is a constant timelike vector with relativistic norm squared, $U^\mu U_\mu = -c^2$. This is clearly the correct expression for a free particle which moves with constant velocity.

Next we consider the Lagrangian of a charged particle in an electromagnetic field. This can be obtained from the free field Lagrangian by simply adding a contribution from the electromagnetic potentials. The Lagrangian has the following Lorentz invariant form

$$L = -mc\sqrt{-g_{\mu\nu}\dot{x}^\mu \dot{x}^\nu} + eA_\mu \dot{x}^\mu \,, \tag{8.79}$$

where the scalar potential Φ and vector potential $\mathbf{A}$ are identified as the components of the electromagnetic *four-potential* $\underline{\mathbf{A}} = (\frac{1}{c}\Phi,\, \mathbf{A})$. It is again a straight forward exercise to check that the corresponding Lagrange's equations give the correct relativistic equations of motion in the relativistic form (8.54).

Let us finally point out that even if covariant expressions have been used in order to construct the Lagrangians with correct relativistic form, there is no problem to re-express them with the coordinate time as parameter, for any chosen inertial frame. After all the parameter λ can be freely chosen and in particular is chosen to coincide with such a coordinate time. The main point is that the action we have found does not depend on the choice of parameter. In particular this means that if the coordinate time is chosen, the action of a free particle should be written as

$$S = -mc^2 \int d\tau = \int L dt \,. \tag{8.80}$$

By use of the time dilation formula $\frac{dt}{d\tau} = \gamma$ this gives the following expression for the Lagrangian, when the coordinate t is chosen as parameter

$$L = -\frac{mc^2}{\gamma} = -mc^2\sqrt{1-\frac{v^2}{c^2}} \,. \tag{8.81}$$

We note that this expression, even if there is some resemblance, is not identical to the energy $E = \gamma mc^2$. In this regard it is different from the non-relativistic Lagrangian.

For a charged particle in an electromagnetic field the corresponding Lagrangian is

$$L = -mc^2\sqrt{1-\frac{v^2}{c^2}} - e\Phi + e\mathbf{v}\cdot\mathbf{A} \,. \tag{8.82}$$

The expressions (8.81) and (8.82) are not Lorentz invariant, but nevertheless correct expressions for the relativistic Lagrangians, as our derivation has shown.

8.5 Exercises

Problem 8.1

Two photons in the laboratory system have frequencies ν_1 and ν_2. The angle between the propagation directions is θ.

a) Find the total energy and the absolute value of the total momentum of the photons in the laboratory system, expressed in terms of the frequencies ν_1, ν_2 and the angle θ.

b) Find the photons' frequency in the center-of-mass system.

c) Is it always possible to find a center-of-mass system for the photons?

Problem 8.2

We send a photon towards an electron at rest.

a) What is the minimum energy of the photon required for the following process to take place?

$$\gamma + e^- \to e^- + e^- + e^+ \,. \tag{8.83}$$

The particles e^- and e^+ have the same mass m_e.

b) Show that the process

$$\gamma \to e^- + e^+ \tag{8.84}$$

is impossible.

Problem 8.3

Fig. 8.5 shows a particle, which in the laboratory frame S is moving with velocity v towards another particle, which is at rest in S. The two particles have the same mass m. First we will assume that the particles collide in such a way that they form one particle after the collision (totally inelastic collision.)

a) Determine the compound particle's energy E, momentum P, velocity V and mass M.

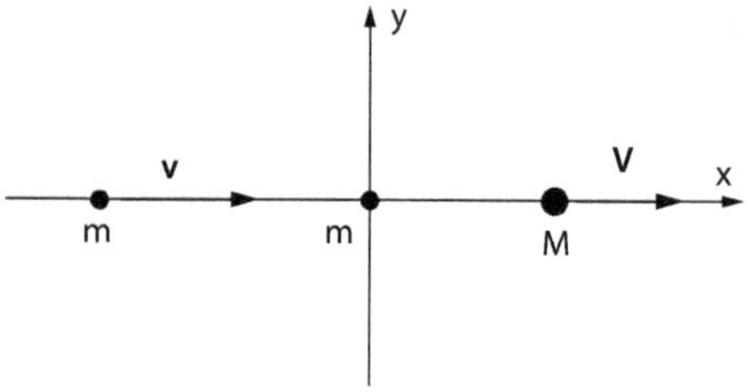

Fig. 8.5 A totally inelastic collision of two particles with equal masses m. In the collision they merge into a single particle of mass M.

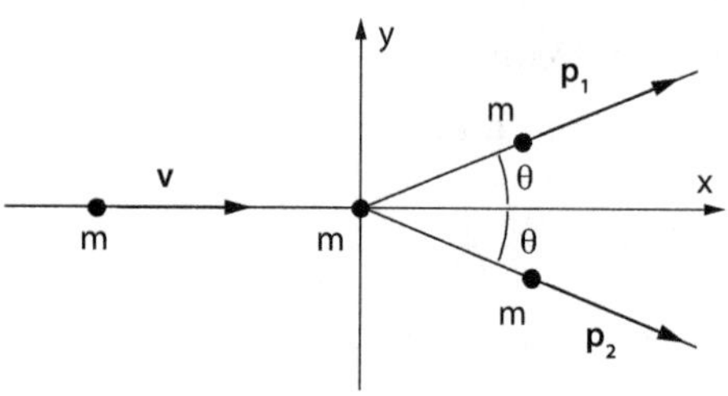

Fig. 8.6 Elastic collision between the two particles with equal masses m.

In the rest of the exercise we will assume that the situation before the collision is as described earlier, but that the particles now collide elastically, i.e. after the collision the two particles are the same as before the collision, with no change in their masses. The collision happens in such a way that the velocities of the particles after the collision make the same angle, θ, with the x-axis in the lab frame S (see Fig. 8.6).

b) Show that after the collision the particles have the same momentum ($|\mathbf{p}_1| = |\mathbf{p}_2|$) and energy ($E_1 = E_2$).

c) Determine $E \equiv E_1 = E_2$ and $p \equiv |\mathbf{p}_1| = |\mathbf{p}_2|$.

d) Determine the angle θ. Find θ in the limiting cases, $\gamma \to 1$ and $\gamma \to \infty$, where γ is the relativistic gamma factor of the moving particle, before the collision. Show that, for all velocities of the particle, we have the inequality $\theta < \pi/4$ satisfied.

Problem 8.4

A photon with energy $E_{ph} = 100\,\text{keV}$ is scattered on a free electron, which is, before the scattering, at rest in the laboratory frame. After the scattering the energy of the photon is E'_{ph}, and the direction of propagation makes an angle θ relative to the direction of the incoming photon. The rest energy of the electron is $E_e = m_e c^2 = 0.51\,\text{MeV}$, and after the scattering it has an energy which we denote by E'_e.

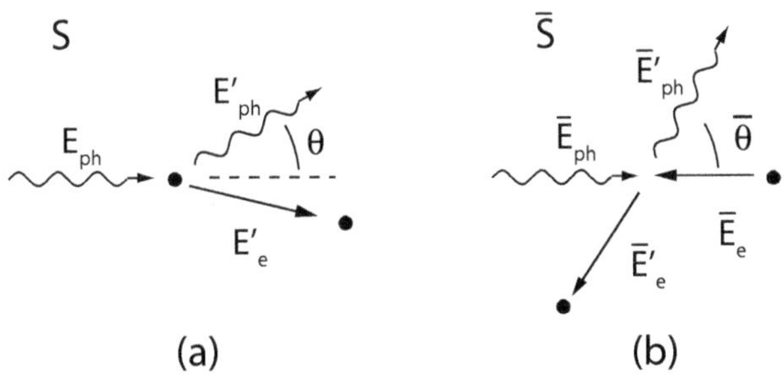

Fig. 8.7 Compton scattering of a photon on a free electron. In (a) the collision is seen in the reference frame where the electron is at rest before the collision, and in (b) it is seen in the center-of-mass system of the photon and the electron.

We examine this process both in the lab frame S (Fig. 8.7a) and in the center of mass system $\bar{S}$ (Fig. 8.7b). In the center of mass system all variables are marked with a "bar", for example with $\bar{E}_{ph}$ as the energy of the incoming photon.

a) What is meant by the center-of-mass system? Use the transformation formulas for energy and momentum to determine the relative velocity between the lab system and the center-of-mass system.

b) Explain why the energy of the incoming and outgoing photon is the same in the center of mass system and find this energy.

c) If $\theta = 90°$ what is the energy of the outgoing photon in the lab system? What is the corresponding energy of the outgoing electron?

Problem 8.5

A Lambda particle (Λ) has momentum $\bar{\mathbf{p}}_\Lambda$ along the x-axis in the laboratory frame $\bar{S}$. The energy of the particle in $\bar{S}$ is $\bar{E}_\Lambda = 3\,\text{GeV}(= 3000\,\text{MeV})$. The mass of Λ is $m_\Lambda = 1116\,\text{MeV/c}^2$.

a) In its rest frame S, the life time of Λ is $\tau_\Lambda = 2.63 \times 10^{-10}s$. What is the corresponding life time $\bar{\tau}_\Lambda$ in the lab frame $\bar{S}$? What is the distance d, which the particle travels in the lab frame, if we assume that it lives exactly the time τ_Λ in its rest frame S?

b) The Λ particle decays to a nucleon N and a pion π. They have masses $m_N = 940\,\text{MeV}$ and $m_\pi = 140\,\text{MeV}$, respectively. Determine the energies, E_N and E_π of the two particles in the rest frame S of Λ.

c) Assume that the pion, in reference frame S, is emitted in a direction which makes the angle 45° with the x-axis. Find in this case the energies $\bar{E}_\pi$ and $\bar{E}_N$ in the lab frame $\bar{S}$.

d) What are the angles $\bar{\theta}_\pi$ and $\bar{\theta}_N$ of the particle's momenta, relative to the x-axis in the lab frame $\bar{S}$?

Problem 8.6

Show that Lagrange's equation, with the Lagrangian given by (8.82), reproduces the relativistic equation of motion for a charged particle in an electromagnetic field, with the electric and magnetic fields given by

$$\mathbf{E} = -\frac{\partial \mathbf{A}}{\partial t} - \boldsymbol{\nabla}\Phi\,, \quad \mathbf{B} = \boldsymbol{\nabla} \times \mathbf{A}\,. \tag{8.85}$$

Problem 8.7

An electron, with charge e, moves in a constant electric field $\mathbf{E}$. The motion is determined by the relativistic Newton's equation

$$\frac{d}{dt}\mathbf{p} = e\mathbf{E}\,, \tag{8.86}$$

where $\mathbf{p}$ denotes the relativistic momentum $\mathbf{p} = m_e\gamma\mathbf{v}$, with m_e as the electron mass, $\mathbf{v}$ as the velocity and $\gamma = 1/\sqrt{1-(v/c)^2}$ as the relativistic gamma factor. We assume the electron to move along the field lines, that is, there is no velocity component orthogonal to $\mathbf{E}$.

a) Show that if $v = 0$ at time $t = 0$, then γ depends on time t as

$$\gamma = \sqrt{1 + \kappa^2 t^2}\,, \tag{8.87}$$

and determine κ.

b) The proper time τ is related to the coordinate time t by the formula $\frac{dt}{d\tau} = \gamma$. Show that if we write $\gamma = \cosh \kappa\tau$ then this relation is satisfied.

c) For linear motion we have the following relation between the proper acceleration a_0 and the acceleration a measured in a fixed inertial reference frame, $a_0 = \gamma^3 a$. Use this to show that the electron has a constant proper acceleration, given by $\mathbf{a}_0 = e\mathbf{E}/m_e$.

Problem 8.8

A particle with charge q and mass m moves with relativistic speed through a region $0 < x < L$ where a constant electric field $\mathbf{E}$ is directed along the y-axis, as indicated in the figure. The particle enters the field at $x = 0$ with momentum $\mathbf{p}_0$ in the direction orthogonal to the field. The relativistic energy at this point is denoted $\mathcal{E}_0$.

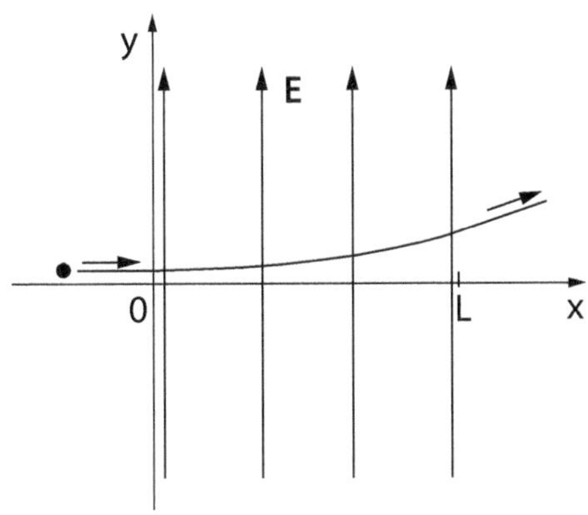

Fig. 8.8 Relativistic motion of a charged particle in an electric field. The velocity of the particle is initially directed orthogonally to the electric field vector $\mathbf{E}$.

a) Use the equation of motion for a charged particle in an electric field to determine the time dependent momentum $\mathbf{p}(t)$ and relativistic energy $\mathcal{E}(t)$ (not including the potential energy) of the particle inside in the electric field. What is the relativistic gamma factor $\gamma(t)$ expressed as a function of coordinate time t?

b) Find the velocity components $v_x(t)$ and $v_y(t)$ and explain the relativistic effect that the velocity in the x-direction decreases with time even if there is no force acting in this direction.

c) Show that the proper time $\Delta\tau$ spent by the particle on the transit through the region $0 < x < L$ is proportional to the length L, $\Delta\tau = \alpha L$, and determine α.

d) What is the transit time Δt through the region when measured in coordinate time?

We remind about the integration formula $\int dx \frac{1}{\sqrt{1+x^2}} = \operatorname{arc\,sinh} x + C$.

Summary

We have in this part studied the basic elements of the special theory of relativity. The starting point has been the fundamental space-time symmetries expressed in the form of Lorentz transformations. They define the transition between Cartesian coordinates of different inertial frames, and the basic difference between these transformations and the Galilean symmetry transformations of non-relativistic physics is the mixing of space and time coordinates in the relativistic case. An important consequence of this is that distance in non-relativistic theory, in the form of the length of the three-vector $\Delta\mathbf{r}$, is replaced by another invariant, which also includes the time difference, $\Delta s^2 = \Delta\mathbf{r}^2 - c^2\Delta t^2$. Invariance of this quantity is directly related to the basic property of Lorentz transformation, that the speed of light does not depend on the choice of reference frame.

The change from Galilean to Lorentz transformations as the fundamental symmetry transformations has many important consequences for relativistic kinematics and dynamics, as first demonstrated by Albert Einstein. We have here derived the kinematical effects of length contraction and time dilation and have stressed the important point that measurements should be performed at *simultaneity* in the reference frame of the observer who makes the measurement. For the time dilation effect this understanding is applied to the twin paradox, which is resolved by taking into account the change in definition of simultaneous events that is performed by one of the twins during his space-time journey.

An introduction to the formalism of four-vectors and tensors has been given, and we have discussed how to apply this formalism when defining covariant relativistic equations. The formalism has been used at several places to derive the relativistic expressions that correspond to known non-relativistic quantities. The idea is to seek covariant expressions, which secures that they are valid in any inertial frame, and to impose the condition that the expressions have the correct non-relativistic limit. This formal approach indeed produces correctly the relativistic extensions of the non-relativistic expressions for the physical quantities and equations. In particular we have introduced the four-vector description of velocity and acceleration, and we have discussed the meaning of proper acceleration. As an example we have studied the so-called hyperbolic motion of a spaceship with constant proper acceleration and effects of the time dilation for time registered on the spaceship and on earth.

The definition of the conserved four-momentum has been shown to have important consequences. The energy and three-vector momentum are there combined into a four-component object, and Lorentz invariance imposes a particular form to the relation between energy and momentum for a moving object. This involves in particular a conversion formula between mass and energy, which is Einstein's famous equation $E = mc^2$. By considering the case of inelastic collisions between particles we demonstrate that this relation is not only a curious coincidence, but it shows that mass can be converted to energy in real physical processes, with a large conversion factor between mass and energy. This points towards the well-known and dramatic effects, where huge amounts of energy are released in nuclear processes.

In the chapter on Relativistic Dynamics we have examined how to update Newton's second law to a relativistic equation and to give meaning to the four-vector force. As a particular application we have examined how to give the equation of motion for a charged particle in an electromagnetic field the correct covariant form. Finally we have discussed how to bring relativistic equations into Lagrangian form and have shown how to resolve the problem which appears when using proper time as the time parameter. The approach has been illustrated by deriving Lagrangians for a free particle and for a charged particle in an electromagnetic field, both in the covariant and the non-covariant forms.

PART 3
Electrodynamics

Introduction

Electric and magnetic phenomena had been known and studied for centuries as separate natural phenomena, before, in the second half of the nineteenth century, a complete theory of electromagnetism was introduced by James Clerk Maxwell (1831–1879). There is a long list of scientists who contributed to the development that led to this important event, with well-known names from the history of physics, such as Charles Augustin de Coulomb (1736–1806), André-Marie Ampère (1775–1836), Hans Christian Ørsted (1777–1851), Carl Friedrich Gauss (1777–1855), and Michael Faraday (1791–1867). Maxwell collected the partial knowledge included in these contributions, and he reformulated them in the form of a consistent set of electromagnetic equations, which we know as Maxwell's equations.

As a crucial element in this process, Maxwell modified what we know as Ampère's equation by adding a term, which implies that a variation in the electric field gives rise to a magnetic field. This is similar to the situation in Faraday's law of induction, where a variation in the magnetic field gives rise to an electric field. In combination these two effects have the important consequence of giving the fields independent dynamics in the form of electromagnetic waves. The existence of such waves was shortly after proven experimentally, and that light is waves of electromagnetic origin was realized by Maxwell already when working on the electromagnetic equations.

Maxwell's formulation of the unified electromagnetic theory is regarded as one of the most important achievements in the history of physics, and its influence on the later development of physics and technology can hardly be overestimated. Modern society is completely dependent on the use of electromagnetism, for information and communication, for transport and energy transfer, and it may be difficult for people today to understand to what extent this scientific achievement about 150 years ago has influenced us all.

Part 3 is devoted to the study of some of the basic phenomena described by electromagnetic theory. Our approach is similar to that of the first two parts. Thus, the aim is to derive our results from the basic principles of the theory, which here means from Maxwell's equations. We begin by summarizing the equations in their non-covariant form, and then showing how to bring the equations into relativistic, covariant form. The use of electromagnetic potentials is important for the discussions and for the following applications. We find the transformation formulas for the electromagnetic fields under Lorentz transformations, and derive the expressions for the energy and momentum densities of the electromagnetic field. Solutions of the free wave equations are first discussed, with focus on the different types of polarization, and solutions with stationary as well as time dependent sources are then derived. Expansions in terms of multipoles are introduced as a method to find expressions for the fields far from the sources. We examine in particular the radiation phenomena, and show in the derivation the close relation between electromagnetic radiation and the relativistic retardation effect.

Chapter 9

Maxwell's equations

In this chapter we formulate and study the fundamental electromagnetic equations. Historically they were developed by studying the different forms of electric and magnetic phenomena and first formulated as independent laws. These phenomena include the creation of electric fields by charges and time dependent magnetic fields, and the creation of magnetic fields by electric currents. We will first recall the form of each of these individual laws and next follow the important step of Maxwell by collecting these in a set of coupled equations for the electromagnetic phenomena. Maxwell's equations gain their most attractive form when written in relativistic, covariant form. We use this formulation to study the Lorentz transformations of electric and magnetic fields and of electric charges and currents. The covariant formulation of Maxwell's equations is also the starting point for the further discussion in later chapters, where we examine different types of solutions to Maxwell's equations.

9.1 Charge conservation

Electric and magnetic fields are produced by electric charges and currents. The charges satisfy the important law of *charge conservation*, which seems to be strictly satisfied in nature. The carriers of electric charge, at the microscopic level these are the elementary particles, may be created and may disappear, but in these processes the total charge is always preserved.

With Q as the total electric charge within a given volume, charge conservation may simply be expressed as

$$\frac{dQ}{dt} = 0 \,. \tag{9.1}$$

However, this equation is correct only when there is no charge passing through the boundary surface of the selected volume. A more general expression is therefore

$$\frac{dQ}{dt} = -I\,, \tag{9.2}$$

where I is the current through the boundary. This equation for the integrated charge and current can be reformulated in terms of the local charge density ρ and current density $\mathbf{j}$, defined by

$$Q(t) = \int_V \rho(\mathbf{r}, t) dV\,, \quad I(t) = \int_S \mathbf{j}(\mathbf{r}, t) \cdot d\mathbf{S}\,. \tag{9.3}$$

In these expressions V is the (arbitrarily) chosen volume with S as the cor-

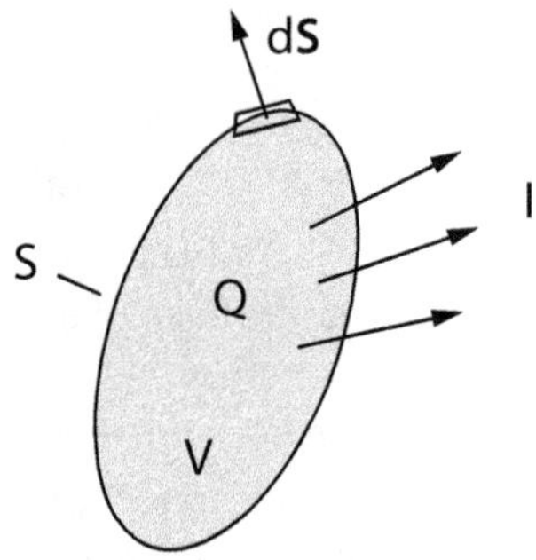

Fig. 9.1 Charge conservation: The change in the charge Q within a volume V is caused by a current I through the boundary surface S.

responding boundary surface, dV is the three-dimensional volume element and $d\mathbf{S}$ is the surface element, with direction orthogonal to the surface. Charge conservation then gets the form

$$\frac{d}{dt} \int_V \rho(\mathbf{r}, t)\, dV + \int_S \mathbf{j}(\mathbf{r}, t) \cdot d\mathbf{S} = 0\,. \tag{9.4}$$

The last term can be rewritten as a volume integral by use of Gauss' theorem, and this gives the following integral form of the equation,

$$\int_V \left(\frac{\partial \rho}{\partial t}(\mathbf{r}, t) + \boldsymbol{\nabla} \cdot \mathbf{j}(\mathbf{r}, t) \right) dV = 0\,. \tag{9.5}$$

Since charge conservation in the form (9.5) is valid at any time t and for an arbitrarily small volume centered at any chosen point $\mathbf{r}$, it can be reformulated as the following local condition on the charge and current densities

$$\frac{\partial \rho}{\partial t} + \boldsymbol{\nabla} \cdot \mathbf{j} = 0\,. \tag{9.6}$$

This form of the condition of charge conservation, as a *continuity equation*, we will later apply repeatedly.

When expressed in the form of *densities*, we have a view of charge and current as being continuously distributed in space. However, we know that at the microscopic level charge has a granular structure, since it is carried by small (point-like) particles. We may take the view that the continuum description is based on a macroscopic approximation, where the local charge is averaged over a volume that is small on a macroscopic scale, but sufficiently large on the microscopic scale to smoothen the granular distribution. In most cases this will be sufficient for our purpose. However, the description of charged particles can also be included by use of Dirac's delta function. For a system of point-like particles the charge and current densities then take the form

$$\begin{aligned} \rho(\mathbf{r},t) &= \sum_i e_i\,\delta(\mathbf{r}-\mathbf{r}_i(t))\,, \\ \mathbf{j}(\mathbf{r},t) &= \sum_i e_i\mathbf{v}_i(t)\,\delta(\mathbf{r}-\mathbf{r}_i(t))\,. \end{aligned} \tag{9.7}$$

In these expressions the label i identifies a particle in the system, with charge e_i, time dependent position $\mathbf{r}_i(t)$ and velocity $\mathbf{v}_i(t)$.

In general, when the motion of the charged particles can be described by a (smooth) velocity field $\mathbf{v}(\mathbf{r},t)$, we have the following relation between current and charge densities,

$$\mathbf{j}(\mathbf{r},t) = \mathbf{v}(\mathbf{r},t)\rho(\mathbf{r},t)\,. \tag{9.8}$$

Note, however, for currents in a conductor there are two independent contributions, from the electrons and from the ions, and these move with different velocities, $\mathbf{v}_e$ and $\mathbf{v}_a$, so that the total current has the form

$$\mathbf{j}(\mathbf{r},t) = \mathbf{v}_e(\mathbf{r},t)\rho_e(\mathbf{r},t) + \mathbf{v}_a(\mathbf{r},t)\rho_a(\mathbf{r},t)\,. \tag{9.9}$$

For the usual situation, with total charge neutrality and with the ions sitting at rest, the expressions for total charge and current densities are

$$\rho(\mathbf{r},t) = 0\,, \quad \mathbf{j}(\mathbf{r},t) = \mathbf{v}_e(\mathbf{r},t)\rho_e(\mathbf{r},t)\,. \tag{9.10}$$

9.2 Gauss' law

This law expresses how electric charge acts as a source for the electric field. Like all of the other electromagnetic equations it can be given both an integral and a differential form. In integral form it relates the flux of the electric field through any given closed surface S to the total charge Q within the surface,

$$\oint_S \mathbf{E} \cdot d\mathbf{S} = \frac{Q}{\epsilon_0} . \tag{9.11}$$

In this equation ϵ_0 is the vacuum permittivity, with the value

$$\epsilon_0 = 8.85 \cdot 10^{-12}\,\mathrm{C}^2/\mathrm{Nm}^2 . \tag{9.12}$$

Equation (9.11) can be rewritten in terms of volume integrals as

$$\int_V \boldsymbol{\nabla} \cdot \mathbf{E}\, dV = \int_V \frac{\rho}{\epsilon_0} dV , \tag{9.13}$$

where on the left hand side Gauss' theorem has been used to rewrite the surface integral as a volume integral. Since this equality should be satisfied for any chosen volume V, the integrands should be equal, and that gives Gauss' law in differential form

$$\boldsymbol{\nabla} \cdot \mathbf{E} = \frac{\rho}{\epsilon_0} . \tag{9.14}$$

Gauss' law is the fundamental equation of *electrostatics*, where the basic problem is to determine the electric field from a given, static charge distribution, with specified boundary conditions satisfied by the field. In its simplest form the problem is to determine the field from an isolated point charge in the open space, in which case Gauss' law in integral form can easily be solved under the assumption of rotational symmetry. Thus, with the charge located at $\mathbf{r} = 0$ and with the electric field of the form $\mathbf{E} = E\mathbf{r}/r$, Gauss' law gives

$$4\pi r^2 E = \frac{q}{\epsilon_0} , \tag{9.15}$$

with q as the charge. This gives the expression for the Coulomb field of a stationary point charge

$$\mathbf{E}(\mathbf{r}) = \frac{q}{4\pi\epsilon_0 r^2} \frac{\mathbf{r}}{r} . \tag{9.16}$$

Due to the fact that Gauss' law gives a linear differential equation for the electric field, the solution for a point charge can be extended to the full solution for a charge distribution. We shall return to the discussion of stationary solutions of Maxwell's equations later on.

9.3 Ampère's law

This law describes how electric currents produce magnetic fields. The integral form is

$$\oint_C \mathbf{B} \cdot d\mathbf{s} = \mu_0 I + \frac{1}{c^2}\frac{d}{dt}\int_S \mathbf{E} \cdot d\mathbf{S}\,, \tag{9.17}$$

and it shows that the line integral of the magnetic field around any closed curve C gets two contributions, one from the total electric current I passing through C and the other from the "displacement current", which is defined by the time derivative of the electric flux through a surface S, with C as boundary. In this equation the vacuum permeability has been introduced. The value of this constant is given by

$$\frac{\mu_0}{4\pi} = 10^{-7}\,\mathrm{N/A^2}\,. \tag{9.18}$$

To rephrase Eq. (9.17) in differential form, the left-hand side is rewritten

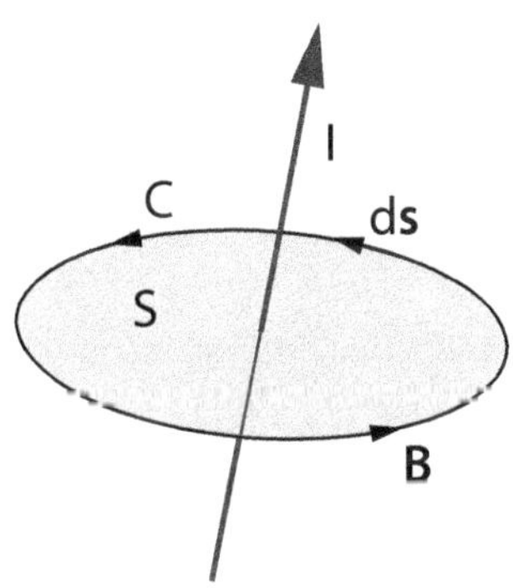

Fig. 9.2 Ampère's law: The circulation of the magnetic field **B** around a closed curve C is determined by the current and the time derivative of the electric flux through the loop.

as a surface integral by use of Stokes' theorem and the current is expressed as a surface integral of the current density,

$$\int_S (\boldsymbol{\nabla} \times \mathbf{B}) \cdot d\mathbf{S} = \int_S \mu_0 \mathbf{j} \cdot d\mathbf{S} + \frac{1}{c^2}\frac{d}{dt}\int_S \mathbf{E} \cdot d\mathbf{S}\,. \tag{9.19}$$

Since this should be satisfied for an arbitrarily chosen surface S there is equality between the integrands, which gives Ampere's law in differential form

$$\boldsymbol{\nabla} \times \mathbf{B} - \frac{1}{c^2}\frac{\partial}{\partial t}\mathbf{E} = \mu_0 \mathbf{j}\,. \tag{9.20}$$

Ampere's law shows that an electric current gives rise to a magnetic field that circulates the current, but it also shows that a changing electric field produces a magnetic field. The origin of the time derivative of the electric field in the equation may not be so obvious, but this term, which was introduced by Maxwell, is needed for the full set of electromagnetic equations to be consistent. An important consequence of this is that the equations have solutions in the form of propagating waves. This follows from the dynamical coupling between the electric and magnetic fields, where time variations in the electric field $\mathbf{E}$ produce a magnetic field $\mathbf{B}$ (Ampere's law), and the time variations in $\mathbf{B}$ give rise to a modified electric field $\mathbf{E}$ (Faraday's law).

An interesting point to notice is that without the contribution from the time derivative of $\mathbf{E}$, (9.20) would be in conflict with the conservation of electric charge. This is seen by taking the divergence of the equation, which would without the electric term give rise to the equation $\boldsymbol{\nabla} \cdot \mathbf{j} = 0$ for the current. However, by comparison with the continuity equation for the charge current, one sees that this is correct only if the charge density is not changing with time. The form of the electric term is in fact precisely what is needed to reproduce the continuity equation, provided there is a specific connection between the constants ϵ_0 and μ_0. To demonstrate this we take the divergence of Eq.(9.20) and apply Gauss' law,

$$\mu_0 \boldsymbol{\nabla} \cdot \mathbf{j} = -\frac{1}{c^2}\frac{\partial}{\partial t}\boldsymbol{\nabla} \cdot \mathbf{E} = -\frac{1}{c^2 \epsilon_0}\frac{\partial \rho}{\partial t} \,. \tag{9.21}$$

This is identical to the continuity equation, provided

$$\epsilon_0 \mu_0 = \frac{1}{c^2} \,, \tag{9.22}$$

which is indeed a correct relation. This shows that conservation of electric charge is not a condition that should be viewed as being independent of the electromagnetic equations. It can be *derived* from the laws of Gauss and Ampère, and can therefore be seen as a consistency requirement for these two electromagnetic equations.

9.4 Gauss' law for the magnetic field and Faraday's law of induction

An important property of the magnetic field is that there exists no isolated magnetic pole. This means that the total magnetic flux through any closed surface S vanishes,

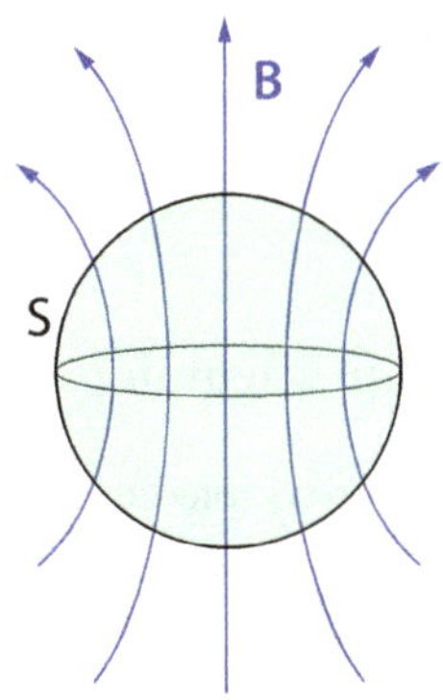

Fig. 9.3 Gauss' law for magnetic fields: The total magnetic flux through any closed surface S is zero. Therefore magnetic flux lines have no end points.

$$\oint_S \mathbf{B} \cdot d\mathbf{S} = 0\,, \tag{9.23}$$

and in differential form this gives

$$\boldsymbol{\nabla} \cdot \mathbf{B} = 0\,. \tag{9.24}$$

It has a similar form as Gauss' law for the electric field, but in this case there is no counterpart to the electric charge density. Expressed in terms of field lines, this means that magnetic field lines are always closed, whereas the electric field lines will generally be open, with end points on the electric charges.

Finally, Faraday's law of induction states that the integral of the electric field around a closed curve C is determined by the time derivative of the magnetic flux through a surface S with C as boundary,

$$\oint_C \mathbf{E} \cdot d\mathbf{s} = -\frac{d}{dt}\int_S \mathbf{B} \cdot d\mathbf{S}\,. \tag{9.25}$$

There is an obvious similarity between this equation and Ampere's law, with the electric field interchanged with the magnetic field. By use of the same method as in the discussion of Ampere's law, we rewrite the equation in differential form as

$$\boldsymbol{\nabla} \times \mathbf{E} + \frac{\partial}{\partial t}\mathbf{B} = 0\,. \tag{9.26}$$

The main difference is that there is no counterpart to the electric current in this equation. We also note from this equation that the electric field will in general not be a conservative field.

Faraday's law of induction describes the important phenomenon of induction of an electric field by a variable magnetic field. This effect is the basis for electromagnetic generators, where mechanical work is transformed into electric energy.

9.5 Maxwell's equations in vacuum

Maxwell's equations are the four coupled equations for the electromagnetic field that we have introduced separately,

$$\begin{aligned} &a)\quad \boldsymbol{\nabla}\cdot\mathbf{E}=\frac{\rho}{\epsilon_0}\,,\\ &b)\quad \boldsymbol{\nabla}\times\mathbf{B}-\frac{1}{c^2}\frac{\partial}{\partial t}\mathbf{E}=\mu_0\mathbf{j}\,,\\ &c)\quad \boldsymbol{\nabla}\cdot\mathbf{B}=0\,,\\ &d)\quad \boldsymbol{\nabla}\times\mathbf{E}+\frac{\partial}{\partial t}\mathbf{B}=0\,. \end{aligned} \tag{9.27}$$

These equation show how electric and magnetic fields are produced by electric charges and currents, and how the two fields interact. As a supplement to these equations we have the continuity equation for charge,

$$\frac{\partial\rho}{\partial t}+\boldsymbol{\nabla}\cdot\mathbf{j}=0\,, \tag{9.28}$$

which however, as we have seen, does not appear as an independent equation, but rather as a consistency condition for Maxwell's equations. To form a complete set, Maxwell's equations should also be supplemented with an equation which shows how the electromagnetic fields act back on charges and currents, here represented by the equation of motion of a charged point particle,

$$\frac{d\mathbf{p}}{dt}=e(\mathbf{E}+\mathbf{v}\times\mathbf{B})\,, \tag{9.29}$$

where $\mathbf{p}$ is the (mechanical) momentum of the particle. Together these equations form a closed set that describes the complete dynamics of the physical system of electromagnetic fields and charged particles.

Maxwell's equations have several interesting symmetry properties. One of these is the symmetry under Lorentz transformations. Maxwell's equations define in fact a fully relativistic theory, even if this theory was developed before Einstein formulated the theory of relativity. This symmetry is seen most clearly when the equations are formulated in the language of four-vectors and tensors.

Another symmetry, which is clearly seen in Maxwell's equations (9.27), is the symmetry under an interchange of electric and magnetic fields. In fact, without the source terms, *i.e.*, with vanishing charge and current densities, equations a) and b) are transformed into equations c) and d) (and *vice versa*) by the following change in the fields, $\mathbf{E} \to c\mathbf{B}$ and $c\mathbf{B} \to -\mathbf{E}$.[1] Even with sources there are symmetries between the electric and magnetic equations, and this can be exploited when solving problems in electrostatics and magnetostatics.

There have been speculations in the past whether the symmetry in Maxwell's equations between $\mathbf{E}$ and $\mathbf{B}$ should be extended to the general form of the equations, by including source terms also for the equations c) and d). The lack of sources for these equations in (9.27) can be understood as reflecting the lack of *magnetic monopoles* in nature. However, the existence of magnetic monopoles in the form of magnetic charges carried by new types of elementary particles cannot be fully excluded. To take this possibility into account in Maxwell's equations would mean to include source terms also in equations c) and d), in the form of *magnetic* charge and current densities. In that case there would be two different types of sources for the electromagnetic field, electric charges and currents, and magnetic charges and currents. There have been performed several experimental searches for elementary particles with magnetic charge, but so far with negative results. We shall here proceed in the usual way, by assuming that no magnetic charges and currents exist, and therefore by keeping Maxwell's equations in the standard form (9.27).

Electromagnetic potentials

When the possibility of magnetic charges has been excluded, and equations c) and d) therefore are homogeneous (source free), the electric and magnetic fields can be expressed in terms of electromagnetic potentials. These are referred to as the scalar potential Φ and the vector potential $\mathbf{A}$, and the electromagnetic fields expressed in terms of these are

$$\mathbf{E} = -\boldsymbol{\nabla}\Phi - \frac{\partial \mathbf{A}}{\partial t}, \quad \mathbf{B} = \boldsymbol{\nabla} \times \mathbf{A}. \tag{9.30}$$

When $\mathbf{E}$ and $\mathbf{B}$ are expressed in terms of the potentials in this way, the two homogeneous Maxwell's equations are in fact satisfied as identities,

[1] This transformation, which is referred to as a *duality transformation*, is a special case of field rotations of the form $\mathbf{E} \to \cos\theta\ \mathbf{E} + \sin\theta\ c\mathbf{B}$ and $\mathbf{B} \to \cos\theta\ \mathbf{B} - \sin\theta\ \mathbf{E}/c$. Without the source terms ρ and $\mathbf{j}$, Maxwell's equations are invariant under general transformations of this form.

as one can readily check. The use of electromagnetic potentials therefore effectively reduces Maxwell's equations to the two inhomogeneous equations a) and b). In addition to reducing the number of field equations, the use of potentials is helpful when solving the field equations.

Expressed in terms of the potentials Maxwell's equations get the form

$$
\begin{aligned}
&a)\ \ \nabla^2\Phi + \frac{\partial}{\partial t}\nabla\cdot\mathbf{A} = -\frac{\rho}{\epsilon_0}\,,\\
&b)\ \ \nabla^2\mathbf{A} - \nabla(\nabla\cdot\mathbf{A}) - \frac{1}{c^2}\frac{\partial}{\partial t}\nabla\Phi - \frac{1}{c^2}\frac{\partial^2}{\partial t^2}\mathbf{A} = -\mu_0\mathbf{j}\,. \qquad (9.31)
\end{aligned}
$$

These equations can be simplified further by imposing a *gauge condition* on the potentials.

Gauge transformations are transformations of the potentials that leave the electromagnetic fields unchanged. They have the form

$$
\begin{aligned}
\Phi \to \Phi' &= \Phi - \frac{\partial\chi}{\partial t}\,,\\
\mathbf{A} \to \mathbf{A}' &= \mathbf{A} + \nabla\chi\,, \qquad (9.32)
\end{aligned}
$$

where $\chi = \chi(\mathbf{r}, t)$ is a differentiable function of the space and time coordinates. It is straight forward to check that such a transformation does not change $\mathbf{E}$ or $\mathbf{B}$,

$$
\begin{aligned}
\mathbf{E} \to \mathbf{E}' &= -\nabla\Phi' - \frac{\partial\mathbf{A}'}{\partial t} = \mathbf{E} + \nabla\frac{\partial\chi}{\partial t} - \frac{\partial}{\partial t}\nabla\chi = \mathbf{E}\,,\\
\mathbf{B} \to \mathbf{B}' &= \nabla\times\mathbf{A}' = \mathbf{B} + \nabla\times\nabla\chi = \mathbf{B}\,. \qquad (9.33)
\end{aligned}
$$

The usual understanding is that gauge transformations do not correspond to any physical operation, since they leave $\mathbf{E}$ or $\mathbf{B}$ unchanged. They only reflect a certain freedom in the choice of the electromagnetic potentials that represent a given electromagnetic field configuration. This freedom can be exploited by making specific gauge choices in the form of conditions that the potentials should satisfy. Two commonly used gauge conditions are the following,

$$1)\ \ \nabla\cdot\mathbf{A} = 0 \qquad \text{Coulomb gauge}\,, \qquad (9.34)$$

$$2)\ \ \partial_\mu A^\mu = 0 \qquad \text{Lorentz gauge}\,, \qquad (9.35)$$

where the abbreviation $\partial_\mu = \frac{\partial}{\partial x^\mu}$, introduced in Chapter 6. The Lorentz gauge condition has a covariant form when expressed in terms of the four-vector potential with components A^μ. This potential is defined by $\underline{\mathbf{A}} = (\frac{1}{c}\Phi, \mathbf{A})$ so that the scalar potential is (up to a factor $1/c$) the time component, and the vector potential is the space component of the four-potential. This gauge condition is often used when it is important to keep the relativistic form of the equations. The Coulomb gauge condition, on the other hand, is often used when the charged particles, such as electrons in atoms, move with non-relativistic velocities, and the relativistic form of the equations therefore is less important. Also other types of gauge conditions can be imposed in order to simplify the electromagnetic equations, but it is important that the constraints they impose on the potentials should not introduce any unphysical constraint on the electromagnetic fields **E** and **B**.

Coulomb gauge

For the Coulomb and Lorentz gauge conditions one can show explicitly that these only affect the choice of potentials, but do not constrain the electromagnetic fields **E** and **B** in any way. Let us consider how this can be demonstrated for the Coulomb gauge. Assume **A** is an arbitrary vector potential, which does *not* satisfy the Coulomb gauge condition. We will change this to a vector potential $\mathbf{A}'$ that does satisfy the condition $\boldsymbol{\nabla}\cdot\mathbf{A}' = 0$, and since the two potentials should be equivalent in the sense that they represent the same magnetic field **B**, they should be related by a gauge transformation, $\mathbf{A}' = \mathbf{A} + \boldsymbol{\nabla}\chi$. The Coulomb gauge condition then implies that the function χ should satisfy the equation

$$\boldsymbol{\nabla}^2\chi = -\boldsymbol{\nabla}\cdot\mathbf{A}\,. \tag{9.36}$$

This equation in fact has the same form as Gauss' law in the static case, where the electric field is determined by the electrostatic potential, $\mathbf{E} = -\boldsymbol{\nabla}\Phi$. Expressed in terms of the electrostatic potential Gauss' law gets the form $\boldsymbol{\nabla}^2\Phi = -\rho/\epsilon_0$, and this equation we know, for an arbitrary charge distribution ρ, to have a well defined solution for Φ, as the sum of the Coulomb potentials from all the parts of the distribution. (We shall later discuss the electrostatic case explicitly.) The solution of (9.36) should then have the same form as the solution of the Coulomb problem, with ρ/ϵ_0 replaced by $\boldsymbol{\nabla}\cdot\mathbf{A}$. This shows that, for any electromagnetic field configuration, one can make a gauge transformation of the vector potential to a form which satisfies the Coulomb gauge condition.

With the Coulomb gauge condition satisfied, $\boldsymbol{\nabla}\cdot\mathbf{A} = 0$, Maxwell's equations take the form

$$\begin{aligned} &a)\quad \boldsymbol{\nabla}^2\Phi = -\frac{\rho}{\epsilon_0}\,,\\ &b)\quad \boldsymbol{\nabla}^2\mathbf{A} - \frac{1}{c^2}\frac{\partial^2}{\partial t^2}\mathbf{A} = -\mu_0\mathbf{j}_\perp\,, \end{aligned} \tag{9.37}$$

where in the second equation we have introduced the *transverse* current density, defined by

$$\mathbf{j}_\perp = \mathbf{j} - \epsilon_0 \frac{\partial}{\partial t} \boldsymbol{\nabla} \Phi \,. \tag{9.38}$$

The word "transverse" here means that it is divergence free, $\boldsymbol{\nabla} \cdot \mathbf{j}_\perp = 0$. This follows by applying Eq. (9.37a) and the continuity equation for charge,

$$\begin{aligned} \boldsymbol{\nabla} \cdot \mathbf{j}_\perp &= \boldsymbol{\nabla} \cdot \mathbf{j} - \epsilon_0 \frac{\partial}{\partial t} \boldsymbol{\nabla}^2 \Phi \\ &= \boldsymbol{\nabla} \cdot \mathbf{j} + \frac{\partial \rho}{\partial t} \\ &= 0 \,. \end{aligned} \tag{9.39}$$

Equation (9.38) can therefore be re-interpreted as a standard (*Helmholtz*) decomposition of the vector field, in a divergence-free (transverse) and a curl-free (longitudinal) component,

$$\mathbf{j} = \mathbf{j}_\perp + \mathbf{j}_\parallel \,, \tag{9.40}$$

and Eq.(9.37b) then shows that only the divergence-free component contributes to the equation.

One should also note that Eq. (9.37a) is non-dynamical in the sense that it involves no time derivative. It can thus be solved like the electrostatic equation, to give the potential Φ expressed in terms of the charge distribution ρ. This is the case even if Φ is time dependent. This means that the dynamical evolution of the electromagnetic field, in the Coulomb gauge, is described by the vector potential $\mathbf{A}$ alone, while the scalar potential Φ is uniquely defined as the Coulomb potential of the charge distribution at any given time.

9.6 Maxwell's equations in covariant form

The covariant form of Maxwell's equations is based on the use of the electromagnetic field tensor. Written as a 4×4 matrix the field tensor takes the form (see Sect. 8.3)

$$(F^{\mu\nu}) = \begin{pmatrix} 0 & E_1/c & E_2/c & E_3/c \\ -E_1/c & 0 & B_3 & -B_2 \\ -E_2/c & -B_3 & 0 & B_1 \\ -E_3/c & B_2 & -B_1 & 0 \end{pmatrix} . \tag{9.41}$$

The reason for the electric and magnetic fields to be arranged into a common object $F^{\mu\nu}$ is that the two fields are mixed under Lorentz transformations.

Such a mixing is implicit both in Maxwell's equations and in the equation of motion for a charged particle (9.29). In the latter case this is obvious, since a reference frame can be chosen where the particle is instantaneously at rest. In such a reference frame there is no contribution to the force from the magnetic field, and therefore the effect in this frame of the electric field alone must be equivalent to the effect of both the electric and magnetic fields in another frame, where the particle is moving. It is an interesting fact that the mixing of the **E** and **B** fields is correctly expressed by combining them in a linear way into the antisymmetric, rank two tensor $F^{\mu\nu}$.

Maxwell's equations, in the form (9.27), get a simplified, compact form when expressed in terms of the electromagnetic field tensor. The two first equations can be rewritten as

$$\begin{aligned} a) \quad & \frac{\partial}{\partial x_k} F^{0k} = \frac{1}{c\epsilon_0}\rho\,, \\ b) \quad & \frac{\partial}{\partial x^l} F^{kl} + \frac{\partial}{\partial x^0} F^{k0} = \mu_0 j^k\,, \end{aligned} \tag{9.42}$$

and these two equations can now be merged into a single covariant equation

$$\partial_\nu F^{\mu\nu} = \mu_0 j^\mu\,. \tag{9.43}$$

In the equation we have introduced the four-vector current density j^μ, which is composed by the charge and current densities in the following way,

$$(j^\mu) = (c\rho, \mathbf{j})\,, \tag{9.44}$$

so that the original three-vector $\mathbf{j}$ is extended to a four-vector $\underline{\mathbf{j}}$ by taking $c\rho$ as the time component of the current.

The continuity equation for charge can also be written in covariant form when the four-vector current is introduced. The covariant form is

$$\partial_\mu j^\mu = 0\,, \tag{9.45}$$

as we can readily verify by separating the time derivative from the space derivative and using the fact that the time component of the 4-current is the charge density (up to a factor c). We have already noticed that charge conservation is needed if Maxwell's equations should be consistent. This is seen very clearly from the covariant equation (9.43), where the continuity equation of the current follows from the antisymmetry of the electromagnetic tensor.

We continue with bringing Maxwell's equations c) and d) into covariant form. To do so we focus on the symmetry between the inhomogeneous equations a) and b), and the homogeneous equations c) and d), under the duality transformation

$$\frac{1}{c}\mathbf{E} \rightarrow -\mathbf{B}\,, \quad \mathbf{B} \rightarrow \frac{1}{c}\mathbf{E}\,. \tag{9.46}$$

It is convenient to define the *dual* field tensor $\tilde{F}^{\mu\nu}$, which is related to $F^{\mu\nu}$ by the same transformation. The matrix form of this tensor is thus

$$(\tilde{F}^{\mu\nu}) = \begin{pmatrix} 0 & -B_1 & -B_2 & -B_3 \\ B_1 & 0 & E_3/c & -E_2/c \\ B_2 & -E_3/c & 0 & E_1/c \\ B_3 & E_2/c & -E_1/c & 0 \end{pmatrix}, \tag{9.47}$$

where the $\mathbf{E}$ and $\mathbf{B}$-fields of the original field tensor have been interchanged according to the substitution (9.46). The relation to the electromagnetic field tensor can be expressed in compact form as

$$\tilde{F}_{\mu\nu} = \frac{1}{2}\epsilon_{\mu\nu\rho\sigma}F^{\rho\sigma}\,, \tag{9.48}$$

where we have introduced the four-dimensional Levi-Civita tensor $\epsilon_{\mu\nu\rho\sigma}$. This is fully antisymmetric under interchange of any pair of indices, and further satisfies $\epsilon_{0klm} = \epsilon_{klm}$.

Based on the symmetry of the field part of Maxwell's equations under the interchange (9.46) of $\mathbf{E}$ and $\mathbf{B}$, it is now clear that the four Maxwell's equations can be written as two compact, covariant equations

$$\begin{aligned} \partial_\nu F^{\mu\nu} &= \mu_0 j^\mu\,, \\ \partial_\nu \tilde{F}^{\mu\nu} &= 0\,. \end{aligned} \tag{9.49}$$

In this form the (partial) symmetry of the equations under duality transformation, $F^{\mu\nu} \rightarrow \tilde{F}^{\mu\nu}$, is seen very clearly. Also the lack of full symmetry is clear, with a "magnetic current" missing in the second equation.

9.7 The electromagnetic four-potential

The absence of a magnetic current in Maxwell's equations makes the symmetry between the electric and magnetic fields not fully complete. However, for the same reason the field tensor can be expressed in terms of the *electromagnetic four-potential* A^μ in the following way,

$$F^{\mu\nu} = \partial^\mu A^\nu - \partial^\nu A^\mu\,. \tag{9.50}$$

As previously discussed, the four-potential is composed of the non-relativistic potentials in such a way that the time component is $A^0 = \Phi/c$

with Φ as the original scalar potential, and the space part of A^μ is identical to the vector potential $\mathbf{A}$. When the field tensor is expressed in terms of the four-potential, the second of the two covariant Maxwell equations is satisfied as an identity, as one now can easily verify by expressing $\tilde{F}^{\mu\nu}$ in terms of A^μ. This means that Maxwell's equations are reduced to one four-vector equation, which is

$$\partial_\nu \partial^\nu A^\mu - \partial^\mu \partial_\nu A^\nu = -\mu_0 j^\mu \,. \tag{9.51}$$

As a last step to simplify the equation we again make use of the freedom to change the potential by a gauge transformation. In covariant form such a transformation is

$$A^\mu \to A'^\mu = A^\mu + \partial^\mu \chi \,, \tag{9.52}$$

where χ is an unspecified differentiable function of the space-time coordinates. In the covariant formulation it is straight forward to check that such a transformation of the four-potential will not change the field tensor. The freedom to change the potential in this way can be used to bring it to the form where the covariant Lorentz gauge condition is satisfied,

$$\partial_\mu A^\mu = 0 \,. \tag{9.53}$$

When this condition is satisfied we find Maxwell's equation reduced to its simplest form

$$\partial_\nu \partial^\nu A^\mu = -\mu_0 j^\mu \,. \tag{9.54}$$

The differential operator, referred to as the d'Alembertian, is often assigned the symbol $\Box$ (or alternatively $\Box^2$),

$$\Box \equiv \partial_\nu \partial^\nu = \boldsymbol{\nabla}^2 - \frac{1}{c^2}\frac{\partial^2}{\partial t^2} \,. \tag{9.55}$$

9.8 Lorentz transformations of the electromagnetic field

When the electric and magnetic fields are collected in the electromagnetic field *tensor*, this means that the correct transformation of $\mathbf{E}$ and $\mathbf{B}$ under Lorentz transformations have been implicitly assumed. This is of course not simply postulated, it is based on the assumption that Maxwell's equations (as well as the equation of motion of charged particles) have the same form in all inertial reference frames, and this is in turn a well established fact based on experimental tests. We will here take the tensor properties of $F^{\mu\nu}$ as the starting point, and show from this how Lorentz transformations mix the electric and magnetic components.

We consider first field transformations under a simple boost in the x-direction. This Lorentz transformation will mix only the time and x-coordinates, and is given by a transformation matrix of the form

$$L = \begin{pmatrix} \gamma & -\beta\gamma & 0 & 0 \\ -\beta\gamma & \gamma & 0 & 0 \\ 0 & 0 & 1 & 0 \\ 0 & 0 & 0 & 1 \end{pmatrix}, \tag{9.56}$$

which means that the only non-vanishing matrix elements are

$$\begin{aligned} L^0{}_0 &= L^1{}_1 = \gamma\,, \\ L^0{}_1 &= L^1{}_0 = -\beta\gamma\,, \\ L^2{}_2 &= L^3{}_3 = 1\,. \end{aligned} \tag{9.57}$$

The tensor properties of the electromagnetic field tensor implies that the transformed field is related to the original field by the equation

$$F'^{\mu\nu} = L^\mu{}_\rho L^\nu{}_\sigma F^{\rho\sigma}\,. \tag{9.58}$$

We extract from this formula the transformation equations for the components of the electric and magnetic fields, in the case where the matrix elements of the Lorentz transformation are given by (9.57). The x-component of the electric field is

$$\begin{aligned} E_1' &= cF'^{01} \\ &= cL^0{}_0 L^1{}_1 F^{01} + cL^0{}_1 L^1{}_0 F^{10} \\ &= c(L^0{}_0 L^1{}_1 - L^0{}_1 L^1{}_0) F^{01} \\ &= \gamma^2(1-\beta^2)E_1 \\ &= E_1\,, \end{aligned} \tag{9.59}$$

which shows that the component in the direction of the boost is unchanged. In the orthogonal directions the components of the transformed field are

$$\begin{aligned} E_2' &= cF'^{02} \\ &= cL^0{}_0 L^2{}_2 F^{02} + cL^0{}_1 L^2{}_2 F^{12} \\ &= \gamma E_2 - \gamma\beta c B_3 \\ &= \gamma(E_2 - vB_3)\,, \end{aligned} \tag{9.60}$$

and

$$\begin{aligned} E_3' &= cF'^{03} \\ &= cL^0{}_0 L^3{}_3 F^{03} + cL^0{}_1 L^3{}_3 F^{13} \\ &= \gamma E_3 + \gamma\beta c B_2 \\ &= \gamma(E_3 + vB_2)\,. \end{aligned} \tag{9.61}$$

These expressions can be written in a form which is independent of the choice of coordinate axes by introducing the *parallel* and *transverse* components of the electric field,

$$E_{\|} = E_1\,, \quad \mathbf{E}_{\perp} = E_2\,\mathbf{e}_y + E_3\,\mathbf{e}_z\,, \tag{9.62}$$

with parallel and transverse here referring to parallel or orthogonal to the boost velocity $\mathbf{v}$. The transformation formulas then are

$$E'_{\|} = E_{\|}\,, \quad \mathbf{E}'_{\perp} = \gamma(\mathbf{E}_{\perp} + \mathbf{v}\times\mathbf{B})\,, \tag{9.63}$$

which shows that the component of the field in the direction of the boost velocity $\mathbf{v}$ is unchanged, while the components orthogonal to $\mathbf{v}$ are mixtures of the original orthogonal components of the electric and magnetic fields.

The transformation of the magnetic components of the field tensor can be found in the same way, and the result is

$$B'_{\|} = B_{\|}\,, \quad \mathbf{B}'_{\perp} = \gamma\left(\mathbf{B}_{\perp} - \frac{\mathbf{v}}{c^2}\times\mathbf{E}\right). \tag{9.64}$$

The transformation formulas for $\mathbf{E}$ and $\mathbf{B}$ have almost the same form, and they are related by the duality transformation already discussed,

$$\frac{1}{c}\mathbf{E} \to -\mathbf{B}\,, \quad \mathbf{B} \to \frac{1}{c}\mathbf{E}\,.$$

The symmetry under this transformation, which gives a mapping between the two field tensors $F^{\mu\nu}$ and $\tilde{F}^{\mu\nu}$, can in fact be used to *derive* the transformation formula for the $\mathbf{B}$-field directly from the transformation formula for the $\mathbf{E}$-field.

Example

Lorentz transformations of a constant magnetic field

As a simple example we assume that in the reference frame S there is no electric field, and a constant magnetic field, $\mathbf{B} = B_0\,\mathbf{e}_z$, directed along the z-axis. The moving frame S' has a velocity $\mathbf{v}$ in the x-direction. We split the fields in parallel and orthogonal components,

$$E_{\|} = \mathbf{E}_{\perp} = 0\,, \quad B_{\|} = 0\,, \; \mathbf{B}_{\perp} = B_0\,\mathbf{e}_z\,. \tag{9.65}$$

For the parallel components of the transformed fields we find

$$E'_{\|} = E_{\|} = 0\,, \quad B'_{\|} = B_{\|} = 0\,, \tag{9.66}$$

and for the orthogonal components

$$\mathbf{E}'_\perp = \gamma(\mathbf{E}_\perp + \mathbf{v} \times \mathbf{B}) = \gamma \mathbf{v} \times \mathbf{B} = -\gamma v B_0 \, \mathbf{e}_y \,,$$
$$\mathbf{B}'_\perp = \gamma \left(\mathbf{B}_\perp - \frac{\mathbf{v}}{c^2} \times \mathbf{E} \right) = \gamma \mathbf{B}_\perp = \gamma B_0 \, \mathbf{e}_z \,. \tag{9.67}$$

Collecting these terms we find that the fields in the reference frame S' are

$$\mathbf{E}' = -\gamma v B_0 \, \mathbf{e}_y \,, \quad \mathbf{B}' = \gamma B_0 \, \mathbf{e}_z \,. \tag{9.68}$$

Also in this reference frame the magnetic field points in the z-direction, but it is stronger than in S due to the factor γ which is larger than 1. In addition there is an electric field in the direction orthogonal to both the velocity of the transformation and the magnetic field.

Lorentz invariants

From the electromagnetic field tensor $F^{\mu\nu}$ we can construct two Lorentz invariant quantities. These are combinations of the electric and magnetic field strengths that take the same value in all inertial frames. For a general tensor $T^{\mu\nu}$ the trace $T^{\mu}{}_{\mu}$ is such an invariant, but in the present case the trace vanishes since $F^{\mu\nu}$ is antisymmetric. This means that there is no invariant that is *linear* in the components of $\mathbf{E}$ and $\mathbf{B}$. However, there are two quadratic expressions that are Lorentz invariants. These are

$$I_1 = \frac{1}{2} F^{\mu\nu} F_{\mu\nu} = \mathbf{B}^2 - \frac{1}{c^2} \mathbf{E}^2 \,,$$
$$I_2 = \frac{1}{4} F^{\mu\nu} \tilde{F}_{\mu\nu} = \frac{1}{c} \mathbf{E} \cdot \mathbf{B} \,. \tag{9.69}$$

It is easy to check that for the example just discussed we get the same expression for the two invariants, whether we evaluate them in reference frame S or S',

$$I_1 = B_0^2 \,, \quad I_2 = 0 \,. \tag{9.70}$$

We note in particular that even if $\mathbf{E}$ and $\mathbf{B}$ get mixed by the Lorentz transformation, the fact that $\mathbf{E}$ dominates $\mathbf{B}$ ($\mathbf{E}^2 > c^2 \mathbf{B}^2$) or $\mathbf{B}$ dominates $\mathbf{E}$ ($\mathbf{E}^2 < c^2 \mathbf{B}^2$) can be stated without reference to any particular inertial frame.

Example

The electromagnetic fields of a linear electric current

In this example we consider the situation where a constant current is running in a straight conducting wire, as shown in Fig. 9.4. We will use the transformation formulas for fields and currents to study these in two different inertial frames. The first one is the reference frame S, which is stationary with respect to the conductor, and the other is a reference frame S', which moves with the electrons in the current. In S we assume the current to take the value I and the conductor to be electrically neutral. In this reference frame the magnetic field circulates the current and outside the conductor the field strength $\mathbf{B}$ is determined by Ampère's law as

$$\oint_C \mathbf{B} \cdot d\mathbf{s} = \mu_0 I \,. \tag{9.71}$$

Assuming the conductor to be rotationally symmetric this determines the field to be

$$\mathbf{B} = \frac{\mu_0 I}{2\pi r} \mathbf{e}_\phi \,, \tag{9.72}$$

with r as the distance from the centre of the conductor and $\mathbf{e}_\phi$ as unit vectors circulating the current. With charge neutrality the electric field orthogonal to the current vanishes, but there is an electric field inside the conductor that drives the current. The current density is

$$\mathbf{j} = \sigma \mathbf{E} \,, \tag{9.73}$$

with σ as the conductivity. We assume electric field to have a constant value, $\mathbf{E}$, inside the conductor, with the same value also outside, close to the conductor.

In reference frame S, where the ions of the conducting material are at rest, the current density gets only contribution from the electrons, with $j = v_e \rho_e$, where v_e is the average velocity and ρ_e the average charge density of the electrons. The current can then be written as

$$I = Aj = A v_e \rho_e \,, \tag{9.74}$$

where A is the cross section area of the conductor. (For simplicity we assume the current density to be constant over the cross section.)

The fields in the reference frame S, when decomposed in the parallel and orthogonal components along the conductor, are given by

$$E_\parallel = E \,, \quad \mathbf{E}_\perp = 0 \,, \quad B_\parallel = 0 \,, \quad \mathbf{B}_\perp = \frac{\mu_0 I}{2\pi r} \mathbf{e}_\phi \,. \tag{9.75}$$

The corresponding fields in S', which moves with velocity $\mathbf{v}_e$ relative to S, are determined by the transformation formulas (9.63) and (9.64), as

$$\begin{aligned} E'_\parallel &= E_\parallel = E \,, \quad & \mathbf{E}'_\perp &= \gamma(\mathbf{E}_\perp + \mathbf{v}_e \times \mathbf{B}) = -\gamma v_e \frac{\mu_0 I}{2\pi r} \mathbf{e}_r \,, \\ B'_\parallel &= B_\parallel = 0 \,, \quad & \mathbf{B}'_\perp &= \gamma \left(\mathbf{B}_\perp - \frac{\mathbf{v}_e}{c^2} \times \mathbf{E} \right) = \gamma \frac{\mu_0 I}{2\pi r} \mathbf{e}_\phi \,. \end{aligned} \tag{9.76}$$

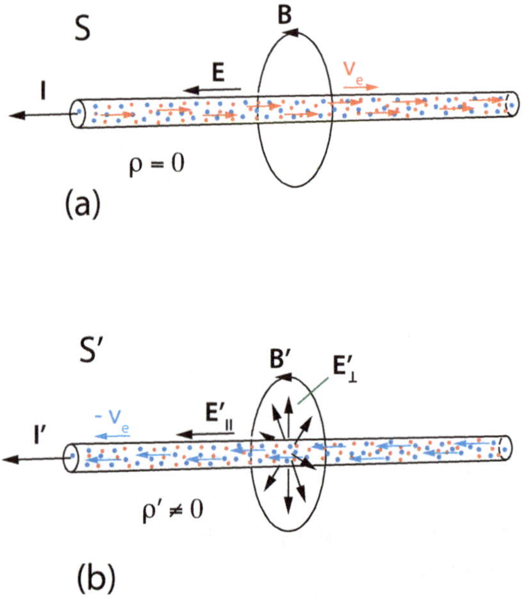

Fig. 9.4 Electromagnetic fields of a linear current. In figure (a) the directions of the electric and magnetic fields are indicated as seen in the rest frame S of the conductor, which is assumed to be charge neutral. The blue points represent charged ions, with positions which are fixed relative to the conductor, while the red points represent moving electrons, which create a current I. The electric field $\mathbf{E}$, which drives the current, is directed along the conductor. In figure (b) the fields are viewed in the (average) rest frame S' of the electrons. In this reference frame there is a non-vanishing charge density ρ' which can be viewed as due to the mixing of charge and current by the Lorentz transformation between S and S'. This implies the presence of a radial component $\mathbf{E}'_\perp$ of the electric field in reference frame S', in addition to the component $\mathbf{E}'_{\parallel}$, which is directed along the conductor.

The magnetic field is also in this reference frame circulating the current, but now it is stronger, enhanced by the factor γ. The electric field we note to have, in addition to the parallel component E, a normal component that is radially directed, out from the conductor. This normal component is somewhat unexpected, since it indicates that the conducting wire in reference frame S' is not charge neutral. Thus, a charge density is needed along the wire in order to create a radially directed electric field. We will check whether this result is consistent with Maxwell's equations by evaluating the charge and current densities in the transformed reference frame.

In reference frame S the charge density is $\rho = 0$ and current density is $j = v_e\rho_e$. To find the corresponding quantities in reference frame S' we use the fact that charge and current densities together form a four-vector $\underline{\mathbf{j}} = (c\rho, \mathbf{j})$. The standard transformation formula for four-vectors gives for the charge and current densities in S',

$$\rho' = \gamma\left(\rho - \frac{v_e}{c^2}j\right) = -\gamma\frac{v_e^2}{c^2}\rho_e = -\gamma\beta^2\rho_e\,,$$
$$j' = \gamma(j - v_e\rho) = \gamma v_e\rho_e\,. \tag{9.77}$$

The charge density in S' is indeed different from zero and the current density is modified by a factor γ. We check now that the expressions given above for the transformed charge and current densities are consistent with what we have found for the transformed fields. We note that the enhancement of the current in S', by the factor γ, is consistent with the corresponding enhancement of the transformed magnetic field. To check the consistency of the transformation of the charge density and the electric field, we consider Gauss' law in reference frame S'. We denote by $\Delta Q'$ the charge in a piece of the conducting wire of length $\Delta L'$ in S', so that $\Delta Q' = \rho'\,\Delta L' A$. According to Gauss' law this charge should create a radially directed electric field given by

$$2\pi r\,\Delta L'\,E'_\perp = \frac{\Delta Q'}{\epsilon_0}\,, \tag{9.78}$$

which gives

$$E'_\perp = \frac{\rho' A}{2\pi r\epsilon_0} = -\gamma\beta^2\frac{\rho_e A}{2\pi r\epsilon_0} = -\gamma\beta\frac{j_e A}{2\pi r c\epsilon_0} = -\gamma v_e\frac{\mu_0 I}{2\pi r}\,, \tag{9.79}$$

where in the last step we have used the relation $\epsilon_0\mu_0 = 1/c^2$. The expression for the radial component of the electric field found in this way is indeed consistent with the result (9.76) found by applying the transformation formula for the electromagnetic field.

Although it may initially seem strange that the conducting wire is charge neutral in one reference frame, but not in the other, it is a clear consequence of the description of charge and current as components of the same four-vector current. Lorentz transformations will mix the time and space components of the 4-current.

As a final comment we should point out that, in a normal conducting wire, the effect we have discussed is extremely small. It is a relativistic effect determined by the ratio of the drift velocity (average velocity) of the electrons to the speed of light. A typical value of the drift velocity may be $v_e = 3 \cdot 10^{-4}$ m/s, corresponding to $\beta = 10^{-12}$ and $\gamma - 1 = 0.5 \cdot 10^{-24}$, indeed a very small number.

9.9 Exercises

Problem 9.1

A vector potential, in cylindrical coordinates, is given as

$$\mathbf{A}(\rho, \phi, z) = A(\phi\rho\, e^{-(\rho/a)^2}\mathbf{e}_\rho + b\, e^{-(\rho/a)^2}\mathbf{e}_z)\,, \tag{9.80}$$

with A, a, and b as constants.

Determine the corresponding magnetic field $\mathbf{B}$, expressed in cylindrical coordinates, and find the current density $\mathbf{j}$ which creates this field. There is no electric field, $\mathbf{E} = 0$.

Problem 9.2

An inertial reference frame S has a set of Cartesian coordinates (t, x, y, z), and a second inertial reference frame S' has coordinates (t', x', y', z'). S' moves along the x-axis, with relativistic velocity v relative to S. The two sets of coordinates are related by the Lorentz transformations

$$t' = \gamma\left(t - \frac{v}{c^2}x\right)\,, \quad x' = \gamma(x - vt)\,, \quad y' = y\,, \quad z' = z\,. \tag{9.81}$$

A point charge q sits at rest at the origin of S, and the electromagnetic field from the charge is therefore a pure Coulomb field in this reference frame.

a) Make use of the Lorentz transformation properties of the scalar and vector potentials, Φ and $\mathbf{A}$, to find the potentials Φ' and $\mathbf{A}'$ in reference frame S', both expressed in terms of the Cartesian coordinates in this frame.

b) Make use of the results from a) to determine the electric field $\mathbf{E}'$ and the magnetic field $\mathbf{B}'$ in S'.

c) Make a two-dimensional (x', y')-plot (with $t' = z' = 0$), which shows equipotential curves of the function $\Phi'(x', y')$ and field lines of the vector field $\mathbf{E}'(x', y')$, for a convenient value for γ. Compare with corresponding plots for the fields in reference frame S.

Problem 9.3

As shown in Sect. 9.8 there are two quadratic Lorentz invariants that can be formed from the electromagnetic fields,

$$\begin{aligned} I_1 &= \frac{1}{2}F^{\mu\nu}F_{\mu\nu} = \mathbf{B}^2 - \frac{1}{c^2}\mathbf{E}^2\,, \\ I_2 &= \frac{1}{4}F^{\mu\nu}\tilde{F}_{\mu\nu} = \frac{1}{c}\mathbf{E}\cdot\mathbf{B}\,. \end{aligned} \tag{9.82}$$

In principle there are also higher order invariants, but they can be shown either to vanish or to be composed of products of the quadratic invariants. We consider here the following cubic and fourth order invariants,

$$I_3 \equiv F^{\mu\nu} F_{\nu\lambda} F^{\lambda}{}_{\mu}\,, \quad I_4 \equiv F^{\mu\nu} F_{\nu\lambda} F^{\lambda\rho} F_{\rho\mu}\,. \tag{9.83}$$

a) Show that $I_3 = 0$ by exploiting the antisymmetry of the electromagnetic field tensor.

b) Show the identity $I_4 = 2I_1^2 + 4I_2^2$ by re-expressing I_4 in terms of the electric and magnetic fields. Note that the calculation can be simplified by orienting the coordinate frame in a specific way, so that the **B**- and **E**-fields are written as

$$\mathbf{B} = B\mathbf{i}\,, \quad \mathbf{E} = E_1\mathbf{i} + E_2\mathbf{j}\,. \tag{9.84}$$

This implies that the field tensor has only three independent, non-vanishing matrix elements, F^{01}, F^{02}, and F^{23}. Also note that the antisymmetry of the field tensor implies

$$F^{0k} = F_{k0}\,, k = 1, 2\,; \quad F^{23} = -F_{32}\,. \tag{9.85}$$

Problem 9.4

Protons in the accelerator ring LHC at CERN are bent into a near circular orbit by a large number of strong magnets. We consider in this problem the motion of a proton within one of the magnets. A strong magnetic field **B** perpendicular to the plane of the ring will bend the orbit with a bending radius R, as illustrated in Fig. 9.5. We consider the magnetic field inside the magnet to be constant in strength.

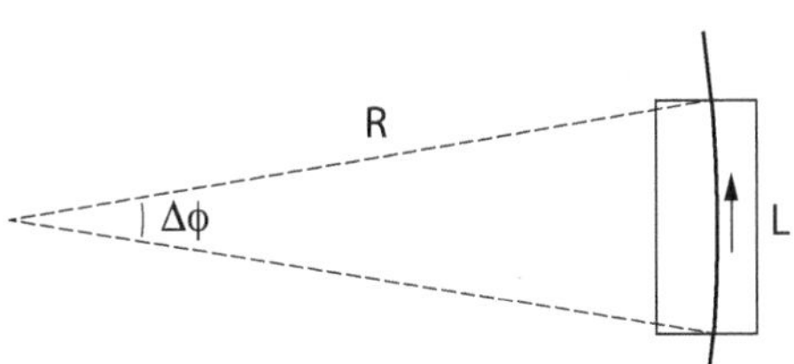

Fig. 9.5 Bending of a proton beam in an accelerator ring.

For the accelerator we have the following information. The proton momentum is $p = 7.0\,\mathrm{TeV/c}$ (or $pc = 7.0 \cdot 10^{12}\,\mathrm{eV}$), the bending radius of the magnet is $R = 2804\,\mathrm{m}$, and the strength of the magnetic field is $B = 8.33\,\mathrm{T}$. The proton mass is $m = 938\,\mathrm{MeV/c^2}$, and the speed of light is $c = 3.0 \cdot 10^8\,\mathrm{m/s}$.

a) Show that we have the following relation between the strength of the magnetic field, the momentum of the proton and the bending radius,

$$eBR = p \tag{9.86}$$

and check that this is satisfied for the given data.

b) Find the relativistic gamma factor γ of the proton, and the acceleration a of the particle within the magnet, both determined in the laboratory frame, where the accelerator ring is at rest.

c) We consider the same situation in the instantaneous rest frame of the proton. What is the strength and orientation of the magnetic field $\mathbf{B}'$ and the electric field $\mathbf{E}'$ in this reference frame, and what is the proper acceleration a_0 of the proton?

Chapter 10

Electromagnetic field dynamics

Maxwell's equations show that the electromagnetic field has its own dynamics. It can propagate as waves through empty space and it can carry energy and momentum. This was realized by Maxwell, and since the propagation velocity is identical to the speed of light, that convinced him that light is such an electromagnetic wave phenomenon. In this chapter we first discuss the wave solutions of Maxwell's equations with particular focus on *polarization* of electromagnetic waves and next examine how Maxwell's equation determine the energy and momentum densities of the electromagnetic field.

10.1 Electromagnetic waves

We consider the source free case, with $j^\mu = 0$, and with the four-potential restricted by the Lorentz gauge condition, $\partial_\mu A^\mu = 0$. The field equation then is

$$\partial_\nu \partial^\nu A^\mu = 0 \,, \tag{10.1}$$

where the d'Alembertian

$$\partial_\nu \partial^\nu = \boldsymbol{\nabla}^2 - \frac{1}{c^2}\frac{\partial^2}{\partial t^2} \,, \tag{10.2}$$

has the form of a *wave operator* in three dimensional space. The wave equation (10.1), which is a *linear* differential equation, has a complete set of *normal modes* as solutions. In the open space, without any physical boundaries, a natural choice for such a set of normal modes is the set of monochromatic plane waves, of the form

$$A^\mu(x) = A^\mu(0)\, e^{ik_\mu x^\mu} \,, \tag{10.3}$$

with $A^\mu(0)$ as the amplitude of field component μ. The four-vector $\underline{\mathbf{k}}$, with components k^μ, is decomposed as

$$\underline{\mathbf{k}} = \left(\frac{\omega}{c}, \mathbf{k}\right) , \tag{10.4}$$

where ω is the angular frequency, and $\mathbf{k}$ is the *wave vector* of the monochromatic plane wave. The plane wave solution (10.3) is a complex solution of Maxwell's equation. Such a complex form is often convenient to use, since it makes the expressions more compact. However, one should keep in mind that the *physical* field is real, and should be identified with the real (or imaginary) part of the solution.

To check that plane waves of the form (10.3) are solutions of the wave equation (10.1) is straight forward. We find

$$\partial_\nu \partial^\nu A^\mu = -k_\nu k^\nu A^\mu \,, \tag{10.5}$$

which shows that the function (10.3) is a solution provided

$$k_\nu k^\nu = 0 \,. \tag{10.6}$$

This means that the four-vector $\underline{\mathbf{k}}$ is a lightlike vector (also called a *null vector*), and this gives rise to the well-known linear relation between frequency and wave number for electromagnetic waves,

$$\omega = c|\mathbf{k}| \,. \tag{10.7}$$

The Lorentz gauge condition further demands the two four-vectors $\underline{\mathbf{k}}$ and $\underline{\mathbf{A}}$ to be orthogonal in the relativistic sense

$$k_\mu A^\mu = 0 \,. \tag{10.8}$$

One should note that the Lorentz gauge condition does not fix uniquely the four-potential for a given electromagnetic field. This is readily seen by assuming A^μ to be a general potential, which satisfies no particular gauge condition. By a gauge transformation

$$A^\mu \to A'^\mu = A^\mu + \partial^\mu \chi \,, \tag{10.9}$$

it can be brought to a form which does satisfy the Lorentz gauge condition $\partial_\mu A'^\mu = 0$, provided χ satisfies the equation

$$\partial_\mu \partial^\mu \chi = -\partial_\mu A^\mu \,. \tag{10.10}$$

χ is not uniquely determined by the equation, since to any particular solution of this differential equation one can add a general solution of the homogeneous (wave) equation $\partial_\mu \partial^\mu \chi = 0$. In the present case, with A^μ satisfying the homogeneous equation (10.1), one can use this freedom to set the time component of the potential to zero.

We therefore assume in the following $A^0 = 0$, with the remaining vector part satisfying the Coulomb gauge condition $\boldsymbol{\nabla} \cdot \mathbf{A} = 0$. The wave equation for $\mathbf{A}$ is

$$\left(\boldsymbol{\nabla}^2 - \frac{1}{c^2} \frac{\partial^2}{\partial t^2} \right) \mathbf{A}(\mathbf{r}, t) = 0 \,. \tag{10.11}$$

Thus, the three components of the vector potential satisfy three identical, uncoupled wave equations, with plane wave solutions

$$\mathbf{A}(\mathbf{r},t) = \mathbf{A}_0 e^{i(\mathbf{k}\cdot\mathbf{r}-\omega t)}\,, \tag{10.12}$$

where the amplitude $\mathbf{A}_0$ is a complex vector orthogonal to $\mathbf{k}$,

$$\mathbf{k}\cdot\mathbf{A}_0 = 0\,, \tag{10.13}$$

in order to satisfy the Coulomb gauge condition.

The general solution to the electromagnetic wave equation (10.11) can now be written as a superposition of plane waves,

$$\mathbf{A}(\mathbf{r},t) = \int d^3k\,\mathbf{A}(\mathbf{k})\,e^{i(\mathbf{k}\cdot\mathbf{r}-\omega t)}\,,\quad \mathbf{k}\cdot\mathbf{A}(\mathbf{k}) = 0\,, \tag{10.14}$$

where each Fourier component $\mathbf{A}(\mathbf{k})$ has to satisfy the transversality condition.

We have in this discussion of electromagnetic waves assumed that they propagate in the open, infinite space. The plane waves then define a complete set of normal modes of the field. If the situation instead corresponds to wave propagation inside some given boundary, for example inside a *wave guide* the normal modes are not the infinite plane waves but solutions that are adjusted to the given boundary conditions. To find the normal modes of the electromagnetic field may then be more demanding, but the general solution is again a linear superposition of these modes.

10.2 Polarization

The plane wave solution (10.12) for the electromagnetic potential gives related expressions for the electric and magnetic fields. For the electric field we find

$$\mathbf{E}(\mathbf{r},t) = -\frac{\partial\mathbf{A}}{\partial t} = i\omega\mathbf{A}_0 e^{i(\mathbf{k}\cdot\mathbf{r}-\omega t)} = i\omega\mathbf{A}(\mathbf{r},t)\,, \tag{10.15}$$

and for the magnetic field

$$\mathbf{B}(\mathbf{r},t) = \boldsymbol{\nabla}\times\mathbf{A} = i\mathbf{k}\times\mathbf{A}_0 e^{i(\mathbf{k}\cdot\mathbf{r}-\omega t)} = i\mathbf{k}\times\mathbf{A}(\mathbf{r},t)\,. \tag{10.16}$$

We note that both these fields satisfy the transversality condition

$$\mathbf{k}\cdot\mathbf{E} = \mathbf{k}\cdot\mathbf{B} = 0\,, \tag{10.17}$$

and they are related by

$$\mathbf{B} = \frac{1}{c}\mathbf{n}\times\mathbf{E}\,,\quad \mathbf{E} = -c\mathbf{n}\times\mathbf{B}\,, \tag{10.18}$$

with $\mathbf{n} = \mathbf{k}/k$ as the unit vector in the direction of propagation of the plane wave. Thus the triplet $(\mathbf{k}, \mathbf{E}, \mathbf{B})$ forms a right handed, orthogonal set of vectors. We further note that for a monochromatic plane wave the two electromagnetic Lorentz invariants previously discussed both vanish

$$\mathbf{E}^2 - c^2\mathbf{B}^2 = 0\,, \quad \mathbf{E} \cdot \mathbf{B} = 0\,. \tag{10.19}$$

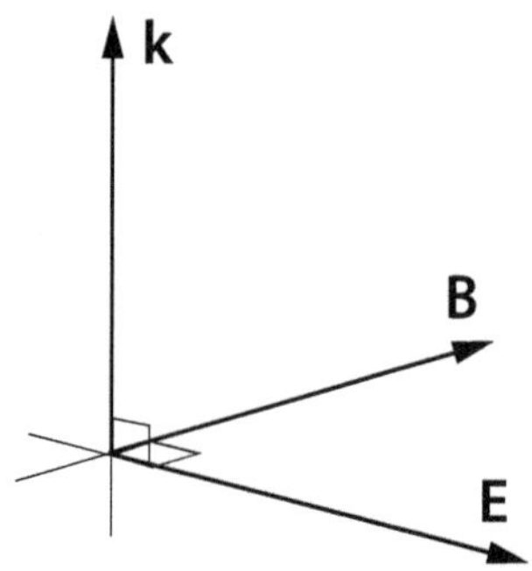

Fig. 10.1 Field vectors of a plane wave. The wave vector **k** together with the two field vectors **E** and **B** form a right-handed set of orthogonal vectors.

The monochromatic plane wave is, as we see, specified on one hand by the wave vector **k**, which gives the direction of propagation and the frequency of the wave, and on the other hand by the orientation of the electric field vector **E** in the plane orthogonal to **k**. The degree of freedom specified by the *direction* of **E** we identify as defining the *polarization* of the electromagnetic wave. We shall take a closer look at the description of different types of polarization. As follows from Eq. (10.18) it is sufficient to focus on the electric field **E**, since the magnetic field **B** is uniquely determined by **E**.

Written in complex form the electric field strength of the plane wave has the form

$$\mathbf{E}(\mathbf{r}, t) = \mathbf{E}_0\, e^{i(\mathbf{k}\cdot\mathbf{r} - \omega t)}\,, \tag{10.20}$$

where the amplitude $\mathbf{E}_0$ is in general a complex vector. We consider the real part of the field (10.20) as the physical field. When decomposed on two arbitrarily chosen orthogonal real unit vectors $\mathbf{e}_1$ and $\mathbf{e}_2$ in the plane orthogonal to **k**, the real field gets the general form

$$\mathbf{E}(\mathbf{r}, t) = E_{10}\,\mathbf{e}_1 \cos(\mathbf{k}\cdot\mathbf{r} - \omega t + \phi_1) + E_{20}\,\mathbf{e}_2 \cos(\mathbf{k}\cdot\mathbf{r} - \omega t + \phi_2)\,. \tag{10.21}$$

The vectors $\mathbf{e}_1$ and $\mathbf{e}_2$ are referred to as polarization vectors, and the triplet of orthogonal unit vectors $(\mathbf{e}_1, \mathbf{e}_2, \mathbf{n})$ are often chosen to form a right-handed reference frame. The two amplitudes E_{10} and E_{20} may be different, and that is also the case for the two phases ϕ_1 and ϕ_2. Note, however, that only the *relative* phase $\phi \equiv \phi_1 - \phi_2$ is physically relevant, since the sum of the phases can be changed by a shift in the definition of the time coordinate t. In the following we shall therefore use this freedom to set $\phi_1 = 0$ and therefore $\phi_2 = -\phi$.

The different types of polarization can be analyzed by considering the orbit described by the real vector $\mathbf{E}(\mathbf{r}, t)$ in the two-dimensional plane when the time coordinate t changes for a fixed point $\mathbf{r}$ in physical space. We consider first some special cases.

Linear polarization

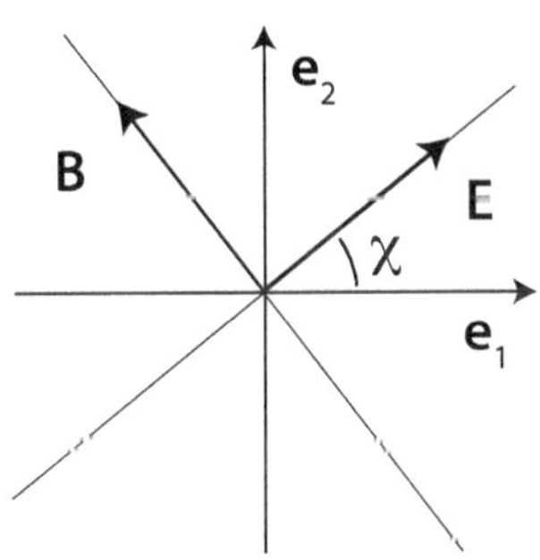

Fig. 10.2 Linear polarization. The electric field vector $\mathbf{E}$ oscillates in a fixed direction orthogonal to the wave vector $\mathbf{k}$, while the magnetic field vector $\mathbf{B}$ oscillates in the direction orthogonal to both $\mathbf{k}$ and $\mathbf{E}$. The vector $\mathbf{k}$ is in this figure directed out of the plane, towards the reader.

This corresponds to the case where the two orthogonal components of the real electric field oscillate *in phase*, with $\phi = 0$ (or $\phi = \pi$). The real-valued electric field then has the form

$$\mathbf{E}(\mathbf{r}, t) = E_0 \cos(\mathbf{k} \cdot \mathbf{r} - \omega t)\mathbf{e}', \tag{10.22}$$

with the rotated unit vector $\mathbf{e}'$ defined by

$$E_0 \mathbf{e}' = E_{10}\mathbf{e}_1 + E_{20}\mathbf{e}_2 . \tag{10.23}$$

The $\mathbf{E}$-field thus oscillates along a fixed axis orthogonal to $\mathbf{k}$, and the $\mathbf{B}$-field oscillates in the direction orthogonal to both $\mathbf{k}$ and $\mathbf{E}$. The axis of oscillation of $\mathbf{E}$, together with the axis defined by $\mathbf{k}$ span a two dimensional plane, often referred to as the *polarization plane* of the electromagnetic field.

Circular polarization

In this case the two orthogonal components of the $\mathbf{E}$-field are 90° out of phase, so that $\phi = \pi/2$ or $\phi = -\pi/2$, while the amplitudes of these components are equal, $E_{10} = E_{20} = E_0/\sqrt{2}$. The electric field then takes the form

$$\mathbf{E}(\mathbf{r},t) = \frac{E_0}{\sqrt{2}}[\cos(\mathbf{k}\cdot\mathbf{r}-\omega t)\mathbf{e}_1 \pm \sin(\mathbf{k}\cdot\mathbf{r}-\omega t)\mathbf{e}_2]\,, \tag{10.24}$$

where the sign $\pm$ determines whether the field vector rotates in the positive or negative direction when t is increasing. The two possibilities are referred to as right-handed or left-handed, circular polarization.

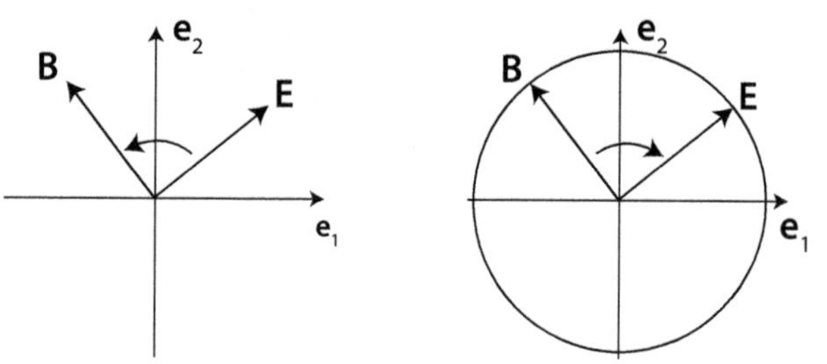

Fig. 10.3 Circular polarization. The electric field vector now rotates in the plane orthogonal to $\mathbf{k}$, and the magnetic field $\mathbf{B}$ also rotates, but 90° out of phase with $\mathbf{E}$. The direction of $\mathbf{k}$ is also here out of the plane, towards the reader. Two cases are shown, corresponding to right handed and left-handed circular polarization.

Elliptic polarization

Next we consider the case where the two orthogonal components are still 90° out of phase, but where the amplitudes of the two components now are different,

$$\begin{aligned}\mathbf{E}(\mathbf{r},t) &= E_{10}\cos(\mathbf{k}\cdot\mathbf{r}-\omega t)\mathbf{e}_1 + E_{20}\sin(\mathbf{k}\cdot\mathbf{r}-\omega t)\mathbf{e}_2\\ &\equiv E_1(\mathbf{r},t)\mathbf{e}_1 + E_2(\mathbf{r},t)\mathbf{e}_2\,.\end{aligned} \tag{10.25}$$

The expression shows that the two components of the field satisfy the ellipse equation

$$\frac{E_1^2}{E_{10}^2} + \frac{E_2^2}{E_{20}^2} = 1\,. \tag{10.26}$$

This means that when we consider the field for a fixed point $\mathbf{r}$ in space, the time dependent vector $\mathbf{E}$ will trace out an ellipse in the plane orthogonal to the direction of propagation of the wave, $\mathbf{n}$. The symmetry axes of the ellipse are in this case along the directions of the real unit vectors $\mathbf{e}_1$ and $\mathbf{e}_2$ and the half axes of the ellipse are given by E_{10} and E_{20}. This is a case of elliptic polarization.

The general case

The case of elliptic polarization discussed above seems not to be the most general one, since we have fixed the relative phase of the two orthogonal components of the polarization vector to be $\pi/2$. However, the most general case does in fact correspond to elliptic polarization, with the only modification that the ellipse is rotated relative to the axes defined by the two chosen real unit vectors $\mathbf{e}_1$ and $\mathbf{e}_2$. To demonstrate this we start with the general expression for the electric field, now written as

$$\mathbf{E}(\mathbf{r},t) = E_{10}\,\mathbf{e}_1 \cos(\mathbf{k}\cdot\mathbf{r} - \omega t + \phi_1) + E_{20}\,\mathbf{e}_2 \sin(\mathbf{k}\cdot\mathbf{r} - \omega t + \phi_2)\,. \tag{10.27}$$

The above expression is the same as (10.21), except for the redefinition $\phi_2 \to \phi_2 - \pi/2$. By expanding the trigonometric functions, we rewrite the expression as

$$\begin{aligned}\mathbf{E}(\mathbf{r},t) &= (E_{10}\cos\phi_1\,\mathbf{e}_1 + E_{20}\sin\phi_2\,\mathbf{e}_2)\cos(\mathbf{k}\cdot\mathbf{r} - \omega t)\\ &\quad - (E_{10}\sin\phi_1\,\mathbf{e}_1 - E_{20}\cos\phi_2\,\mathbf{e}_2)\sin(\mathbf{k}\cdot\mathbf{r} - \omega t)\,.\end{aligned} \tag{10.28}$$

The next step is to write this in a form similar to (10.25),

$$\mathbf{E}(\mathbf{r},t) = E'_{10}\cos(\mathbf{k}\cdot\mathbf{r} - \omega t)\mathbf{e}'_1 + E'_{20}\sin(\mathbf{k}\cdot\mathbf{r} - \omega t)\mathbf{e}'_2\,, \tag{10.29}$$

where the new amplitudes and unit vectors are defined by

$$\begin{aligned}E'_{10}\mathbf{e}'_1 &= E_{10}\cos\phi_1\,\mathbf{e}_1 + E_{20}\sin\phi_2\,\mathbf{e}_2\,,\\ E'_{20}\mathbf{e}'_2 &= -E_{10}\sin\phi_1\,\mathbf{e}_1 + E_{20}\cos\phi_2\,\mathbf{e}_2\,.\end{aligned} \tag{10.30}$$

However, orthogonality of the new unit vectors then demands

$$E_{10}^2\sin\phi_1\cos\phi_1 = E_{20}^2\sin\phi_2\cos\phi_2\,. \tag{10.31}$$

The important point to note is that we have not, in the above expression, exploited the freedom that lies in the definition of the two phases ϕ_1 and ϕ_2. As already noted there is a free variable in the definition of the phases, since only the relative phase has physical significance. The above equation can therefore be satisfied, simply by identifying it as a condition which fixes the free phase variable. The expressions for the new amplitudes we find by exploiting the normalization of the unit vectors,

$$\begin{aligned} {E'_{10}}^2 &= E_{10}^2 \cos^2\phi_1 + E_{20}^2 \sin^2\phi_2 \,, \\ {E'_{10}}^2 &= E_{10}^2 \sin^2\phi_1 + E_{20}^2 \cos^2\phi_2 \,. \end{aligned} \tag{10.32}$$

This shows that the electric field vector (10.27), when re-expressed, as in (10.29), in terms of the rotated unit vectors $\mathbf{e}'_1$ and $\mathbf{e}'_2$, has the same form (10.25) as previously discussed for elliptically polarized plane waves. Fig. 10.4 illustrates a case of elliptic polarization where the electric field vector and the magnetic field vector trace out two orthogonal ellipses under the time evolution.

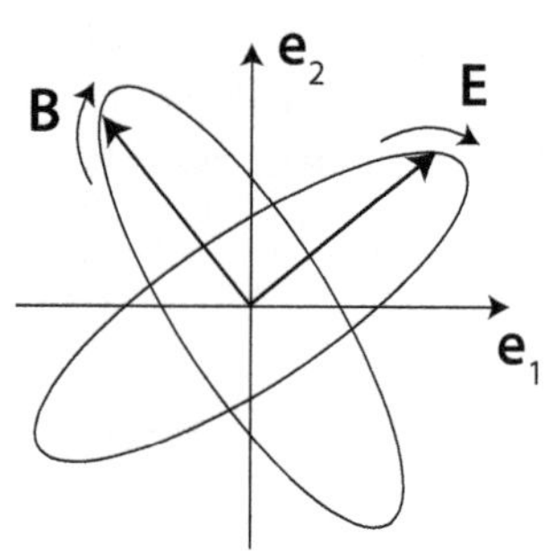

Fig. 10.4 Elliptic polarization. The time dependent electric field vector **E** describes an ellipse in the plane orthogonal to **k**, while the magnetic field vector **B** describes a similar ellipse, rotated by 90°. Also here the direction of **k** is out of the plane, towards the reader.

One should note that all the effects of polarization that we have discussed in this section can be seen as consequences of linear *superposition*. Thus, the monochromatic plane waves can, in all cases, be viewed as superpositions of two linearly polarized plane waves, with the same wave vector **k**, and with polarization along two arbitrarily chosen orthogonal directions. The different types of polarization are then produced by varying the relative amplitudes and relative phases of these two partial waves.

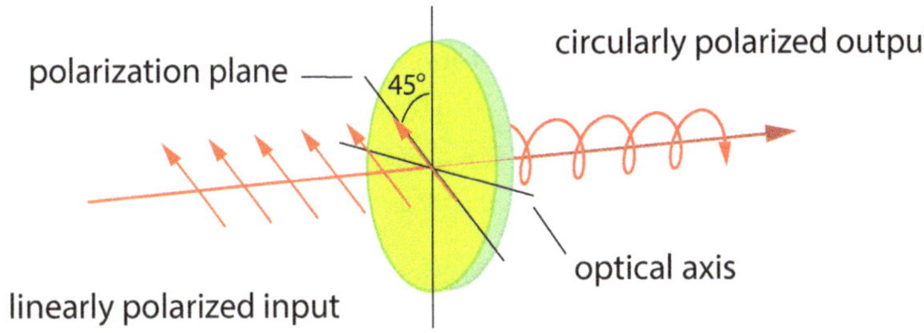

Fig. 10.5 Transforming linear to circular polarization in a birefringent crystal. The crystal is cut into a plate, with the orientation of the cut made so that the *optical axis* of the crystal is parallel to the surfaces of the plate. Plane waves with polarization in this direction propagate with a different speed than waves with polarization in the orthogonal direction. In the figure the incoming light is linearly polarized with the polarization plane tilted 45° relative to the optical axis. The width of the crystal is adjusted to give a quarter-wavelength relative phase shift of the two polarization components during the propagation in the crystal. This gives rise to a circularly polarized output.

Electromagnetic waves can propagate not only in vacuum, but also in transparent media. The speed of propagation in such a medium is generally slower than in vacuum, as a consequence of the fact that the effective values of the permittivity ϵ and permeability μ are different from their values in vacuum. A particularly interesting medium is found in birefringent crystals where the speed of propagation is not unique, but depends on the polarization of the electromagnetic wave. This is exploited in so-called *wave plates*, which have the property of changing the polarization of transient waves. There are for example wave plates that can change linear polarization of the incoming wave to circular polarization of the outgoing wave, as illustrated in Fig. 10.5.

10.3 Electromagnetic energy and momentum

Maxwell's equations describe how moving charges give rise to electromagnetic fields and the Lorentz force describes how the fields act back on the charges. Since the field acts with forces on charged particles, this implies that energy and momentum is transferred between the field and the particles, and consequently the electromagnetic field has to be a carrier of energy and momentum. The expressions for the energy and momentum densities of the field can be determined from Maxwell's equations and the Lorentz force, under the assumption of *conservation of energy* and *conservation of momentum.*

To demonstrate this we examine the change in energy and momentum of a point-like test particle, which interacts with the electromagnetic field. The charge and momentum density in this case can be expressed as

$$\begin{aligned} \rho(\mathbf{r}, t) &= q\, \delta(\mathbf{r} - \mathbf{r}(t))\,, \\ \mathbf{j}(\mathbf{r}, t) &= q\, \mathbf{v}(t)\, \delta(\mathbf{r} - \mathbf{r}(t))\,, \end{aligned} \tag{10.33}$$

with q as the charge of the particle, $\mathbf{r}(t)$ as the time dependent position vector, and $\mathbf{v}(t)$ as the velocity of the particle. The Lorentz force which acts on the particle is

$$\mathbf{F} = q(\mathbf{E} + \mathbf{v} \times \mathbf{B})\,. \tag{10.34}$$

This force will in general change the energy of the particle, and the time derivative of the energy is

$$\frac{d}{dt}\mathcal{E}_{part} = \mathbf{F} \cdot \mathbf{v} = q\mathbf{v} \cdot \mathbf{E}\,. \tag{10.35}$$

Energy conservation means that the total energy of both field and particle is left unchanged. With $\mathcal{E}_{field}$ as the field energy within a finite volume V, with boundary surface Σ, energy conservation takes the form

$$\frac{d}{dt}(\mathcal{E}_{field} + \mathcal{E}_{part}) = -\int_\Sigma \mathbf{S} \cdot d\mathbf{A}\,, \tag{10.36}$$

where $\mathbf{S}$ is the energy current density of the field and $d\mathbf{A}$ is the area element of the surface. The right hand side of the equation describes the energy loss in V caused by the energy current through the boundary surface. The time derivative of the field energy can then be written

$$\begin{aligned} \frac{d}{dt}\mathcal{E}_{field} &= -q\mathbf{v} \cdot \mathbf{E} - \int_\Sigma \mathbf{S} \cdot d\mathbf{A} \\ &= -\int_V \mathbf{j} \cdot \mathbf{E}\, dV - \int_\Sigma \mathbf{S} \cdot d\mathbf{A}\,, \end{aligned} \tag{10.37}$$

where the expression for the time derivative of the particle energy has been rewritten by use of expression (10.33) for the current density. In the last form the equation is in fact valid for arbitrary charge configurations within the volume V.

By use of Ampere's law the current density can be replaced by the electric and and magnetic fields in the following way,

$$\mathbf{j} = \frac{1}{\mu_0}\boldsymbol{\nabla} \times \mathbf{B} - \epsilon_0 \frac{\partial \mathbf{E}}{\partial t}\,, \tag{10.38}$$

and this gives for the volume integral in (10.37)

$$\int_V \mathbf{j} \cdot \mathbf{E}\, dV = \int_V \left[\frac{1}{\mu_0}\mathbf{E} \cdot (\boldsymbol{\nabla} \times \mathbf{B}) - \epsilon_0 \mathbf{E} \cdot \frac{\partial \mathbf{E}}{\partial t}\right] dV\,. \tag{10.39}$$

We further modify the integrand of the first term by using field identities and Faraday's law of induction,

$$\mathbf{E}\cdot(\boldsymbol{\nabla}\times\mathbf{B}) = \nabla\cdot(\mathbf{B}\times\mathbf{E}) + \mathbf{B}\cdot(\boldsymbol{\nabla}\times\mathbf{E})$$

$$= -\nabla\cdot(\mathbf{E}\times\mathbf{B}) - \mathbf{B}\cdot\frac{\partial\mathbf{B}}{\partial t}\,. \tag{10.40}$$

This gives

$$\begin{aligned}\int_V \mathbf{j}\cdot\mathbf{E}\,dV &= -\int_V \Big[\epsilon_0\mathbf{E}\cdot\frac{\partial\mathbf{E}}{\partial t} + \frac{1}{\mu_0}\mathbf{B}\cdot\frac{\partial\mathbf{B}}{\partial t} + \frac{1}{\mu_0}\boldsymbol{\nabla}\cdot(\mathbf{E}\times\mathbf{B})\Big]dV \\ &= -\frac{d}{dt}\int_V \frac{1}{2}\Big[\epsilon_0 E^2 + \frac{1}{\mu_0}B^2\Big]dV - \frac{1}{\mu_0}\int_\Sigma(\mathbf{E}\times\mathbf{B})\cdot d\mathbf{A}\,,\end{aligned} \tag{10.41}$$

where in the last step a part of the volume integral has been rewritten as a surface integral by use of Gauss' theorem.

Writing the field energy as a volume integral, $\mathcal{E}_{field} = \int_V u\,dV$, with u as the energy density of the field, and separating the volume and surface integrals, we get from Eq. (10.37)

$$\frac{d}{dt}\int_V\left(u - \frac{1}{2}\left(\epsilon_0\mathbf{E}^2 + \frac{1}{\mu_0}\mathbf{B}^2\right)\right)dV = -\int_\Sigma\left(\mathbf{S} - \frac{1}{\mu_0}(\mathbf{E}\times\mathbf{B})\right)\cdot d\mathbf{A}\,, \tag{10.42}$$

which should be satisfied for an arbitrarily chosen volume and for general field configurations. Assuming the volume integral and the surface integral should vanish independently, this determines the energy density,

$$u = \frac{1}{2}\Big[\epsilon_0 E^2 + \frac{1}{\mu_0}B^2\Big]\,, \tag{10.43}$$

and the energy current density,

$$\mathbf{S} = \frac{1}{\mu_0}\mathbf{E}\times\mathbf{B}\,. \tag{10.44}$$

These are the standard expressions for the energy density and the energy current density of the electromagnetic field, and the derivation shows that the field equations combined with energy conservation leads to these expressions. The vector $\mathbf{S}$ is called the *Poynting vector.*[1]

[1] John Henry Poynting (1852–1914) was a British physicist, working mainly on problems in electromagnetic theory.

The expression for the momentum density of the electromagnetic field can be derived in the same way. We start with the expression for the time derivative of the particle momentum,

$$\frac{d}{dt}\mathbf{P}_{part} = \mathbf{F} = q(\mathbf{E} + \mathbf{v} \times \mathbf{B}), \tag{10.45}$$

Since we in the present case will focus only the momentum density of the field, we are not explicit about the surface contributions and write the momentum conservation equation simply as

$$\frac{d}{dt}(\mathbf{P}_{field} + \mathbf{P}_{part}) = \text{surf. terms}. \tag{10.46}$$

We follow the same approach as for the field energy, first by applying Maxwell's equations to replace the charge and current densities with field variables,

$$\begin{aligned}\frac{d}{dt}\mathbf{P}_{field} &= -\int \Big[\rho\mathbf{E} + \mathbf{j} \times \mathbf{B}\Big]\, dV + \text{surf. terms} \\ &= -\int \Big[\epsilon_0 \mathbf{E}\,(\boldsymbol{\nabla} \cdot \mathbf{E}) + \frac{1}{\mu_0}\left(\boldsymbol{\nabla} \times \mathbf{B} - \frac{1}{c^2}\frac{\partial \mathbf{E}}{\partial t}\right) \times \mathbf{B}\Big]\, dV \\ &\quad + \text{surf. terms}. \end{aligned} \tag{10.47}$$

We further manipulate the expression, and separate the terms that contribute to the volume part and to the surface part of the integral.

For the magnetic field we rewrite the expression $(\boldsymbol{\nabla}\times\mathbf{B})\times\mathbf{B}$ by resolving the double cross product (see Appendix C.2),

$$\mathbf{B} \times (\boldsymbol{\nabla} \times \mathbf{B}) = \mathbf{B}(\boldsymbol{\nabla} \cdot \mathbf{B}) - \sum_i \mathbf{e}_i \boldsymbol{\nabla} \cdot \left(B_i \mathbf{B} - \frac{1}{2}\mathbf{e}_i \mathbf{B}^2\right), \tag{10.48}$$

with $\{\mathbf{e}_i, i = 1, 2, 3\}$ as a set of normalized unit vectors. We notice that the first term is identically zero, since the magnetic field is divergence free, and the second term will under integration only give a surface contribution. For the electric field we have a similar expression,

$$\mathbf{E} \times (\boldsymbol{\nabla} \times \mathbf{E}) = \mathbf{E}(\boldsymbol{\nabla} \cdot \mathbf{E}) - \sum_i \mathbf{e}_i \boldsymbol{\nabla} \cdot \left(E_i \mathbf{E} - \frac{1}{2}\mathbf{e}_i \mathbf{E}^2\right), \tag{10.49}$$

where the second term, also here, only gives surface terms, while the first term gives a non-vanishing volume contribution. We have, as follows from Faraday's law,

$$\mathbf{E} \times (\boldsymbol{\nabla} \times \mathbf{E}) = -\mathbf{E} \times \frac{\partial \mathbf{B}}{\partial t}, \tag{10.50}$$

and by introducing this in (10.47), we then obtain

$$\begin{aligned}\frac{d}{dt}\mathbf{P}_{field} &= \int \left[\epsilon_0\mathbf{E}\times\frac{\partial\mathbf{B}}{\partial t} + \epsilon_0\frac{\partial\mathbf{E}}{\partial t}\times\mathbf{B}\right] dV + \text{surf. terms} \\ &= \frac{d}{dt}\int \epsilon_0\mathbf{E}\times\mathbf{B}\,dV + \text{surf. terms}\,. \end{aligned} \qquad (10.51)$$

This result gives the following expression for the field momentum density,

$$\mathbf{g} = \epsilon_0\mathbf{E}\times\mathbf{B}\,, \qquad (10.52)$$

and we note that, up to a factor $1/c^2 = \epsilon_0\mu_0$, it is identical to the energy current density $\mathbf{S}$.

In the relativistic formulation the energy density and the momentum density are combined in the symmetric *energy-momentum tensor*,

$$T^{\mu\nu} = \frac{1}{\mu_0}\left(F^{\mu\rho}F^{\nu}{}_{\rho} - \frac{1}{4}g^{\mu\nu}F_{\rho\sigma}F^{\rho\sigma}\right)\,. \qquad (10.53)$$

The energy density corresponds here to the component T^{00}, the components of the energy current density and momentum density are proportional to $T^{0i} = T^{i0}$, $i = 1, 2, 3$, and the momentum current density is described by the components T^{kl} of the tensor.

Example

Energy and momentum density of a plane wave

We consider a plane wave with the electric and magnetic field vectors related by

$$\mathbf{B} = \frac{1}{c}\mathbf{n}\times\mathbf{E}\,, \quad \mathbf{E} = -c\mathbf{n}\times\mathbf{B}\,, \qquad (10.54)$$

where $\mathbf{n} = \mathbf{k}/k$ is a unit vector in the direction of propagation of the wave. This gives $\mathbf{B}^2 = \mathbf{E}^2/c^2$ and therefore the energy density of the field is

$$u = \frac{1}{2}\left(\epsilon_0\mathbf{E}^2 + \frac{1}{\mu_0}\mathbf{B}^2\right) = \epsilon_0\mathbf{E}^2\,, \qquad (10.55)$$

with equal contributions from the electric and magnetic fields.

Poynting's vector, which determines the energy current and momentum densities of the plane wave, is

$$\mathbf{S} = \frac{1}{\mu_0}\mathbf{E}\times\mathbf{B} = \epsilon_0 c\mathbf{E}^2\mathbf{n} = u\,c\mathbf{n}\,. \qquad (10.56)$$

It is directed along the direction of propagation of the wave, and the last expression in (10.56) is consistent with the interpretation that the field energy is transported in the direction of the propagating wave with the speed of light.

Field energy and potential energy

Let us consider two static point charges q_1 and q_2 at relative position $\mathbf{r} = \mathbf{r}_1 - \mathbf{r}_2$. The Coulomb energy of the system,

$$U(r) = \frac{q_1 q_2}{4\pi\epsilon_0 r}\,, \tag{10.57}$$

is usually considered as the potential energy of the two charges. However, in the preceding discussion we have found an expression for the local energy density of the electromagnetic field, which should also apply to this static situation. This raises the question of how the potential energy of the charges is related to the electromagnetic field energy. An important point to notice is that we should *not* consider the two energies as something we should add in order to obtain the total energy of the system of charges and fields. Instead the integrated field energy is identical to the *total* electromagnetic energy of the charges and fields, and the potential energy can be extracted as the part of this energy that depends on the position of the static charges. We demonstrate this by calculating the integrated field energy of the two charges.

The integrated field energy is

$$\mathcal{E} = \frac{1}{2}\epsilon_0 \int \mathbf{E}(\mathbf{r}')^2 d^3r'\,, \tag{10.58}$$

where the electrostatic field $\mathbf{E}$ is the superposition of the Coulomb field from the two charges,

$$\mathbf{E}(\mathbf{r}') = \mathbf{E}_1(\mathbf{r}') + \mathbf{E}_2(\mathbf{r}') = \frac{q_1}{4\pi\epsilon_0}\frac{\mathbf{r}'-\mathbf{r}_1}{|\mathbf{r}'-\mathbf{r}_1|^3} + \frac{q_2}{4\pi\epsilon_0}\frac{\mathbf{r}'-\mathbf{r}_2}{|\mathbf{r}'-\mathbf{r}_2|^3}\,. \tag{10.59}$$

The field energy then has a natural separation into three parts,

$$\mathcal{E} = \mathcal{E}_1 + \mathcal{E}_2 + \mathcal{E}_{12}\,, \tag{10.60}$$

where the first two parts are the contributions from the Coulomb energies of each of the two charges, disregarding the presence of the other,

$$\begin{aligned}
\mathcal{E}_1 &= \frac{1}{2}\epsilon_0 \int \mathbf{E}_1(\mathbf{r}')^2 d^3r' = \frac{{q_1}^2}{32\pi^2\epsilon_0}\int \frac{d^3r'}{r'^4} = \frac{{q_1}^2}{8\pi\epsilon_0}\int_0^\infty \frac{dr'}{r'^2}\,,\\
\mathcal{E}_2 &= \frac{1}{2}\epsilon_0 \int \mathbf{E}_2(\mathbf{r}')^2 d^3r' = \frac{{q_2}^2}{32\pi^2\epsilon_0}\int \frac{d^3r'}{r'^4} = \frac{{q_1}^2}{8\pi\epsilon_0}\int_0^\infty \frac{dr'}{r'^2}\,.
\end{aligned} \tag{10.61}$$

We note that these two terms are independent of the positions of the charges. They are referred to as the *self energies* of the charges, and these energies are in a sense always bound to the charges in the Coulomb field surrounding each of them. Except for the different charge factors the self energy of the two charges are the same, but we note that for *point charges* the integrated self energy diverges in the limit $r' \to 0$. This is a separate point to discuss, and we shall return to this question briefly.

The third contribution to the field energy comes from the superposition of the Coulomb fields of the two charges,

$$\mathcal{E}_{12}(r) = \epsilon_0 \int \mathbf{E}_1(\mathbf{r}') \cdot \mathbf{E}_2(\mathbf{r}') d^3r' \,. \tag{10.62}$$

As indicated in the equation it depends on the distance between the two charges. To calculate this term it is convenient to introduce the Coulomb potential of one of the charges $\mathbf{E}_1 = -\boldsymbol{\nabla}\Phi_1$. We restrict the volume integral to a finite volume V and extract, by partial integration, a surface term as the integral over the boundary surface S of the volume V,

$$\begin{aligned}
\mathcal{E}_{12} &= -\epsilon_0 \int_V \boldsymbol{\nabla}\Phi_1 \cdot \mathbf{E}_2 \, d^3r' \\
&= -\epsilon_0 \int_V \boldsymbol{\nabla} \cdot (\Phi_1 \mathbf{E}_2) \, d^3r' + \epsilon_0 \int_V \Phi_1 \boldsymbol{\nabla} \cdot \mathbf{E}_2 \, d^3r' \\
&= -\epsilon_0 \oint_S \Phi_1 \mathbf{E}_2 \cdot d\mathbf{S} + \int_V \Phi_1(\mathbf{r}') \, q_2 \, \delta(\mathbf{r}' - \mathbf{r}_2) \, d^3r' \,.
\end{aligned} \tag{10.63}$$

In the last term Gauss' law for the electromagnetic field has been applied to rewrite the divergence of the electric field as a charge density. Since we consider point charges this density is proportional to a Dirac delta function.

Let us now consider the limit the volume extends to infinity. We note that in this limit the surface integral tends to zero, since far from the charges the product $\Phi_1 E_2$ falls off with distance as $1/r'^3$. We are then left with the volume integral, which is easy to evaluate due to the presence of the delta function,

$$\mathcal{E}_{12}(r) = q_2 \Phi_1(\mathbf{r}_2) = q_2 \frac{q_1}{4\pi\epsilon_0 |\mathbf{r}_1 - \mathbf{r}_2|} = \frac{q_1 q_2}{4\pi\epsilon_0 r} = U(r) \,. \tag{10.64}$$

This shows that the Coulomb energy $U(r)$ can be identified as the part of the total field energy that depends on the distance between the charges, and is due to the overlap of the electric fields of the two charges.

We return now to the question of how to understand the expression for the self energy terms. For an isolated point charge q located at the origin the energy of the Coulomb field is

$$\mathcal{E} = \frac{1}{2}\epsilon_0 \int \mathbf{E}^2 d^3r = \frac{q^2}{8\pi\epsilon_0} \int_0^\infty \frac{dr}{r^2} \,, \tag{10.65}$$

and this energy is obviously infinite due to the divergence of the integral as $r \to 0$. A reasonable assumption is that there is nothing wrong with the expression for the field energy, but that the idealization of treating the charge as being located at a mathematical point is the origin of the problem. This is a problem which exists not only in the classical theory; also in the quantum description of particles and fields there are infinities associated with the electromagnetic self energies that have to be taken care of by the theory.

One way to handle the problem with the divergent self energy integral is to assume that the charge has a finite size a, with this as an effective cutoff of the integral,

$$\mathcal{E}_a = \frac{q^2}{8\pi\epsilon_0}\int_a^\infty \frac{dr}{r^2} = \frac{q^2}{8\pi\epsilon_0 a}\,. \tag{10.66}$$

This gives a finite result, and at least formally, solves the problem with the infinite energy. However, to make a consistent picture of physical charged particles, like electrons, as small charged bodies is not so simple.

A way to get around the problem is instead to focus on the fact that the self energy is bound to each individual charge and therefore is not important for the interactions between the charges. One may therefore avoid the problem of a precise theory of point-like charges by simply assuming that the energy carried by the field is finite, and that the only physical effect of this energy is to change the mass of the charged particle according to Einstein's relation

$$\Delta mc^2 = \mathcal{E}\,. \tag{10.67}$$

The physical mass of the charged particle can then be written as a sum

$$m = m_b + \Delta m\,, \tag{10.68}$$

where m_b is the so called bare mass, which is the (imagined) mass of the particle without the Coulomb field. The method of absorbing the self energy of the electromagnetic energy in the physical mass is referred to as *mass renormalization*, and the effect it has is to remove the otherwise divergent self energy contributions in the description of interacting charged particles.

Finally, let us use this interpretation of the self energy to give an estimate of the value of the corresponding length parameter a for an electron, given in (10.66). We know that $\Delta m \leq m_e$, with m_e as the physical electron mass and with equality meaning that *all* the electron mass is due to the electromagnetic energy of its Coulomb field. In this limit we get

$$\frac{e^2}{8\pi\epsilon_0 a} = m_e c^2 \quad \Rightarrow \quad a = \frac{e^2}{8\pi\epsilon_0 m_e c^2}\,. \tag{10.69}$$

The numerical factor of the length parameter a in (10.69) is based on a simple, classical model of the electron as a charged shell, with the electron mass m_e being completely of electromagnetic origin. Other charge distributions and models of the electron mass would lead to somewhat different values. Thus, the precise value is not important, but the idea that there must be an effective cutoff in the electromagnetic field energy contribution to the electron mass, at a length scale of the order of (10.69), remains. Conventionally the *classical electron radius* is defined by this expression enhanced by the factor 2,

$$r_e = \frac{e^2}{4\pi\epsilon_0 m_e c^2} \,. \tag{10.70}$$

Its numerical value is

$$r_e = 2.818 \times 10^{-15}\,\mathrm{m}\,, \tag{10.71}$$

which shows that it is indeed a very small radius, comparable to the radius of an atomic nucleus.

Example

Field contribution to the momentum of a charged particle in a constant magnetic field

For a charged particle moving in a constant, homogeneous magnetic field the system is symmetric both under rotations around the magnetic field lines, and under translations in all three space directions. One should from this expect that the momentum $m\dot{\mathbf{r}}$ of the particle and the component along the $\mathbf{B}$ field of the angular momentum $m\mathbf{r} \times \dot{\mathbf{r}}$ would be conserved quantities. However, that is not the case, but if we add some specific "correction" terms, we find indeed constants of motion that correspond to these symmetries. These additional terms can be interpreted as field contributions to the usual (mechanical) momentum and angular momentum of the particle, and we will here show explicitly that this interpretation is correct. For simplicity we assume non-relativistic motion.

We first recall how the conserved momentum is derived from the equation of motion

$$m\ddot{\mathbf{r}} = e\mathbf{v} \times \mathbf{B}\,, \tag{10.72}$$

with e as the charge of the particle. The equation can clearly be rewritten as

$$\frac{d}{dt}(m\dot{\mathbf{r}} - e\mathbf{r} \times \mathbf{B}) = 0\,, \tag{10.73}$$

which means that $m\dot{\mathbf{r}} - e\mathbf{r}\times\mathbf{B}$ is a constant of motion. What we will show is that the $\mathbf{r}\times\mathbf{B}$ term can be re-written as a field contribution, given by the integral of the field momentum density $\mathbf{g} = \epsilon_0\mathbf{E}\times\mathbf{B}$.

It is convenient to make, as we have done before, the following gauge choice $\mathbf{A} = -\frac{1}{2}\mathbf{r}\times\mathbf{B}$ for the vector potential of the constant magnetic field. It satisfies, as it should, the condition $\mathbf{B} = \boldsymbol{\nabla}\times\mathbf{A}$, but it has a further useful symmetry, $\partial_i A_j = -\partial_j A_i$, which one can easily check. This is not satisfied for general vector potentials of the constant magnetic field. With $\mathbf{E}$ as the electric field from the charged particle and $\mathbf{B}$ as the constant magnetic field we now derive the expression for the field momentum

$$\begin{aligned}
\int g_i dV &= \epsilon_0 \int [\mathbf{E}\times\mathbf{B}]_i\, dV \\
&= \epsilon_0 \int [\mathbf{E}\times(\boldsymbol{\nabla}\times\mathbf{A})]_i\, dV \\
&= \epsilon_0 \epsilon_{ijk}\epsilon_{klm} \int E_j \partial_l A_m\, dV \\
&= \epsilon_0(\delta_{il}\delta_{jm} - \delta_{im}\delta_{jl}) \int E_j \partial_l A_m\, dV \\
&= \epsilon_0 \int (E_j\partial_i A_j - E_j\partial_j A_i)\, dV \\
&= -2\epsilon_0 \int E_j \partial_j A_i\, dV \\
&= 2\epsilon_0 \int \boldsymbol{\nabla}\cdot\mathbf{E}\, A_i\, dV - 2\epsilon_0 \oint A_i\, \mathbf{E}\cdot d\mathbf{S}\,,
\end{aligned} \tag{10.74}$$

where the last integral is a surface integral, which we may assume to be taken on a sphere far from the charged particle. The antisymmetry of $\mathbf{A}$ under inversion $\mathbf{r}\to-\mathbf{r}$ then implies that there is a cancellation of contributions from opposite directions on the integration surface.

With the surface contribution omitted and applying Gauss' law, we find for the integrated field momentum

$$\begin{aligned}
\int \mathbf{g} dV &= 2\epsilon_0 \int \mathbf{A}\, \boldsymbol{\nabla}\cdot\mathbf{E}\, dV \\
&= -\epsilon_0 \int (\mathbf{r}'\times\mathbf{B})\, \frac{e}{\epsilon_0}\delta(\mathbf{r}-\mathbf{r}')\;\; dV' \\
&= -e\mathbf{r}\times\mathbf{B}\,,
\end{aligned} \tag{10.75}$$

where we in the expression above have introduced $\mathbf{r}'$ as the integration variable to distinguish it from the position vector $\mathbf{r}$ of the charged particle.

The result of our investigation confirms the claim that the additional term in the conserved momentum of the particle can be interpreted as momentum contributions from the overlapping **E** and **B** fields.

10.4 Exercises

Problem 10.1

The monochromatic plane wave solutions discussed in this chapter have the electric field **E** perpendicular to the magnetic field **B**. Consider the following case where the fields instead are parallel

$$\begin{aligned}\mathbf{E}(\mathbf{r},t) &= E_0 \sin\omega t(\sin kx\,\mathbf{e}_y + \cos kx\,\mathbf{e}_z)\,,\\ \mathbf{B}(\mathbf{r},t) &= \frac{1}{c}E_0 \cos\omega t(\sin kx\,\mathbf{e}_y + \cos kx\,\mathbf{e}_z).\end{aligned} \tag{10.76}$$

a) Show that this is a standing wave solution of Maxwell's equations.

b) Determine the momentum and energy densities of the electromagnetic wave.

c) Show that this wave solution can be written as a superposition of two propagating plane waves, and check that the field components **E** and **B** are orthogonal for each of these. Specify the polarization of the two plane waves.

d) The electromagnetic wave can be viewed as a superposition of plane polarized waves, with polarization (direction of the **E** field) along the orthogonal directions $\mathbf{e}_y$ and $\mathbf{e}_z$. Find these components of the wave and check that they all satisfy the orthogonality relation $\mathbf{E}\cdot\mathbf{B} = 0$.

Problem 10.2

We consider a situation where a monochromatic plane wave of light is sent through a birefringent crystal in the z-direction. The wave can be decomposed in linearly polarized components, where polarization in the x-direction corresponds to a wave with a faster propagation velocity, c_f, than the velocity c_s, of a wave which is polarized in the y-direction. We assume that the plane wave enters the crystal at $z = 0$. The electric component of the wave we assume inside the crystal to have the form

$$\mathbf{E}(z,t) = \frac{E_0}{\sqrt{2}}[\cos(k_f z - \omega t)\mathbf{e}_x + \cos(k_s z - \omega t)\mathbf{e}_y]\,, \tag{10.77}$$

with $k_f = \omega/c_f$, $k_s = \omega/c_s$, and E_0 as the amplitude of the oscillating field.

a) When entering the crystal, at $z = 0$, the plane wave is linearly polarized. Show this and determine its direction of polarization.

b) We introduce the following variables, $k_0 = (k_f + k_s)/2$ and $\Delta k = k_s - k_f$. Show that inside the crystal, with $z > 0$, the electric field (10.77) can be written in the form

$$\begin{aligned}\mathbf{E}(z,t) &= E_{10}(z)\cos(k_0 z - \omega t)\mathbf{e}_1 + E_{20}(z)\sin(k_0 z - \omega t)\mathbf{e}_2 \\ &\equiv E_1(z,t)\mathbf{e}_1 + E_2(z,t)\mathbf{e}_2\,, \end{aligned} \tag{10.78}$$

and determine the amplitudes E_{10} and E_{20}, and the orthogonal unit vectors $\mathbf{e}_1$ and $\mathbf{e}_2$.

c) Explain in what sense the expression shows that the wave, for general z, is elliptically polarized, and describe qualitatively the change in polarization as the wave propagates through the crystal. Determine for what values of z the wave is linearly and circularly polarized.

Problem 10.3

A point charge q is placed in a constant magnetic field $\mathbf{B}$. Assume Cartesian coordinates (x, y, z) are chosen with the magnetic field in the z-direction and the charge position at the origin $x = y = z = 0$. Show, by use of the expression for Poynting's vector, that the superposition of the electric field from the charge and the magnetic field gives rise to field energy and momentum currents that circulate around the z-axis.

Chapter 11

Maxwell's equations with stationary sources

We return to the original form of Maxwell's equations in the Lorentz gauge,

$$\partial_\nu \partial^\nu A^\mu = -\mu_0 j^\mu \,, \tag{11.1}$$

and assume the 4-current j^μ, and therefore both the charge and current densities, to be independent of time,

$$\rho = \rho(\mathbf{r}) \,, \quad \mathbf{j} = \mathbf{j}(\mathbf{r}) \,. \tag{11.2}$$

Note that this is the case only in a preferred inertial frame (the lab frame). With time independent sources we may also assume the electromagnetic potential A^μ, as a solution of (11.1), to be time independent. This means that the Lorentz gauge condition again reduces to the Coulomb gauge condition, $\nabla \cdot \mathbf{A} = 0$, and Maxwell's equation has a natural decomposition in two independent equations,

$$\boldsymbol{\nabla}^2 \Phi = -\frac{\rho}{\epsilon_0} \,, \tag{11.3}$$

$$\boldsymbol{\nabla}^2 \mathbf{A} = -\mu_0 \mathbf{j} \,, \tag{11.4}$$

where the scalar potential $\Phi = A^0/c$ determines the electric field and vector potential $\mathbf{A}$ determines the magnetic field.

Since there is no coupling between the two equations (11.3) and (11.4), they can be studied separately. Equation (11.3) is the basic equation in *electrostatics*, where static charges give rise to a time independent electric field, while Eq. (11.4) is the basic equation in *magnetostatics* where stationary currents give rise to a time independent magnetic field. As differential equations they are of the same type, known as the *Poisson equation*, and even if there are some differences, the methods of finding the electrostatic and magnetostatic fields with given sources, are much the same. We examine the two cases separately.

11.1 The electrostatic equation

Since the electrostatic equation (11.3) is a linear differential equation, the solution can be seen as a linear superposition of contributions from point-like parts of the charge distribution. For a single point charge q located at the origin, the solution to Gauss' law is given by the Coulomb field,

$$\mathbf{E}(\mathbf{r}) = \frac{q}{4\pi\epsilon_0 r^2}\frac{\mathbf{r}}{r}\,, \tag{11.5}$$

with the corresponding Coulomb potential as

$$\Phi(\mathbf{r}) = \frac{q}{4\pi\epsilon_0 r}\,. \tag{11.6}$$

For a charge distribution $\rho(\mathbf{r})$, which is no longer point-like, the potential can be written directly as a sum (or integral) over the Coulomb potential from all parts of the distribution,

$$\Phi(\mathbf{r}) = \frac{1}{4\pi\epsilon_0}\int \frac{\rho(\mathbf{r}')}{|\mathbf{r}-\mathbf{r}'|}d^3r'\,. \tag{11.7}$$

The corresponding electric field strength is

$$\mathbf{E}(\mathbf{r}) = -\boldsymbol{\nabla}\Phi = \frac{1}{4\pi\epsilon_0}\int \frac{\rho(\mathbf{r}')}{|\mathbf{r}-\mathbf{r}'|^3}(\mathbf{r}-\mathbf{r}')d^3r'\,. \tag{11.8}$$

In reality the above solution is a *particular* solution of the differential equation. A general solution will therefore be of the form

$$\Phi(\mathbf{r}) = \Phi_c(\mathbf{r}) + \Phi_0(\mathbf{r})\,, \tag{11.9}$$

where Φ_c denotes the solution given above and Φ_0 is a general solution of the source free *Laplace equation*

$$\boldsymbol{\nabla}^2\Phi_0 = 0\,. \tag{11.10}$$

The solution (11.7) written above *implicitly* assumes the boundary condition that is natural in the open, infinite space, namely that the potential falls to zero at infinity.

When we consider the electric field in a finite region V of space, with given boundary conditions on the boundary surface S, the contribution from Φ_0 will generally be important. This contribution to the potential will correct the contribution from the integrated Coulomb potential so that the total potential satisfies the boundary conditions. As a particular situation we may consider the electric field produced by a charge within a cavity of an electric conductor. Since the boundary surface of the conductor is an *equipotential* surface, the function Φ_0 is determined as a solution of

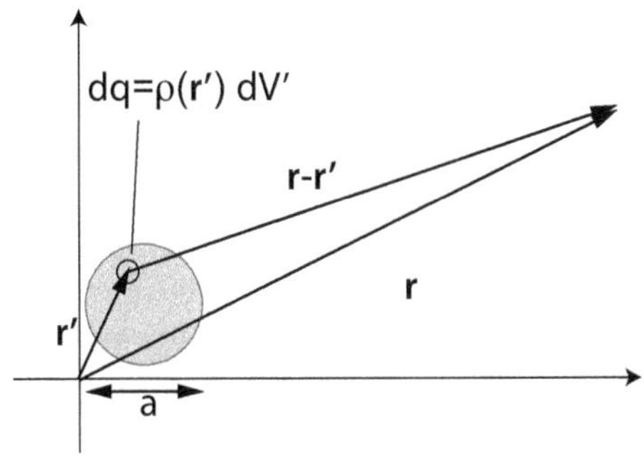

Fig. 11.1 The electrostatic potential from a charge distribution. The potential at a point $\mathbf{r}$ is determined as a linear superposition of contributions from small pieces dq of the charge, located at points $\mathbf{r}'$ in the charge distribution. The multipole expansion, expressed as an expansion in powers of r'/r, converges rapidly when the distance r to the point where the potential is evaluated, is much larger than the typical length scale a of the distribution.

the Laplace equation (11.10) with the following boundary condition on the surface of the conductor,

$$\Phi_0(\mathbf{r}) = -\Phi_c(\mathbf{r}) = -\frac{1}{4\pi\epsilon_0}\int_V \frac{\rho(\mathbf{r}')}{|\mathbf{r}-\mathbf{r}'|}d^3r' \quad \mathbf{r}\in S\,. \tag{11.11}$$

More generally we often make a distinction between two types of boundary conditions where either the potential Φ is specified on the boundary (Dirichlet condition) or the electric field $\mathbf{E} = -\boldsymbol{\nabla}\Phi$ is specified (Neumann condition). To determine the potential that satisfies the correct boundary conditions is generally a non-trivial problem, and several methods have been developed to solve the problem for different types of boundary condition. This we shall not discuss further, but instead assume the simple form (11.7) of the potential in the open, infinite space to be valid.

The integral expression for $\Phi(\mathbf{r})$ in a sense gives the solution to the electrostatic problem. But for any specified charge distribution the problem remains how to solve the integral in order to determine the electrostatic potential. However, when far from the charges this problem can be simplified by use of *multipole expansion*, and we will next study how such an expansion can be applied to give useful approximations to the electrostatic potential.

Multipole expansion

This expansion is based on the assumption that the point $\mathbf{r}$ where we are interested in determining the potential and the electric field is at some distance from the charge distribution. To be more precise, let us assume that the charge distribution has a finite extension of linear dimension a, as illustrated in Fig. 11.1. Our assumption is that the point where the potential should be determined lies at a distance from the charges which is much larger than a. With the origin chosen to lie close to the charges we may write this assumption as

$$r >> r' \approx a\,, \tag{11.12}$$

where $\mathbf{r}'$ is the variable in the integration over the charge distribution. In the integral formula for the potential we may introduce the small vector $\boldsymbol{\xi} \equiv \mathbf{r}'/r$, and make a Taylor expansion in powers of the vector.

The inverse distance between the points $\mathbf{r}'$ and $\mathbf{r}$, when expressed in terms of $\boldsymbol{\xi}$ is

$$\begin{aligned}\frac{1}{|\mathbf{r}-\mathbf{r}'|} &= (r^2 + r'^2 - 2\mathbf{r}\cdot\mathbf{r}')^{-\frac{1}{2}} \\ &= \frac{1}{r}\left[1 + \left(\frac{r'}{r}\right)^2 - 2\frac{\mathbf{r}}{r}\cdot\frac{\mathbf{r}'}{r}\right]^{-\frac{1}{2}} \\ &= \frac{1}{r}\left[1 - 2\frac{\mathbf{r}}{r}\cdot\boldsymbol{\xi} + \xi^2\right]^{-\frac{1}{2}} \\ &\equiv \frac{1}{r} f(\boldsymbol{\xi})\,. \end{aligned} \tag{11.13}$$

We make a Taylor expansion of the function $f(\boldsymbol{\xi})$ introduced at the last step,

$$\begin{aligned} f(\boldsymbol{\xi}) &= f(0) + \sum_i \xi_i \frac{\partial f}{\partial \xi_i}(0) + \frac{1}{2}\sum_{ij} \xi_i\xi_j \frac{\partial^2 f}{\partial\xi_i\partial\xi_j}(0) + \ldots \\ &= 1 + \frac{\mathbf{r}}{r}\cdot\boldsymbol{\xi} + \sum_{ij}\frac{1}{2}\left(3\frac{x_i x_j}{r^2} - \delta_{ij}\right)\xi_i\xi_j + \ldots\,, \end{aligned} \tag{11.14}$$

and reintroduce the integration variable $\mathbf{r}'$, in the corresponding expansion of the inverse distance

$$\frac{1}{|\mathbf{r}-\mathbf{r}'|} = \frac{1}{r}\left(1 + \frac{\mathbf{r}\cdot\mathbf{r}'}{r^2} + \frac{1}{2}\left(3\frac{(\mathbf{r}\cdot\mathbf{r}')^2}{r^4} - \frac{r'^2}{r^2}\right) + \ldots\right). \tag{11.15}$$

For the electrostatic potential this gives the following expansion

$$\Phi(\mathbf{r}) = \frac{1}{4\pi\epsilon_0 r}\int \rho(\mathbf{r}')\left[1 + \frac{\mathbf{r}\cdot\mathbf{r}'}{r^2} + \frac{1}{2}\left(3\frac{(\mathbf{r}\cdot\mathbf{r}')^2}{r^4} - \frac{r'^2}{r^2}\right) + \ldots\right] d^3r'$$

$$\equiv \Phi_0(\mathbf{r}) + \Phi_1(\mathbf{r}) + \Phi_2(\mathbf{r}) + \ldots\,, \tag{11.16}$$

with Φ_n as the n'th term of the expansion of the potential in powers of ξ.

We consider the first terms in the expansion, beginning with the *monopole* term,

$$\Phi_0(\mathbf{r}) = \frac{1}{4\pi\epsilon_0 r}\int \rho(\mathbf{r}')\, d^3r' = \frac{q}{4\pi\epsilon_0 r}\,, \tag{11.17}$$

with $q = \int \rho(\mathbf{r}')\, d^3r'$ as the total charge of the charge distribution. This shows that the lowest order term of the expansion gives a potential which is the same as if the total charge were collected at the origin of the coordinate system. This first term will give a good approximation to the true potential if the point $\mathbf{r}$ is sufficiently far away and the origin is chosen sufficiently close to the (center of the) charge distribution.

The second term of the expansion is the electric *dipole* term,

$$\Phi_1(\mathbf{r}) = \frac{1}{4\pi\epsilon_0 r^3}\int \rho(\mathbf{r}')(\mathbf{r}\cdot\mathbf{r}')\, d^3r' = \frac{\mathbf{r}\cdot\mathbf{p}}{4\pi\epsilon_0 r^3}\,, \tag{11.18}$$

where we have introduced the electric dipole moment,

$$\mathbf{p} = \int \rho(\mathbf{r})\,\mathbf{r}\, d^3r\,. \tag{11.19}$$

This term gives a correction to the monopole term, and we note that for large r it falls off with distance like $1/r^2$, while the monopole term falls off like $1/r$. As a consequence, the monopole term will always dominate the dipole term for sufficiently large r (unless $q = 0$).

We include one more term of the expansion in our discussion. This is the electric quadrupole term,

$$\Phi_2(\mathbf{r}) = \frac{1}{8\pi\epsilon_0 r^3}\int \rho(\mathbf{r}')(3(\mathbf{n}\cdot\mathbf{r}')^2 - r'^2)\, d^3r' = \frac{Q_{\mathbf{n}}}{8\pi\epsilon_0 r^3}\,, \tag{11.20}$$

with $\mathbf{n} = \mathbf{r}/r$ as the unit vector in direct of the point $\mathbf{r}$ and $Q_{\mathbf{n}}$ as the *quadrupole moment* about the axis $\mathbf{n}$. It can be written as $Q_{\mathbf{n}} = \sum_{ij} Q_{ij} n_i n_j$, with

$$Q_{ij} = \int \rho(\mathbf{r})(3x_i x_j - r^2\delta_{ij})\, d^3r\,, \tag{11.21}$$

as the *quadrupole moment tensor*.

The electric field can now be expanded in the same way,

$$\mathbf{E} = \mathbf{E}_0 + \mathbf{E}_1 + \mathbf{E}_2 + \ldots\,, \tag{11.22}$$

with $\mathbf{E}_n = -\boldsymbol{\nabla}\Phi_n$ for the $n'th$ term in the expansion. We give the explicit expressions for the first two terms. The monopole term is

$$\mathbf{E}_0 = -\boldsymbol{\nabla}\Phi_0 = \frac{q}{4\pi\epsilon_0 r^3}\mathbf{r}\,, \tag{11.23}$$

which is the Coulomb field of a point charge q located in the origin. The next term is

$$\mathbf{E}_1 = -\boldsymbol{\nabla}\Phi_1 = -\boldsymbol{\nabla}\left(\frac{\mathbf{r}\cdot\mathbf{p}}{4\pi\epsilon_0 r^3}\right) = \frac{1}{4\pi\epsilon_0 r^3}(3\mathbf{n}(\mathbf{n}\cdot\mathbf{p}) - \mathbf{p})\,, \tag{11.24}$$

with $\mathbf{n}=\mathbf{r}/\mathrm{r}$ as before. This field is called the *electric dipole field.*

It should be clear from the above construction that the higher the multipole index n is, the faster the corresponding potential and electric field fall off with distance. Thus, for large r the n'th multipole term of the potential falls off like $r^{-(n+1)}$, while the corresponding term in the expansion of the electric field field falls off like $r^{-(n+2)}$. When considering the electric field far from the charges, it is often sufficient to consider only the first terms of the multipole expansion. In particular that is the case when we are interested in electromagnetic radiation far from the source, as we shall soon consider. In that case the field is determined by the time derivatives of the multipole moments. Since the total charge is conserved there will be no contribution from the monopole term, but for large r the main contribution will be from the electric dipole term, unless this term is absent.

Elementary multipoles

Elementary multipole fields can be produced by point charges in the following way. An elementary monopole field is simply the Coulomb field of a point charge located at the origin. This Coulomb field has no higher multipole components. A dipole field is produced by two point charges of opposite sign, $\pm q$, located symmetrically about the origin, at positions $\pm\mathbf{d}/2$. The dipole moment of the charge configuration is $\mathbf{p} = q\mathbf{d}$. This field has no monopole component, and in the limit where $d \to 0$ with qd fixed, all higher multipole components vanish and the electric field is a pure dipole field. Such an electric dipole field, with d finite, is illustrated in Fig. 11.2.

In a similar way a pure quadrupole field can be produced by two dipoles of opposite signs, which have positions with a relative shift $\boldsymbol{l}$. For this charge configuration only the quadrupole component of the electric field survives in the limit $l \to 0$ with $p\,l$ fixed. Such an elementary quadrupole field is shown in Fig. 11.3.

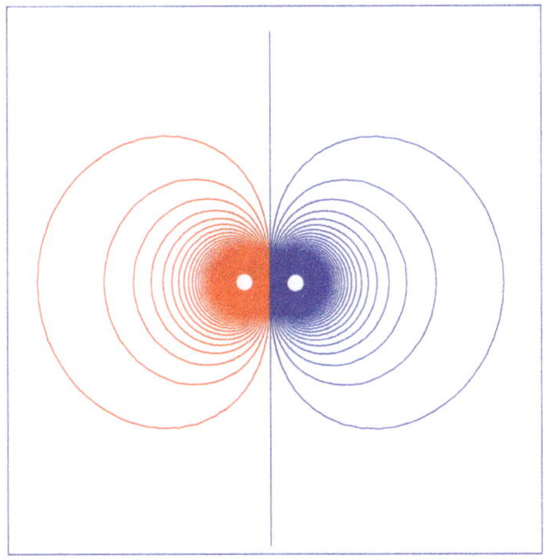

Fig. 11.2 Electric dipole potential. Two electric point charges of opposite sign, but equal magnitude, are place at shifted positions. Equipotential lines are shown for a plane which includes the two charges. In the figure red corresponds to positive potential values and blue to negative values. The potential diverges towards the point charges.

11.2 Magnetic fields from stationary currents

For magnetic fields produced by stationary currents the basic equation is

$$\boldsymbol{\nabla}^2 \mathbf{A} = -\mu_0 \mathbf{j}\,, \tag{11.25}$$

with $\mathbf{j} = \mathbf{j}(\mathbf{r})$ as a time independent current density. We note that the equation has, for each component of $\mathbf{A}(\mathbf{r})$, the same form as the equation for a static electric potential, and the Coulomb field solution can immediately be translated to the following solution of the magnetic equation

$$\mathbf{A}(\mathbf{r}) = \frac{\mu_0}{4\pi} \int \frac{\mathbf{j}(\mathbf{r}')}{|\mathbf{r} - \mathbf{r}'|} d^3 r'\,. \tag{11.26}$$

The corresponding magnetic field is

$$\mathbf{B}(\mathbf{r}) = \boldsymbol{\nabla} \times \mathbf{A}(\mathbf{r}) = \frac{\mu_0}{4\pi} \int \left(\boldsymbol{\nabla} \frac{1}{|\mathbf{r} - \mathbf{r}'|} \right) \times \mathbf{j}(\mathbf{r}')\, d^3 r'\,. \tag{11.27}$$

The gradient in the integrand can easily be calculated by changing temporarily the position of the origin so that $\mathbf{r}' = 0$. We have

$$\boldsymbol{\nabla} \left(\frac{1}{r} \right) = \frac{d}{dr} \left(\frac{1}{r} \right) \boldsymbol{\nabla} r = -\frac{\mathbf{r}}{r^3}\,. \tag{11.28}$$

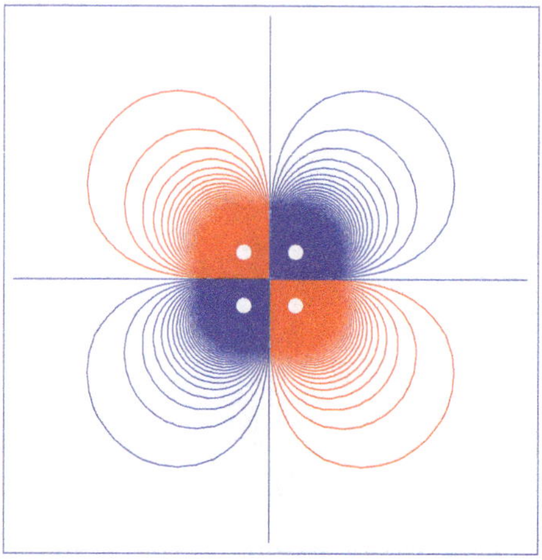

Fig. 11.3 Electric quadrupole potential. In this case four point charges located in a plane, with pairwise opposite signs, produce the potential. The charges form two shifted dipoles of opposite orientation. Equipotential lines are shown for the plane which includes the four charges. In the figure red corresponds to positive potential values and blue to negative values. The potential diverges towards each of the four point charges.

Shifting the origin back to the correct position gives

$$\boldsymbol{\nabla}\frac{1}{|\mathbf{r}-\mathbf{r}'|} = -\frac{\mathbf{r}-\mathbf{r}'}{|\mathbf{r}-\mathbf{r}'|^3}\,, \tag{11.29}$$

and therefore the following expression for the magnetic field

$$\mathbf{B}(\mathbf{r}) = \frac{\mu_0}{4\pi}\int \frac{\mathbf{j}(\mathbf{r}')\times(\mathbf{r}-\mathbf{r}')}{|\mathbf{r}-\mathbf{r}'|^3}\,d^3r'\,. \tag{11.30}$$

The above expression gives the magnetic field from a general stationary current distribution. However, another form is often more useful, and that corresponds to the situation where the magnetic field is produced by a current in a thin conducting cable. When the cross section can be regarded as vanishingly small the volume integral of the current density can be replaced by the line integral of the current along the curve defined by the thin cable. To find this expression we use the following replacement in the integral, as illustrated in Fig. 11.4,

$$\mathbf{j}\,d^3r' \to \mathbf{j}\,\Delta A\,dr' = j\,\Delta A\,d\mathbf{r}' = I d\mathbf{r}'\,. \tag{11.31}$$

The current density is here assumed to be constant over the cross section of the cable, with ΔA is the cross section area, and I as the current running

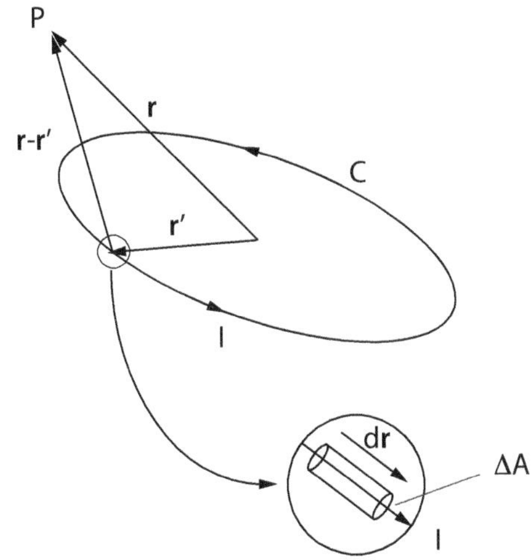

Fig. 11.4 Magnetic field from a current loop. The magnetic field at a given point P is a superposition of contributions from each part of the current loop. The field expressed as a line integral around the loop is derived from the general volume integral of the current density by first integrating over the cross-section of the conductor, with area ΔA.

in the cable. This gives the following line integral representation of the magnetic field

$$\mathbf{B}(\mathbf{r}) = -\frac{\mu_0 I}{4\pi} \int_C \frac{(\mathbf{r} - \mathbf{r}')}{|\mathbf{r} - \mathbf{r}'|^3} \times d\mathbf{r}' , \tag{11.32}$$

with C as the curve that the current follows. This expression for the magnetic field produced by a stationary current is known as *Biot-Savart's law.*[1]

Multipole expansion for the magnetic field

For positions $\mathbf{r}$ far from the current a multipole expansion can be given for the magnetic field, which is similar to that for the electric field. To show this we expand the integrand of (11.26) in powers of r'/r, and for the vector

[1] Jean-Baptiste Biot (1791–1841) was a French physicist, astronomer and mathematician. In physics he made many contributions, notably in optics and on magnetism. Felix Savart (1774–1862) was a French physicist and mathematician. A main interest of his was on acoustics, and he worked with Biot on the theory of magnetism and electric currents.

potential that gives

$$\begin{aligned}\mathbf{A}(\mathbf{r}) &= \frac{\mu_0}{4\pi}\int \frac{\mathbf{j}(\mathbf{r}')}{|\mathbf{r}-\mathbf{r}'|}d^3r' \\ &= \frac{\mu_0}{4\pi r}\int \mathbf{j}(\mathbf{r}')\left[1+\frac{\mathbf{r}\cdot\mathbf{r}'}{r^2}+\frac{1}{2}\left(3\frac{(\mathbf{r}\cdot\mathbf{r}')^2}{r^4}-\frac{r'^2}{r^2}\right)+...\right]d^3r' \\ &\equiv \mathbf{A}_0(\mathbf{r})+\mathbf{A}_1(\mathbf{r})+\mathbf{A}_2(\mathbf{r})+\dots\end{aligned} \tag{11.33}$$

We focus on the first two terms of the expansion, and derive explicit expressions for these. The monopole term is

$$\mathbf{A}_0(\mathbf{r}) = \frac{\mu_0}{4\pi r}\int \mathbf{j}(\mathbf{r}')\,d^3r'\,, \tag{11.34}$$

and the dipole term is

$$\mathbf{A}_1(\mathbf{r}) = \frac{\mu_0}{4\pi r^3}\int (\mathbf{r}\cdot\mathbf{r}')\,\mathbf{j}(\mathbf{r}')d^3r'\,. \tag{11.35}$$

As we shall see the monopole term vanishes identically, and the dipole term can be re-expressed in a more convenient form. This we will show by exploiting the continuity equation for electric charge,

$$\nabla\cdot\mathbf{j}+\frac{\partial\rho}{\partial t}=0\,, \tag{11.36}$$

which is valid for general space- and time-dependent charge and current densities. Obviously this equation implies the following identity, valid for an arbitrary function $f(\mathbf{r})$,

$$\int f(\mathbf{r})\nabla\cdot\mathbf{j}(\mathbf{r},t)d^3r+\frac{d}{dt}\int f(\mathbf{r})\rho(\mathbf{r},t)d^3r=0\,. \tag{11.37}$$

In the following we shall assume the integration volume to be sufficiently large, so that no charge is transported through the surface of the volume. By a partial integration we can then rewrite the identity as

$$-\int[\boldsymbol{\nabla} f(\mathbf{r})]\cdot\mathbf{j}(\mathbf{r},t)d^3r+\frac{d}{dt}\int f(\mathbf{r})\rho(\mathbf{r},t)d^3r=0\,, \tag{11.38}$$

and by choosing f as products of the Cartesian coordinates of $\mathbf{r}$, we can derive relations between moments of the current distribution and the time derivative of moments of the charge distribution.

The simplest, non-trivial case is given by $f(\mathbf{r}) = x_k, k = 1,2,3$, which implies $\boldsymbol{\nabla} f(\mathbf{r}) = \mathbf{e}_k$. In this case Eq. (11.38) gives

$$-\int j_k(\mathbf{r},t)d^3r+\frac{d}{dt}\int x_k\rho(\mathbf{r},t)d^3r=0\,. \tag{11.39}$$

The second term in this expression can be recognized as the time derivative of the k-component of the electric dipole moment, and in vector form the identity given above can then be written as

$$\int \mathbf{j}(\mathbf{r},t)d^3r = \dot{\mathbf{p}}\,. \tag{11.40}$$

In the stationary case, with $\dot{\mathbf{p}} = 0$, this implies that the volume integral of the charge current vanishes, and as a consequence, so does also the magnetic monopole term,

$$\mathbf{A}_0(\mathbf{r}) = 0\,. \tag{11.41}$$

We next examine the case $f(\mathbf{r}) = x_k x_l$; $k, l = 1, 2, 3$. The identity (11.38) now takes the form

$$\int (x_k j_l(\mathbf{r},t) + x_l j_k(\mathbf{r},t))d^3r = \frac{d}{dt}\int x_k x_l \rho(\mathbf{r},t)d^3r\,, \tag{11.42}$$

which in the stationary case, with no contribution from the time derivative, gives

$$\int (x_k j_l(\mathbf{r}) + x_l j_k(\mathbf{r}))d^3r = 0\,. \tag{11.43}$$

We derive from this the following identity

$$\int [\mathbf{r}'\,(\mathbf{r}\cdot\mathbf{j}(\mathbf{r}')) + (\mathbf{r}\cdot\mathbf{r}')\,\mathbf{j}(\mathbf{r}')]d^3r' = 0\,, \tag{11.44}$$

where we have made a change of integration variable. This gives

$$\begin{aligned}\int (\mathbf{r}\cdot\mathbf{r}')\,\mathbf{j}(\mathbf{r}')d^3r' &= \frac{1}{2}\int [(\mathbf{r}\cdot\mathbf{r}')\,\mathbf{j}(\mathbf{r}') - \mathbf{r}'\,(\mathbf{r}\cdot\mathbf{j}(\mathbf{r}'))]d^3r' \\ &= \frac{1}{2}\left[\int \mathbf{r}'\times\mathbf{j}(\mathbf{r}')d^3r'\right]\times\mathbf{r}\,. \end{aligned} \tag{11.45}$$

Introducing this in the expression for the dipole term $\mathbf{A}_1(\mathbf{r})$, we can now write this in the form

$$\mathbf{A}_1(\mathbf{r}) = \frac{\mu_0}{4\pi}\frac{\mathbf{m}\times\mathbf{r}}{r^3}\,, \tag{11.46}$$

where $\mathbf{m}$ is the magnetic dipole moment of the current distribution, defined as

$$\mathbf{m} = \frac{1}{2}\int (\mathbf{r}\times\mathbf{j}(\mathbf{r}))\,d^3r\,. \tag{11.47}$$

The corresponding magnetic dipole field is

$$\begin{aligned}\mathbf{B}_1(\mathbf{r}) &= \boldsymbol{\nabla}\times\mathbf{A}_1(\mathbf{r}) \\ &= \frac{\mu_0}{4\pi}\boldsymbol{\nabla}\times\left(\frac{\mathbf{m}\times\mathbf{r}}{r^3}\right) \\ &= \frac{\mu_0}{4\pi r^3}(3\mathbf{n}(\mathbf{n}\cdot\mathbf{m}) - \mathbf{m})\,, \end{aligned} \tag{11.48}$$

with $\mathbf{n} = \mathbf{r}/r$ as before. We note that the form of the magnetic dipole field is precisely the same as that of the electric dipole field (11.24), with the electric dipole moment $\mathbf{p}$ replaced by the magnetic dipole moment $\mathbf{m}$.

11.3 Force on charge and current distributions

The electric and magnetic multipole moments appear in several ways in electromagnetic theory. One of these cases is when we consider electromagnetic radiation, and we shall discuss that in the next section. Another one is when we consider the electromagnetic force on a body with a non-vanishing charge or current distribution. We consider the last situation here.

Let us first consider a body with a given charge density $\rho(\mathbf{r})$, which is subject to an electric field $\mathbf{E}(\mathbf{r})$ which varies little over the charge distribution. We assume that the origin of our coordinate system ($\mathbf{r} = 0$) is chosen at a central point of the body, and we make an expansion of the field around this point,

$$\mathbf{E}(\mathbf{r}) = \mathbf{E}(0) + \big[(\mathbf{r} \cdot \boldsymbol{\nabla}')\mathbf{E}(\mathbf{r}')\big]_{\mathbf{r}'=0} + \dots . \tag{11.49}$$

The total force that acts on the body is then

$$\begin{aligned}
\mathbf{F}_e &= \int d^3r\, \rho(\mathbf{r})\, \mathbf{E}(\mathbf{r}) \\
&= \int d^3r\, \rho(\mathbf{r})\, \mathbf{E}(0) + \int d^3r\, \rho(\mathbf{r})\, [(\mathbf{r} \cdot \boldsymbol{\nabla}')\mathbf{E}(\mathbf{r}')]_{\mathbf{r}'=0} + \dots \\
&= q\mathbf{E} + (\mathbf{p} \cdot \boldsymbol{\nabla})\mathbf{E} + \dots ,
\end{aligned} \tag{11.50}$$

with q as the total charge of the body and $\mathbf{p}$ as the electric dipole moment. We note in particular the form of the dipole contribution to the force, which depends on the derivatives of the electric field.

The multipole moments also appear in the expansion of the torque,

$$\begin{aligned}
\boldsymbol{\tau}_e &= \int d^3r\, \rho(\mathbf{r})\, \mathbf{r} \times \mathbf{E}(\mathbf{r}) \\
&= \int d^3r\, \rho(\mathbf{r})\, \mathbf{r} \times (\mathbf{E}(0) + \int d^3r\, \rho(\mathbf{r})\, \mathbf{r} \times [(\mathbf{r} \cdot \boldsymbol{\nabla}')\mathbf{E}(\mathbf{r}')]_{\mathbf{r}'=0} + \dots \\
&= \mathbf{p} \times \mathbf{E} + \dots .
\end{aligned} \tag{11.51}$$

In the expressions above one should note that $\mathbf{E}$ is the *external* field acting on the charge distribution. The internal field from one part of the charge distribution to another part does not contribute, since internal forces do not contribute to the total force or torque.

We may describe in a similar way the magnetic force and torque acting on a current distribution. To obtain the expressions in this case, it is necessary to make use of the identity (11.43). We here leave out some details in the derivations, and refer to Problem 11.4, for filling in these details. The magnetic force is

$$\begin{aligned}\mathbf{F}_m &= \int d^3r\,\mathbf{j}\times\mathbf{B}(\mathbf{r})\\ &= \int d^3r\,\mathbf{j}(\mathbf{r})\times\mathbf{B}(0) + \int d^3r\,\mathbf{j}(\mathbf{r})\times[(\mathbf{r}\cdot\boldsymbol{\nabla}')\mathbf{B}(\mathbf{r}')]_{\mathbf{r}'=0} + \dots\,.\end{aligned} \tag{11.52}$$

For a stationary current the first term gives no contribution since $\int \mathbf{j}(\mathbf{r})d^3r = 0$, as previously discussed. The leading term, written in vector component form, is therefore

$$(F_m)_i = \epsilon_{iln}\left(\int d^3r x_k j_l(\mathbf{r})\right)\partial_k B_n(0) + \dots\,. \tag{11.53}$$

By use of (11.43) and the definition of the magnetic moment $\mathbf{m}$, we find the following identity

$$\int d^3r\, x_k j_l(\mathbf{r}) = \epsilon_{kls} m_s\,. \tag{11.54}$$

Introduced in the expression for the force this gives (see Appendix A.2 for product formulas of Levi-Civita symbols),

$$(F_m)_i = m_n\partial_i B_n - m_i\partial_n B_n + \dots\,, \tag{11.55}$$

where the last term vanishes due to the condition $\boldsymbol{\nabla}\cdot\mathbf{B} = 0$. When written in vector form we therefore obtain for the magnetic force

$$\begin{aligned}\mathbf{F}_m &= \boldsymbol{\nabla}(\mathbf{m}\cdot\mathbf{B}) + \dots\\ &= (\mathbf{m}\cdot\boldsymbol{\nabla})\mathbf{B} + \dots\,.\end{aligned} \tag{11.56}$$

The final expression given above is obtained by use of the condition $\boldsymbol{\nabla}\times\mathbf{B} = 0$. As follows from Ampere's law, this is satisfied when the electric field $\mathbf{E}$ is time independent and there is no current. The assumption here is that the currents of the *external* magnetic field are far from the body on which the external fields are acting. Comparison with the leading terms of the electric force acting on the body, we see that there is — as we should expect — no term corresponding to the charge (monopole) component of the electric force, while the dipole components in the two cases have precisely the same form.

The leading term in the expansion of the torque is

$$\boldsymbol{\tau}_m = \int d^3r\,\mathbf{r}\times(\mathbf{j}(\mathbf{r})\times\mathbf{B}(0)) + \dots\,, \tag{11.57}$$

which in component form is

$$(\tau_m)_i = \epsilon_{iks}\epsilon_{sln}\left(\int d^3r\, x_k j_l(\mathbf{r})\right) B_n(0) + \dots . \tag{11.58}$$

Reducing the product of the two Levi-Civita symbols we obtain

$$(\tau_m)_i = \left(\int d^3r\, x_k j_i(\mathbf{r})\right) B_k(0) - \left(\int d^3r\, x_k j_k(\mathbf{r})\right) B_i(0) + \dots . \tag{11.59}$$

Due to the identity (11.54) the second term vanishes, and the first term can be rewritten as

$$(\tau_m)_i = \epsilon_{ijk} m_j B_k + \dots . \tag{11.60}$$

In vector form the leading term of the torque of the magnetic force is therefore

$$\boldsymbol{\tau}_n = \mathbf{m} \times \mathbf{B} + \dots . \tag{11.61}$$

Also in this case we find the same correspondence between the electric and magnetic expressions, where the electric dipole moment and the electric field in the first case is replaced by the magnetic moment and the magnetic field in the second case.

11.4 Exercises

Problem 11.1

Three point charges, two with charge $+q$ and one with charge $-q$ are positioned in the sequence $(+q, -q, +q)$ along the x-axis. The distances between neighboring charges are equal to d, and the middle charge is placed at the origin $x = 0$.

a) Find an expression for the full Coulomb potential of the three charges in the x, y-plane and make a contour plot of the potential (which shows equipotential lines of the potential).

b) Determine the monopole moment (total charge) q_{tot}, dipole moment $\mathbf{p}$, and the quadrupole moment $Q_{\mathbf{n}}$ of the charge distribution, where $\mathbf{n}$ demotes a unit vector in the x, y-plane.

c) Find the corresponding three contributions to the Coulomb potential, and make a contour plot of the sum. Compare this plot with that of the full potential in a).

Problem 11.2

A non-relativistic particle, with electric charge q and mass m moves in a magnetic dipole field, given by the vector potential

$$\mathbf{A} = \frac{\mu_0}{4\pi r^3}(\boldsymbol{\mu} \times \mathbf{r})\,, \tag{11.62}$$

with $\boldsymbol{\mu}$ as the magnetic dipole moment of a static charge distribution centered at the origin. (We use here the notation $\boldsymbol{\mu}$ for the dipole moment to avoid confusion with the particle mass m).

a) Show that the particle's Lagrangian is

$$L = \frac{1}{2}m\mathbf{v}^2 + \frac{q\mu_0}{4\pi m r^3}\,\boldsymbol{\mu}\cdot\boldsymbol{\ell}\,, \tag{11.63}$$

with $\boldsymbol{\ell} = m\,\mathbf{r}\times\mathbf{v}$ as the particle's orbital angular momentum.

We make now the assumption that the magnetic dipole moment is oriented along the z-axis and that the particle moves in the x,y-plane. Choose in the following $r = |\mathbf{r}|$ and the angle ϕ between the x-axis and the position vector $\mathbf{r}$ as coordinates.

b) Show that the Lagrangian of the particle, when expressed in terms of r, ϕ and their time derivatives, takes the form

$$L = \frac{1}{2}m(\dot{r}^2 + r^2\dot{\phi}^2) + \lambda\frac{\dot{\phi}}{r}\,, \tag{11.64}$$

with $\lambda \equiv q\mu_0|\boldsymbol{\mu}|/4\pi$. Find the canonical momentum p_ϕ conjugate to ϕ, and give the physical interpretation of this quantity. Also comment on the consequence of L having no explicit time dependence.

c) Write Lagrange's equation for the coordinate r, expressed in terms of r, $\ddot{r}$ and p_ϕ, and use the equation to find $\dot{r}^2$ as a function of r and p_ϕ. Compare the expression with that of the particle's kinetic energy.

Problem 11.3

Figure 11.5 shows a rectangular current loop ABCD. In the loop's rest frame, S, the loop has length a in the x-direction and length b in the y-direction, the current is I and the charge density is zero. We remind you about the following general definitions of the electric dipole moment $\mathbf{p}$, and the magnetic dipole moment $\mathbf{m}$ of a current distribution are

$$\mathbf{p} = \int \mathbf{r}\rho(\mathbf{r})\,d^3r\,, \quad \mathbf{m} = \frac{1}{2}\int (\mathbf{r}\times\mathbf{j}(\mathbf{r}))d^3r\,, \tag{11.65}$$

where $\rho(\mathbf{r})$ is the charge density and $\mathbf{j}(\mathbf{r})$ is the current density in the wire. For simplicity we may assume the current density to be constant over the cross section of the wire, which gives $I = j\Delta$, with Δ as the area of the cross section.

a) Show that in the rest frame the loop's electric dipole moment is zero and the magnetic moment is $\mathbf{m} = I\mathbf{a} \times \mathbf{b}$.

In the following we will examine how the loop is observed in a reference frame S', where the loop is moving with velocity $\mathbf{v}$ to the right ($\beta = v/c$ and $\gamma = 1/\sqrt{1-\beta^2}$). The Lorentz transformation formulas for charge and current densities may be useful when solving the problems below.

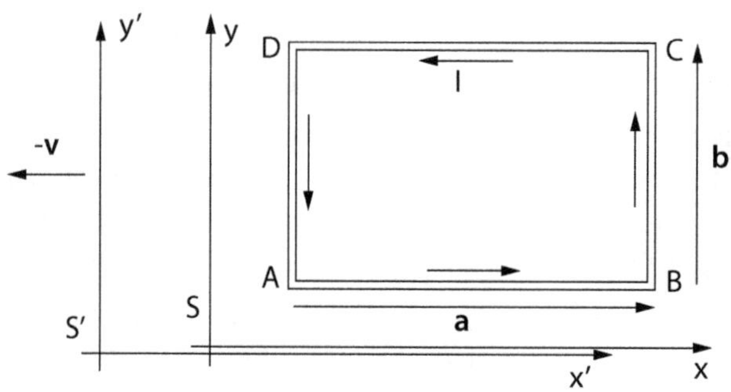

Fig. 11.5 A moving current loop.

b) What is the length and width of the loop in S'?

c) Show that the parts AB and CD of the loop have charges $\pm aIv/c^2$ in S'.

d) Show that in S' the loop's electric dipole moment is $\mathbf{p}' = -\frac{1}{c^2}\mathbf{m} \times \mathbf{v}$, and the magnetic dipole moment is $\mathbf{m}' = (1 - \beta^2/2)\mathbf{m}$.

e) Show that the current is $I\gamma$ in the AB and CD and I/γ in BC and DA.

f) Show that the result in e) is consistent with charge conservation.

Problem 11.4

In this exercise the task is to fill in some details in the derivations of the force and torque on a stationary current in a magnetic field, discussed in Section 11.3. Make use of the identity (11.43), the definition of the magnetic dipole moment $\mathbf{m}$, and the properties of the Levi-Civita symbol, discussed in Appendix A.2.

a) Show that the electric current $\mathbf{j}(\mathbf{r})$ satisfies the following identity

$$\int d^3r x_k\, j_l(\mathbf{r}) = \epsilon_{nkl}\, m_n\,, \tag{11.66}$$

with m_n as components of the magnetic dipole moment.

b) Use this to show that the expression (11.53) for the magnetic force can be reformulated as (11.55), and further, in vector form, as

$$\mathbf{F}_m = \boldsymbol{\nabla}(\mathbf{m}\cdot\mathbf{B}) + \ldots\,, \tag{11.67}$$

where the derivative acts on the $\mathbf{B}$-field, but treats $\mathbf{m}$ as a constant.

c) Assuming the magnetic field satisfies $\boldsymbol{\nabla}\times\mathbf{B} = 0$, show that the expression for $\mathbf{F}_m$ can also be written as

$$\mathbf{F}_m = (\mathbf{m}\cdot\boldsymbol{\nabla})\mathbf{B} + \ldots\,. \tag{11.68}$$

d) Use (11.66) to show that the expression (11.58) for the torque of the magnetic force can be written as

$$\boldsymbol{\tau}_n = \mathbf{m}\times\mathbf{B} + \ldots\,. \tag{11.69}$$

Chapter 12

Electromagnetic radiation

We consider now the full problem of solving Maxwell's equations with time dependent sources. The field equation, in the Lorentz gauge, is as before

$$\partial_\nu \partial^\nu A^\mu = -\mu_0 j^\mu\,, \quad \partial_\mu A^\mu = 0\,, \tag{12.1}$$

where the current may now depend both on space and time coordinates, $j^\mu = j^\mu(\mathbf{r}, t)$. The solution involves a *retardation effect*, since the field at some distance from the source will respond to changes in the source at a delayed time, in accordance with the fact that the speed of wave propagation is finite. We shall, as the next step, examine how this retardation effect gives rise to the phenomenon of electromagnetic radiation.

12.1 Solutions to the time dependent equation

We note that also in the general case, with time dependent sources, the equations for each vector component of A^μ can be solved separately, and the equations are all of the same form. There is a coupling between the components, as follows from the Lorentz gauge condition $\partial_\mu A^\mu = 0$, but this equation is automatically taken care of by the continuity equation $\partial_\mu j^\mu = 0$, for fields that tend to zero at infinity. In non-covariant form the differential equation to be solved is

$$\left(\boldsymbol{\nabla}^2 - \frac{1}{c^2}\frac{\partial^2}{\partial t^2}\right) f(\mathbf{r}, t) = -s(\mathbf{r}, t)\,, \tag{12.2}$$

where f represents one of the components of the potential and s represents the corresponding component of the current density. When discussing solutions of this equation we consider the source term $s(\mathbf{r}, t)$ as a known function, while $f(\mathbf{r}, t)$ is the unknown function, to be determined as a solution of the differential equation.

To proceed we introduce the Fourier transformation of the equation with respect to time. For the function $f(\mathbf{r},t)$ this transformation is

$$f(\mathbf{r},t) = \int_{-\infty}^{\infty} \tilde{f}(\mathbf{r},\omega)e^{-i\omega t}d\omega\,,$$
$$\tilde{f}(\mathbf{r},\omega) = \frac{1}{2\pi}\int_{-\infty}^{\infty} f(\mathbf{r},t)e^{i\omega t}dt\,, \tag{12.3}$$

and the same type of transformation formulas are valid for the source function $s(\mathbf{r},t)$. In the Fourier transformed version time t is then replaced by the frequency variable ω, while the space coordinate $\mathbf{r}$ is left unchanged. Applied to the differential equation (12.2) the transformation gives the following equation for the Fourier transformed fields,

$$\left(\boldsymbol{\nabla}^2 + \frac{\omega^2}{c^2}\right)\tilde{f}(\mathbf{r},\omega) = -\tilde{s}(\mathbf{r},\omega)\,. \tag{12.4}$$

This differential equation, which only includes derivatives with respect to the space coordinates, shows a clear resemblance to the electrostatic equation. However, the presence of the constant $\frac{\omega^2}{c^2}$ makes it different. The differential equation (12.4) is known as the inhomogeneous *Helmholtz equation.*

Even if there is a difference, we may take some inspiration from the Coulomb problem. As we have earlier discussed, the usual way to find the solution of the electrostatic problem is first to find the electrostatic potential of a point charge, and to use this to find a general solution by integrating over the actual charge distribution. For a point charge q the charge distribution is $\rho(\mathbf{r}) = q\delta(\mathbf{r})$, and the electrostatic equation is

$$\boldsymbol{\nabla}^2\Phi = -\frac{\rho}{\epsilon_0} = -\frac{q}{\epsilon_0}\delta(\mathbf{r})\,, \tag{12.5}$$

with the Coulomb potential as solution

$$\Phi = \frac{q}{4\pi\epsilon_0 r}\,. \tag{12.6}$$

This shows that we (formally) have the following identity

$$\boldsymbol{\nabla}^2\left(\frac{1}{r}\right) = -4\pi\delta(\mathbf{r})\,. \tag{12.7}$$

We will show that a similar relation is valid when a constant is added to $\boldsymbol{\nabla}^2$, in the following way

$$(\boldsymbol{\nabla}^2 + \alpha^2)\left(\frac{e^{i\alpha r}}{r}\right) = -4\pi\delta(\mathbf{r})\,, \tag{12.8}$$

which gives the solution of the inhomogeneous Helmholtz equation with a point source.

To this end, we split the function in (12.8) in two parts

$$\frac{e^{i\alpha r}}{r} = \frac{1}{r} - \frac{1 - e^{i\alpha r}}{r}, \tag{12.9}$$

where the first part carries the singularity of the function, which gives rise to the delta-function in (12.8). Thus, the second term is non-singular at $r = 0$, and to find the result of acting with the Laplacian on this part is therefore straight forward. We use the expression for the Laplacian in polar coordinates,

$$\begin{aligned}\boldsymbol{\nabla}^2\left(\frac{1 - e^{i\alpha r}}{r}\right) &= \frac{1}{r^2}\frac{d}{dr}r^2\frac{d}{dr}\left(\frac{1 - e^{i\alpha r}}{r}\right) \\ &= \frac{1}{r^2}\frac{d}{dr}\left(-1 + e^{i\alpha r} - i\alpha r e^{i\alpha r}\right) \\ &= \alpha^2\frac{e^{i\alpha r}}{r}\,. \end{aligned} \tag{12.10}$$

From this follows

$$\begin{aligned}\boldsymbol{\nabla}^2\left(\frac{e^{i\alpha r}}{r}\right) &= \boldsymbol{\nabla}^2\left(\frac{1}{r}\right) - \boldsymbol{\nabla}^2\left(\frac{1 - e^{i\alpha r}}{r}\right) \\ &= -4\pi\delta(\mathbf{r}) - \alpha^2\frac{e^{i\alpha r}}{r}, \end{aligned} \tag{12.11}$$

which reproduces Eq. (12.8).

We immediately rewrite the above equation in a form directly related to the problem we want to solve:

$$\left(\boldsymbol{\nabla}^2 + \frac{\omega^2}{c^2}\right)\left(\frac{e^{\pm i\frac{\omega}{c}r}}{r}\right) = -4\pi\delta(\mathbf{r})\,. \tag{12.12}$$

Note that in this expression we have explicitly made use of the fact that α is only up to a sign determined by α^2. Our interpretation of this equation is now the following. Assume we modify the electrostatic equation by adding the term proportional to ω^2/c^2. This change in the field equation will modify the potential set up by a point charge so it is no longer a Coulomb potential. Actually the modification is not unique, the Coulomb potential can be modified either by the factor $\exp(+i\frac{\omega}{c}r)$ or $\exp(-i\frac{\omega}{c}r)$. However, as we shall soon see, there is a reason for choosing one of these as the *physical* solution.

With the solution established for a point source, the potential for a source distribution is found by integrating over the distribution, in the same way as done for the Coulomb problem (Section 11.1). This gives, with $\tilde{s}(\mathbf{r},\omega)$ as the source term, the following solutions,

$$\tilde{f}_{\pm}(\mathbf{r},\omega) = \frac{1}{4\pi}\int \frac{e^{\pm i\frac{\omega}{c}|\mathbf{r}-\mathbf{r}'|}}{|\mathbf{r}-\mathbf{r}'|}\tilde{s}(\mathbf{r}',\omega)d^3r' \,, \tag{12.13}$$

where the distance r from the point charge is now replaced by the distance $|\mathbf{r}-\mathbf{r}'|$ from the point of integration. The corresponding time dependent solution of the original equation is found as the Fourier integral

$$\begin{aligned} f_{\pm}(\mathbf{r},t) &= \int_{-\infty}^{\infty} \tilde{f}_{\pm}(\mathbf{r},\omega)e^{-i\omega t}d\omega \\ &= \frac{1}{4\pi}\int \left(\int_{-\infty}^{\infty} e^{-i\omega(t\mp\frac{|\mathbf{r}-\mathbf{r}'|}{c})}\tilde{s}(\mathbf{r}',\omega)d\omega\right)\frac{1}{|\mathbf{r}-\mathbf{r}'|}d^3r' \,. \end{aligned} \tag{12.14}$$

We recognize the integral in the brackets as the Fourier integral of the function $s(\mathbf{r}',t\mp\frac{|\mathbf{r}-\mathbf{r}'|}{c})$, and this gives for $f_\pm$ the following expression

$$f_{\pm}(\mathbf{r},t) = \frac{1}{4\pi}\int \frac{s(\mathbf{r}',t\mp\frac{|\mathbf{r}-\mathbf{r}'|}{c})}{|\mathbf{r}-\mathbf{r}'|}d^3r' \,. \tag{12.15}$$

The solutions we have found are similar in form to the Coulomb potential, since the potential $f_\pm$ is determined as the integral of the source term divided by the distance between the source and the point of the potential. But there is one important difference, which has to do with the time dependence. The potential at a given time t is determined by the source at another time $t_\pm = t \pm |\mathbf{r}-\mathbf{r}'|/c$. One of these is earlier than t and the other is later than t. The solution f_- is called the *retarded* solution, since $t_- < t$, and the effect that the source has on the field therefore is delayed in time. Similarly f_+ is called the *advanced* solution, since $t_+ > t$, and the effect that the source has on the potential is advanced in time. For this reason we usually consider the retarded solution f_- as the physical one. Note however that Maxwell's equations accept both these solutions, since they are invariant under time reversal, $t \to -t$. The two types of solutions can be understood as corresponding to two different types of boundary conditions. Usually we specify *initial conditions*, with the solution of Maxwell's equation given as the retarded potential, but it is also possible to specify *final conditions* with the solution given as the advanced potential.

It is of interest to note that the two space-time points $(\mathbf{r},t)$ and $(\mathbf{r}',t_\pm)$ can be connected by a light signal, since we have

$$(\mathbf{r}-\mathbf{r}')^2 - c^2(t-t_\pm)^2 = 0 \,, \tag{12.16}$$

as we can readily check. Thus $(\mathbf{r}',t_-)$ lies on the past light cone relative to $(\mathbf{r},t)$, while $(\mathbf{r}',t_+)$ lies on the future light cone.

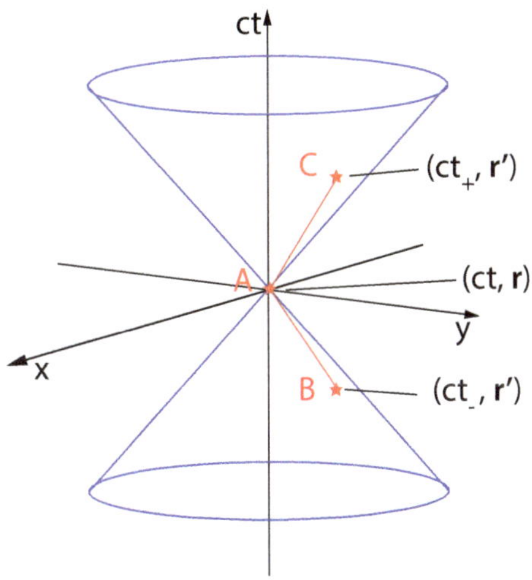

Fig. 12.1 Advanced and retarded space-time points. Given a point A with coordinates $(ct, \mathbf{r})$ a point B with *retarded* time coordinate $t_- = t - |\mathbf{r} - \mathbf{r}'|/c$ is located on the *past light cone* of the point A. This means that a light signal emitted from B can reach the point A. Similarly a point C with *advanced* time coordinate $t_+ = t + |\mathbf{r} - \mathbf{r}'|/c$ is located on the *future light cone* of the point A. A light signal emitted from A is then able to reach the point C.

The retarded potential

We now translate the results we have found to expressions for the electromagnetic potentials. In the following we shall consider only the retarded solutions, which we regard as the physical ones. For the scalar and vector potentials the expressions are

$$\Phi(\mathbf{r}, t) = \frac{1}{4\pi\epsilon_0} \int \frac{\rho(\mathbf{r}', t_-)}{|\mathbf{r} - \mathbf{r}'|} d^3r',$$

$$\mathbf{A}(\mathbf{r}, t) = \frac{\mu_0}{4\pi} \int \frac{\mathbf{j}(\mathbf{r}', t_-)}{|\mathbf{r} - \mathbf{r}'|} d^3r', \tag{12.17}$$

with $t_- = t - |\mathbf{r} - \mathbf{r}'|/c$ referred to as the *retarded time*. It is interesting to note that the potentials we have found have precisely the same form as the potentials previously found with static sources. The only effect of the time dependence sits in the retardation effect, the effect that there is a time delay between the change in the charge and current distributions and the effects measured in the potentials. This means that the volume integrals in the expressions for Φ and $\mathbf{A}$ are not integrals over space at constant t.

Instead they are integrals over the three-dimensional past light cone of the point $(ct, \mathbf{r})$.

Even if the effect of time evolution of the source terms looks simple (and innocent) when we consider the potentials, that is not so when we consider the electromagnetic fields $\mathbf{E}$ and $\mathbf{B}$. This is because the retarded time t_- depends on $\mathbf{r}$ and $\mathbf{r}'$. When the fields are expressed through derivatives of the potentials, this dependence on $\mathbf{r}$ gives rise to new terms in the expressions for $\mathbf{E}$ and $\mathbf{B}$. These terms have an immediate physical interpretation. They describe radiation from the time dependent sources.

12.2 Electromagnetic potentials of a point charge

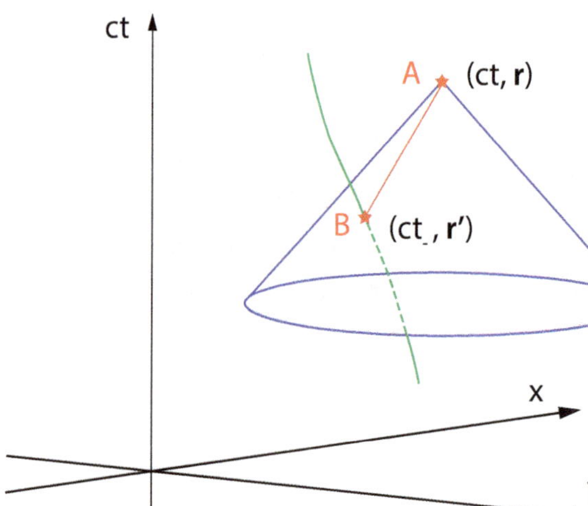

Fig. 12.2 Retarded point on the world line of a point charge. Given a point A with coordinates $(ct, \mathbf{r})$ there is only one point B that is both located on the world line and on the past light cone of A. This means that there is only one point in the volume integral of the retarded electromagnetic potentials that gives a contribution at the point A.

In this case the charge and current densities are expressed as

$$\begin{aligned} \rho(\mathbf{r}, t) &= q\delta(\mathbf{r} - \mathbf{r}(t)), \\ \mathbf{j}(\mathbf{r}, t) &= q\mathbf{v}(t)\delta(\mathbf{r} - \mathbf{r}(t)), \end{aligned} \tag{12.18}$$

with q as the charge, $\mathbf{r}$(t) as the time dependent position of the charge and $\mathbf{v}(t)$ as the velocity. The presence of the delta function means that, in the expressions we have derived for the potentials produced by charges and currents, the integral over the densities will get contributions only from a single point. However, there is a complication due to the retardation effect. We consider first the scalar potential,

$$\Phi(\mathbf{r},t) = \frac{q}{4\pi\epsilon_0}\int \frac{\delta(\mathbf{r}' - \mathbf{r}(t_-))}{|\mathbf{r}-\mathbf{r}'|} d^3r' \,. \tag{12.19}$$

One should note that the retarded time $t_- = t - |\mathbf{r}-\mathbf{r}'|/c$ is a function of the integration variable $\mathbf{r}'$ and this we have to take into account when integrating over the delta function (see Appendix A.1). It is convenient to introduce the argument of the delta function as a new integration variable,

$$\mathbf{r}'' = \mathbf{r}' - \mathbf{r}(t_-)\,, \tag{12.20}$$

where we note that the vector $\mathbf{r}$ in the definition of t_- is a constant under the integration. The change of variable introduces a change in the integration measure given by

$$d^3r'' = Jd^3r'\,, \tag{12.21}$$

where J is the Jacobian of the transformation, which is the determinant of the matrix with elements

$$J_{kl} = \frac{\partial x''_k}{\partial x'_l}\,. \tag{12.22}$$

We find for this matrix element the following expression

$$\begin{aligned} J_{kl} &= \delta_{kl} - \frac{dx_k}{dt_-}\frac{\partial t_-}{\partial x'_l} \\ &= \delta_{kl} + \frac{1}{c}v_k(t_-)\frac{\partial}{\partial x'_l}\sqrt{r^2 + r'^2 - 2\mathbf{r}\cdot\mathbf{r}'} \\ &= \delta_{kl} - \frac{1}{c}v_k(t_-)\frac{x_l - x'_l}{|\mathbf{r}-\mathbf{r}'|}\,. \end{aligned} \tag{12.23}$$

To simplify expressions we introduce $\boldsymbol{\beta}(t) = \mathbf{v}(t)/c$ and $\mathbf{n} = (\mathbf{r}-\mathbf{r}')/|\mathbf{r}-\mathbf{r}'|$. The matrix element of the Jacobian can then be written as

$$J_{kl} = \delta_{kl} - \beta_k n_l\,. \tag{12.24}$$

When calculating the corresponding determinant it is useful temporarily to chose the x-axis in the direction of $\mathbf{n}$, which gives $n_1 = 1,\ n_2 = n_3 = 0$. The result is simply $1 - \beta_1$ which we rewrite in a coordinate independent way as

$$J = 1 - \boldsymbol{\beta}\cdot\mathbf{n}\,. \tag{12.25}$$

The integral in the expression for the potential can now be evaluated,

$$\Phi(\mathbf{r},t) = \frac{q}{4\pi\epsilon_0}\int \frac{\delta(\mathbf{r}'')}{|\mathbf{r}-\mathbf{r}'|}\,\frac{1}{1-\boldsymbol{\beta}\cdot\mathbf{n}}\,d^3r''\,. \tag{12.26}$$

In this integral the effect of the delta function is simply to put $\mathbf{r}''=0$, which is equivalent to $\mathbf{r}'=\mathbf{r}(t_-)$, and the potential can therefore be written as

$$\Phi(\mathbf{r},t) = \frac{q}{4\pi\epsilon_0|\mathbf{r}-\mathbf{r}(t_-)|(1-\boldsymbol{\beta}(t_-)\cdot\mathbf{n}(t_-))}\,. \tag{12.27}$$

To simplify this expression we introduce the relative vector $\mathbf{R}(t)=\mathbf{r}-\mathbf{r}(t)$ and use the label *ret* to indicate that expression should be evaluated at time $t=t_-$,

$$\Phi(\mathbf{r},t) = \frac{q}{4\pi\epsilon_0\,(R-\boldsymbol{\beta}\cdot\mathbf{R})_{ret}}\,. \tag{12.28}$$

The vector potential can be found in precisely the same way, and we simply give the result

$$\mathbf{A}(\mathbf{r},t) = \frac{\mu_0 q}{4\pi}\left(\frac{\mathbf{v}}{R-\boldsymbol{\beta}\cdot\mathbf{R}}\right)_{ret}\,. \tag{12.29}$$

The expressions we have found for the potentials of a moving point charge are called the *Liénard-Wiechert potentials.*[1] We note that these expressions are valid with no restriction on the motion of the charge; it may be at rest, move with constant speed, or be accelerated. Therefore the potentials implicitly contain all electromagnetic effects of a charge in motion, in particular radiation from an accelerated charge.

There is a clear similarity between the expressions found here and that of the Coulomb potential of a stationary point charge. This we see most clearly if we choose as inertial frame the rest frame of the moving charge *at the retarded time* t_-. In this frame the potential are

$$\Phi(\mathbf{r},t) = \frac{q}{4\pi\epsilon_0\,R_{ret}}\,,\qquad \mathbf{A}(\mathbf{r},t)=0\,. \tag{12.30}$$

This is simply the Coulomb potential with the distance to the charge determined by its position at the retarded time.

This gives us a simple picture of how the potential in the surrounding space-time is created by the moving charge. Each point along its trajectory determines the potential on the future light cone from the chosen point, as the Coulomb potential in the rest frame of the charge. If the charge is accelerated the rest frame will change along the path, and this means that the potential is not that of a Coulomb potential in a *fixed* inertial frame.

[1]Alfred-Marie Liénard (1869–1958) was a French physicist and engineer. Emil Johann Wiechert (1861–1928) was a German physicist and geophysicist.

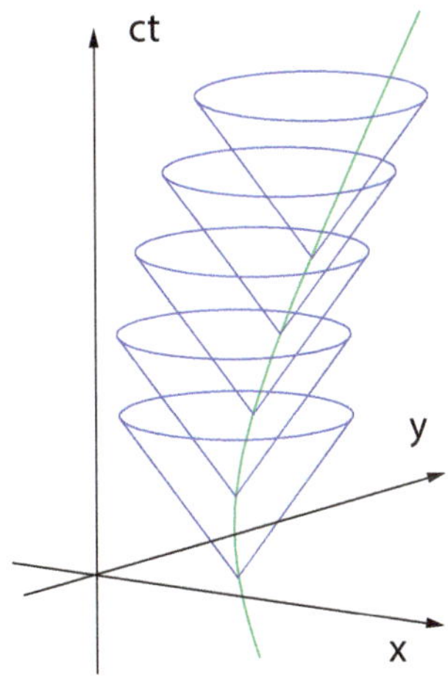

Fig. 12.3 Geometrical interpretation of the electromagnetic potentials produced by an accelerated charged point particle. The potential at any given space-time point is uniquely determined by the position and velocity of the charge at the *retarded* time. This is the time at which a light signal sent from the particle is able to reach the chosen space-time point. As a consequence, for any chosen point on the world line of the particle, the electromagnetic potential on the corresponding future light cone will be determined by the position and velocity of the charge at this point. The figure illustrates how the full space-time dependent potential can be viewed as composed of a sequence of contributions defined on the future light cones (in blue) of the moving charge, here represented by the green curve. Note that even if the potential has no direct dependence on the acceleration of the particle, the electromagnetic fields do, since they depend on the space time derivatives of the potential.

12.3 The general situation: Fields far from the sources

We study now the potentials of a general time-dependent charge and current distribution, far from the sources, where the leading contributions to the potentials fall off with distance as $1/r$. The same approximation technique as used in the multipole expansions of static distributions can then be applied. With $\mathbf{r}$ denoting the position at which the potential is evaluated, and $\mathbf{r}'$ as the integration variable over the charge distribution, we again assume the origin to be chosen close to the charges, so that r'/r is a small quantity which we can use as an expansion parameter. The distance to the charge distribution is as before given by

$$|\mathbf{r} - \mathbf{r}'| = r - \frac{\mathbf{r} \cdot \mathbf{r}'}{r} + \dots . \tag{12.31}$$

The expression for the retarded time can be expanded in a similar way,

$$t_- = t - \frac{r}{c} + \frac{\mathbf{r}\cdot\mathbf{r}'}{rc} + \dots\,, \tag{12.32}$$

where, in the following, we include only terms up to first order in r'/r.

To find the leading contribution to the scalar potential we need to make an expansion of the charge density

$$\begin{aligned}\rho(\mathbf{r}', t_-) &= \rho\left(\mathbf{r}', t - \frac{r}{c}\right) + \frac{\mathbf{r}\cdot\mathbf{r}'}{rc}\frac{\partial\rho}{\partial t}\left(\mathbf{r}', t - \frac{r}{c}\right) + \dots \\ &= \rho(\mathbf{r}', t_r) + \frac{\mathbf{r}\cdot\mathbf{r}'}{rc}\frac{\partial\rho}{\partial t}(\mathbf{r}', t_r) + \dots\,,\end{aligned} \tag{12.33}$$

where we have introduced $t_r = t - r/c$, which is the retarded time, not for a general point $\mathbf{r}'$ of the charge distribution, but rather for the origin $\mathbf{r}' = 0$. This we assume to be a central point of the distribution. From the above expressions we find

$$\frac{\rho(\mathbf{r}', t_-)}{|\mathbf{r} - \mathbf{r}'|} = \frac{1}{r}\rho(\mathbf{r}', t_r) + \frac{\mathbf{r}\cdot\mathbf{r}'}{r^2 c}\frac{\partial\rho}{\partial t}(\mathbf{r}', t_r) + \dots\,, \tag{12.34}$$

where one should note that higher order terms in the time derivative of ρ, which have not been included explicitly, also give contributions of order $1/r$. This expression is now inserted in the integral expression for the potential, which gives

$$\begin{aligned}\Phi(\mathbf{r}, t) &= \frac{1}{4\pi\epsilon_0}\int \frac{\rho(\mathbf{r}', t_-)}{|\mathbf{r} - \mathbf{r}'|} d^3r' \\ &= \frac{1}{4\pi\epsilon_0 r}\int \rho\left(\mathbf{r}', t - \frac{r}{c}\right) d^3r' \\ &\quad + \frac{1}{4\pi\epsilon_0 r^2 c}\int (\mathbf{r}\cdot\mathbf{r}')\frac{\partial\rho}{\partial t}\left(\mathbf{r}', t - \frac{r}{c}\right) d^3r' + \dots \\ &= \frac{q}{4\pi\epsilon_0 r} + \frac{\mathbf{r}\cdot\dot{\mathbf{p}}_{ret}}{4\pi\epsilon_0 r^2 c} + \dots\,,\end{aligned} \tag{12.35}$$

with q as the total charge and $\mathbf{p}$ as the electric dipole moment. In addition to the terms included explicitly in (12.35) there are higher order multipole contributions, which come from higher order time derivatives in the expansion of the charge density (12.33). These are all of order $1/r$, with the number of time derivatives increasing with the degree of the multipole. The dipole contribution is thus first order in time derivative, the quadrupole term is second order etc.

We continue now to analyze the vector potential in the same way. The general expression is

$$\mathbf{A}(\mathbf{r}, t) = \frac{\mu_0}{4\pi}\int \frac{\mathbf{j}(\mathbf{r}', t_-)}{|\mathbf{r} - \mathbf{r}'|} d^3r'\,, \tag{12.36}$$

where a Taylor expansion is introduced for the current density in the same way as done above for the charge density. We find

$$\mathbf{A}(\mathbf{r},t) = \frac{\mu_0}{4\pi r}\int \mathbf{j}(\mathbf{r}',t_r)d^3r' + \frac{\mu_0}{4\pi c r^2}\int (\mathbf{r}\cdot\mathbf{r}')\left[\frac{\partial \mathbf{j}}{\partial t}(\mathbf{r}',t_r)\right]d^3r' + \dots . \tag{12.37}$$

The first term can be expressed in terms of the electric dipole moment as

$$\int \mathbf{j}(\mathbf{r}',t)d^3r' = \dot{\mathbf{p}}(t)\,, \tag{12.38}$$

which is an identity earlier derived in Eq. (11.40).

To rewrite the second term, another identity, derived in Eq. (11.42), is needed

$$\int x_k\, j_l(\mathbf{r},t)\, d^3r = -\int x_l\, j_k(\mathbf{r},t)\, d^3r + \frac{d}{dt}\int x_k x_l\, \rho(\mathbf{r},t)\, d^3r\,. \tag{12.39}$$

This implies

$$\begin{aligned}\int (\mathbf{r}\cdot\mathbf{r}')\mathbf{j}(\mathbf{r}',t)d^3r' &= \frac{1}{2}\left[\int \mathbf{r}'\times\mathbf{j}(\mathbf{r}',t)d^3r'\right]\times\mathbf{r}\\ &\quad+\frac{1}{2}\frac{d}{dt}\int \mathbf{r}'(\mathbf{r}\cdot\mathbf{r}')\rho(\mathbf{r}',t)d^3r'\\ &= \mathbf{m}\times\mathbf{r} + \frac{1}{2}r\dot{\mathbf{D}}_{\mathbf{n}}\,,\end{aligned} \tag{12.40}$$

where the magnetic dipole moment $\mathbf{m}$ and the electric quadrupole *vector* $\mathbf{D}_{\mathbf{n}}$ are defined by

$$\begin{aligned}\mathbf{m} &= \frac{1}{2}\int \mathbf{r}'\times\mathbf{j}(\mathbf{r}',t)d^3r'\,,\\ \mathbf{D}_{\mathbf{n}} &= \int \mathbf{r}'(\mathbf{r}'\cdot\mathbf{n})\rho(\mathbf{r}',t)d^3r'\,,\end{aligned} \tag{12.41}$$

with $\mathbf{n} = \mathbf{r}/r$ as the unit vector in direction of $\mathbf{r}$. By use of these expressions we are able to write vector potential as

$$\mathbf{A}(\mathbf{r},t) = \frac{\mu_0}{4\pi r}\left(\dot{\mathbf{p}} + \frac{1}{c}\dot{\mathbf{m}}\times\mathbf{n} + \frac{1}{2c}\ddot{\mathbf{D}}_{\mathbf{n}} + \dots\right)_{ret}\,,\quad \mathbf{n} = \frac{\mathbf{r}}{r}\,, \tag{12.42}$$

where the subscript *ret* now means that the vectors should be taken at retarded time $t_r = t - r/c$. We note that both the electric and the magnetic dipole moments, as well as higher multipole moments, contribute to the vector potential.

Let us next consider the magnetic field that corresponds to the vector potential that we have found,

$$\begin{aligned}\mathbf{B}(\mathbf{r},t) &= \boldsymbol{\nabla}\times\mathbf{A}(\mathbf{r},t)\\ &= -\frac{\mu_0}{4\pi r^2}\boldsymbol{\nabla} r\times\left(\dot{\mathbf{p}}+\frac{1}{c}\dot{\mathbf{m}}\times\mathbf{n}+\frac{1}{2c}\ddot{\mathbf{D}}_{\mathbf{n}}+...\right)_{ret}\\ &\quad+\frac{\mu_0}{4\pi r}(\boldsymbol{\nabla} t_r)\times\frac{d}{dt}\left(\dot{\mathbf{p}}+\frac{1}{c}\dot{\mathbf{m}}\times\mathbf{n}+\frac{1}{2c}\ddot{\mathbf{D}}_{\mathbf{n}}+...\right)_{ret}\\ &\quad+\ldots\,. \end{aligned}\tag{12.43}$$

One should note the two kinds of contributions, with the first one coming from the explicit dependence of r in the expression (12.42) for $\mathbf{A}(\mathbf{r},t)$, and the second one coming from the r-dependence of the retarded time t_-. The gradient of the retarded time is given by

$$\boldsymbol{\nabla} t_r = \boldsymbol{\nabla}\left(t-\frac{r}{c}\right) = -\frac{1}{c}\mathbf{n}\,,\tag{12.44}$$

and this gives for the magnetic field the following expression

$$\begin{aligned}\mathbf{B}(\mathbf{r},t) &= \frac{\mu_0}{4\pi r^2}\left(\dot{\mathbf{p}}+\frac{1}{c}\dot{\mathbf{m}}\times\mathbf{n}+\frac{1}{2c}\ddot{\mathbf{D}}_{\mathbf{n}}+...\right)_{ret}\times\mathbf{n}\\ &\quad+\frac{\mu_0}{4\pi rc}\left(\ddot{\mathbf{p}}+\frac{1}{c}\ddot{\mathbf{m}}\times\mathbf{n}+\frac{1}{2c}\dddot{\mathbf{D}}_{\mathbf{n}}+...\right)_{ret}\times\mathbf{n}\\ &\quad+\ldots\,. \end{aligned}\tag{12.45}$$

One should note that all the terms that fall off as $1/r$, are obtained by differentiation through the retarded time variable t_r. They dominate, for large r, over the terms that fall off as higher powers of $1/r$. Among these are the static terms of the multipole expansion which we have examined earlier, as well as the terms with $1/r^2$ dependence included above. A similar expression as for the magnetic field (12.45) is found for the electric field.

12.4 Radiation fields

Sufficiently far away from the sources only the field components that fall off with distance as $1/r$ give substantial contributions. They describe the *radiation field* generated by the time dependent sources. As shown by the expression (12.45) the magnetic component of the radiation field is

$$\mathbf{B}_{rad}(\mathbf{r},t) = \frac{\mu_0}{4\pi rc}\left(\ddot{\mathbf{p}}\times\mathbf{n}+\frac{1}{c}(\ddot{\mathbf{m}}\times\mathbf{n})\times\mathbf{n}+\frac{1}{2c}\dddot{\mathbf{D}}_{\mathbf{n}}\times\mathbf{n}+...\right)_{ret}\,.\tag{12.46}$$

To find the corresponding expression for the electric field we may write the **E**-field in terms of the potentials and follow the same procedure as for **B**.

However, we may make a short cut in the following way. In the far-field region, the fields can locally be regarded as describing plane waves, which propagate in the direction $\mathbf{n}$. (They are not necessarily *monochromatic* plane waves, since the Fourier transform of the time dependent multipole moments may contain more than one frequency.) As previously shown, for electromagnetic plane waves we have a simple relation between $\mathbf{E}$ and $\mathbf{B}$, which is not dependent of the frequency of the wave, $\mathbf{E} = -c\mathbf{n} \times \mathbf{B}$. In the present case the electric field therefore takes the form

$$\mathbf{E}_{rad}(\mathbf{r}, t) = \frac{\mu_0}{4\pi r}\left((\ddot{\mathbf{p}} \times \mathbf{n}) \times \mathbf{n} - \frac{1}{c}\ddot{\mathbf{m}} \times \mathbf{n} + \frac{1}{2c}(\dddot{\mathbf{D}}_{\mathbf{n}} \times \mathbf{n}) \times \mathbf{n} + ...\right)_{ret} . \tag{12.47}$$

There are two conditions which should be satisfied if the radiation fields, given by (12.46) and (12.47), should dominate the full electromagnetic field. The first one is $r >> a$, where a gives a measure of the linear size of the charge and current distribution. When this is satisfied, the expansion parameter is small, $r'/r \approx a/r << 1$, and the higher order terms in r'/r can be omitted. The other condition is $r >> \lambda$, with λ as a typical wave length of the radiation. If that is not satisfied, there are contributions to the fields where a smaller number of time derivatives of the multipole moments could compensate for a higher power in $1/r$. In particular this condition is necessary for the second set of terms in (12.45) to dominate the first set of terms. If we furthermore have the following condition satisfied, $\lambda >> a$, then the first terms of the multipole expansions of the radiation field, (12.46) and (12.47), will dominate the later ones, so that the electric dipole contribution will be more important than the electric quadrupole contribution etc.

The electric and magnetic dipole contributions may seem to be giving comparable contributions to the radiation, but under normal conditions that is not the case. The reason for this is that the magnetic moment depends on the charge *currents* and therefore on the velocity of the charges (usually electrons). This implies that the magnetic dipole term is damped by a factor v/c relative to that of the electric dipole term, with v as the (average) velocity of the charges.

Electric dipole radiation will therefore usually be dominant, for example in the radiation from an antenna. This contribution to the radiation is described by the term which depends on $\ddot{\mathbf{p}}$, and it is for short referred to as the $E1$ radiation term. However, under certain conditions this type of radiation may be suppressed so that *magnetic dipole radiation* will be dominant. This is the term depending on $\ddot{\mathbf{m}}$, with the short hand notation

$M1$. Similarly *electric quadrupole radiation*, referred to as $E2$, may also under certain conditions be important, etc.

As an interesting point to stress, the radiation field described by (12.46) and (12.47) appears as a direct consequence of the retardation effects. This has been clearly demonstrated in our derivation of the magnetic field $\mathbf{B}$ in (12.43). It is the derivative with respect to r through the retarded time t_r that gives the field contributions that fall off with distance like $1/r$, while the direct derivation of the potentials with respect to r gives field contributions that fall off like $1/r^2$.

Electric dipole radiation

When the electric dipole term dominates the radiation, the expressions for the radiation field simplify to

$$\begin{aligned}\mathbf{E}_{rad}(\mathbf{r},t) &= \frac{1}{4\pi\epsilon_0 rc^2}\,(\ddot{\mathbf{p}}_{ret}\times\mathbf{n})\times\mathbf{n}\\ \mathbf{B}_{rad}(\mathbf{r},t) &= \frac{\mu_0}{4\pi rc}\,\ddot{\mathbf{p}}_{ret}\times\mathbf{n}\,. \end{aligned}\tag{12.48}$$

Poynting's vector for this field is

$$\begin{aligned}\mathbf{S}(\mathbf{r},t) &= \frac{1}{\mu_0}\mathbf{E}_{rad}\times\mathbf{B}_{rad}\\ &= \frac{c}{\mu_0}\mathbf{B}_{rad}^2\,\mathbf{n}\\ &= \frac{\mu_0}{16\pi^2 r^2 c}(\ddot{\mathbf{p}}_{ret}\times\mathbf{n})^2\,\mathbf{n}\\ &= \frac{\mu_0}{16\pi^2 r^2 c}(\ddot{\mathbf{p}}^2\sin^2\theta)_{ret}\,\mathbf{n}\,, \end{aligned}\tag{12.49}$$

where the angle θ introduced in the last step is the angle between the vectors $\ddot{\mathbf{p}}$ and $\mathbf{n}$. The subscript *ret* is a reminder, that all variables at the source should be taken at the retarded time $t_r = t - r/c$.

Since $\mathbf{S}(\mathbf{r},t)$ gives the energy current density of the electromagnetic field, the above expression shows that the radiation is, as one should expect, directed in the radial direction $\mathbf{n}$ away from the source of the radiation. The radiation power is given as the integral of $\mathbf{S}$ over all angles,

$$\begin{aligned}P &= \frac{\mu_0}{16\pi^2 c}\ddot{\mathbf{p}}_{ret}^2\iint d\phi d\theta\sin^3\theta\\ &= \frac{\mu_0}{8\pi c}\ddot{\mathbf{p}}_{ret}^2\int_{-1}^{1}du(1-u^2)\\ &= \frac{\mu_0}{6\pi c}\ddot{\mathbf{p}}_{ret}^2\,. \end{aligned}\tag{12.50}$$

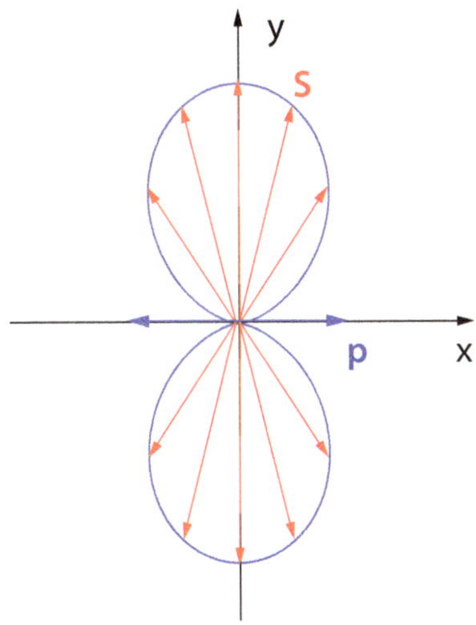

Fig. 12.4 Angular distribution of radiated energy in electric dipole radiation from a linear antenna. The electric dipole moment **p**, which here oscillates along the x-axis, gives a rotationally symmetric radiation pattern about this axis. The magnitude of the Poynting vector **S**, which determines the angular distribution of the radiated power is shown in the figure.

For radiation from a linear antenna the electric dipole moment oscillates with its direction fixed by the direction of the antenna. The angular distribution of the radiation then has the simple form

$$\mathbf{S}(\mathbf{r}, t) = \frac{\mu_0 \, \ddot{p}_{ret}^2}{16\pi^2 r^2 c} \sin^2\theta \, \mathbf{n} \,, \tag{12.51}$$

where θ now is the (time independent) angle between the directions of $\mathbf{n}$ and of the antenna. The angular distribution of the radiated energy is illustrated in Fig. 12.4, and we note in particular that the radiation is maximal in directions perpendicular to the direction of the antenna.

Let us further assume the time variation of the electric dipole moment of the antenna to have a simple harmonic form,

$$p(t) = p_0 \cos \omega t \,, \tag{12.52}$$

with oscillation period $T = 2\pi/\omega$. The expression for the time averaged radiation power from the antenna is then

$$\bar{P} = \frac{1}{T} \int_0^T P(t) dt$$

$$= \frac{\mu_0 p_0^2 \omega^4}{6\pi c T} \int_0^T \cos^2 \omega t \, dt$$

$$= \frac{\mu_0 \, p_0^2 \, \omega^4}{12\pi c} \, . \tag{12.53}$$

We note in particular that, for fixed p_0, the radiated power increases strongly with the frequency of the oscillating dipole moment.

Example

Electric dipole radiation from a linear antenna

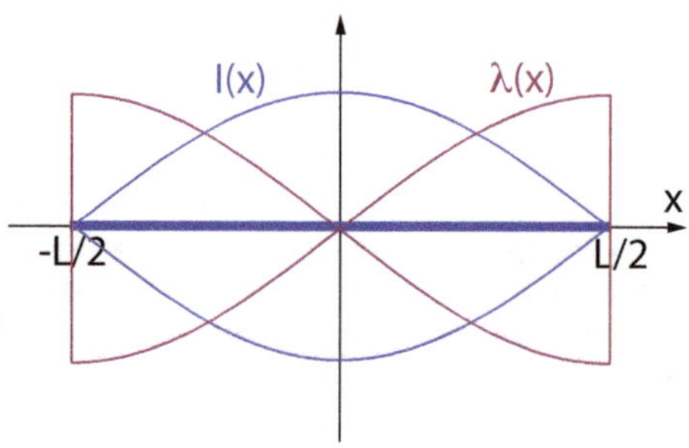

Fig. 12.5 Oscillating current and charge in a linear antenna. The figure shows the current $I(x)$ and charge density $\lambda(x)$ along the antenna, where the current here has a cosine form and the charge density a sine form as functions of x. They both oscillate in time, with a phase shift of $\pi/2$, so that the charge density vanishes when the current has its maximum and *vice versa*.

Let us assume a linear antenna of length L is directed along the x-axis as illustrated in Fig. 12.5. Let us further assume an oscillating current is induced in the antenna, of the form

$$I(x,t) = I_0 \cos\left(\frac{x}{L}\pi\right) \cos \omega t \, . \tag{12.54}$$

The x-dependence of the current shows that it has its maximum at the midpoint of the antenna and that it vanishes, as it should, at the endpoints. Charge conservation now gives a connection between the space variation in current and the time variation in the charge density, which has the form

$$\frac{\partial I}{\partial x} + \frac{\partial \lambda}{\partial t} = 0 \tag{12.55}$$

where λ is the *linear* charge density, *i.e.*, the charge per unit length along the antenna. The equation is the one-dimensional form of the continuity equation for the charge, which we earlier have formulated as an equation in three space dimensions.

The electric dipole moment can, in this case, be expressed as a one dimensional integral along the antenna,

$$p(t) = \int_{-\frac{L}{2}}^{\frac{L}{2}} \lambda(x,t)\, x\, dx\,, \tag{12.56}$$

with direction $\mathbf{p} = p\,\mathbf{i}$ along the x-axis. For the time derivative we find

$$\begin{aligned}
\dot{p} &= \int_{-\frac{L}{2}}^{\frac{L}{2}} \frac{\partial \lambda}{\partial t}\, x\, dx \\
&= -\int_{-\frac{L}{2}}^{\frac{L}{2}} \frac{\partial I}{\partial x}\, x\, dx \\
&= -\int_{-\frac{L}{2}}^{\frac{L}{2}} \frac{\partial}{\partial x}(\, xI)\, dx + \int_{-\frac{L}{2}}^{\frac{L}{2}} I\, dx \\
&= \int_{-\frac{L}{2}}^{\frac{L}{2}} I\, dx \\
&= \frac{2L}{\pi} I_0 \cos\omega t\,.
\end{aligned} \tag{12.57}$$

The corresponding expression for the oscillating dipole moment is

$$p(t) = p_0 \sin\omega t\,, \tag{12.58}$$

with $p_0 = 2L/\pi\omega$. The double time derivative of the dipole moment, which is needed for the radiation formula, is

$$\ddot{p}(t) = -p_0\omega^2 \sin\omega t = \frac{2L}{\pi}\omega I_0 \sin\omega t\,. \tag{12.59}$$

The formula for the radiation power now gives

$$\begin{aligned}
P(t) &= \frac{\mu_0}{6\pi c}\, \ddot{p}_{ret}^2 \\
&= \frac{\mu_0}{6\pi c}\, p_0^2\, \omega^4 \sin^2 \omega t_r\,, \qquad t_r = t - r/c\,,
\end{aligned} \tag{12.60}$$

which, when expressed in terms of the current amplitude, is

$$P(t) = \frac{2\mu_0 L^2\omega^2 I_0^2}{3\pi^3 c}\, \sin^2 \omega t_r\,. \tag{12.61}$$

For the time average of the power this gives

$$\bar{P} = \frac{\mu_0 L^2 \omega^2 I_0^2}{3\pi^3 c}\,, \tag{12.62}$$

since the average value of $\sin^2 \omega t$ is $1/2$. This result is consistent with what has earlier been presented in (12.53), which can easily be shown by use of the relation between the current amplitude I_0 and the electric dipole amplitude p_0.

Let us finally consider the polarization of the radiation as it is measured by a receiver. As we already know, both $\mathbf{E}$ and $\mathbf{B}$ are orthogonal to the direction $\mathbf{n}$ of the wave propagation, from the antenna to the receiver. Since the dipole moment oscillates in strength but not in direction, the general expressions for the electric and magnetic fields produced in electric dipole radiation, (12.48), shows that $\mathbf{B}$ will be oscillating along the fixed line $\mathbf{i} \times \mathbf{n}$ and $\mathbf{E}$ along the fixed line $(\mathbf{i} \times \mathbf{n}) \times \mathbf{n}$. This means that the radiation field will, for any direction $\mathbf{n}$, be *linearly* polarized.

12.5 Larmor's radiation formula

As a last point we shall consider radiation from an accelerated point charge. The fields produced by a moving point charge has previously been given in the form of the Liénard-Wiechert potentials (see (12.28) and (12.29)). We consider the non-relativistic limit of these potentials, which corresponds to $\beta = v/c \to 0$. This gives

$$\Phi(\mathbf{r}, t) = \frac{q}{4\pi\epsilon_0 R_{ret}}\,, \qquad \mathbf{A}(\mathbf{r}, t) = \frac{\mu_0 q}{4\pi}\left(\frac{\mathbf{v}}{R}\right)_{ret}\,, \tag{12.63}$$

where R is the relative distance between the chosen point $\mathbf{r}$ and the position of the point, and $\mathbf{r}_p$ is the velocity of the charge. The retarded time in the expressions is given by

$$t_- = t - \frac{1}{c}|\mathbf{r} - \mathbf{r}_p(t_-)|\,. \tag{12.64}$$

We will be interested in the radiation fields, which depend on R as $1/R$. As earlier discussed, the relevant contributions to the radiation fields come from the retardation effect. We focus first on the magnetic field,

$$\mathbf{B}(\mathbf{r}, t) = \boldsymbol{\nabla} \times \mathbf{A} = \boldsymbol{\nabla} t_- \times \frac{\partial \mathbf{A}}{\partial t_-} + \mathcal{O}(1/R^2)\,, \tag{12.65}$$

where the gradient of t_- is

$$\begin{aligned}\nabla t_- &= \nabla\left(t - \frac{1}{c}|\mathbf{r} - \mathbf{r}_p(t_-)|\right) \\ &= \frac{1}{c}\left(\frac{\mathbf{R}}{R}\right)_{ret} + \left(\frac{\mathbf{R}}{R}\cdot\boldsymbol{\beta}\right)_{ret}\nabla t_- \\ &\approx \frac{1}{c}\mathbf{n}_{ret}\,, \quad \mathbf{n}_{ret} = \left(\frac{\mathbf{R}}{R}\right)_{ret}.\end{aligned} \tag{12.66}$$

In the last step we have made the non-relativistic approximation $\boldsymbol{\beta} \approx 0$. This gives for the magnetic field

$$\begin{aligned}\mathbf{B}(\mathbf{r}, t) &= \frac{1}{c}\frac{\partial \mathbf{A}}{\partial t_-} \times \mathbf{n}_{ret} + \mathcal{O}(1/R^2) \\ &= \frac{\mu_0 q}{4\pi c}\left(\frac{\mathbf{a}\times\mathbf{n}}{R}\right)_{ret} + \mathcal{O}(1/R^2)\,.\end{aligned} \tag{12.67}$$

Since the electric component of the radiation field is determined by the magnetic field as $\mathbf{E} = c\mathbf{B}(\mathbf{r}, t) \times \mathbf{n}$, we find for the radiation fields from the point charge the following expressions

$$\begin{aligned}\mathbf{B}_{rad}(\mathbf{r}, t) &= \left[\frac{q\mu_0}{4\pi Rc}\mathbf{a}\times\mathbf{n}\right]_{ret}, \\ \mathbf{E}_{rad}(\mathbf{r}, t) &= \left[\frac{q\mu_0}{4\pi R}(\mathbf{a}\times\mathbf{n})\times\mathbf{n}\right]_{ret}.\end{aligned} \tag{12.68}$$

In these expressions we have

$$\mathbf{R}(t) = \mathbf{r} - \mathbf{r}_p(t)\,, \quad \mathbf{n} = \frac{\mathbf{R}}{R}\,, \quad \mathbf{a}(t) = \ddot{\mathbf{r}}_p(t)\,, \tag{12.69}$$

with $\mathbf{r}$ as the position where the fields are evaluated, and $\mathbf{r}_p(t)$ as the time dependent position vector of the moving charge. Note that in Eq. (12.68) the retarded time is measured relative to the position of the moving charge,

$$t_- = t - \frac{1}{c}|\mathbf{r} - \mathbf{r}_p(t_-)|\,. \tag{12.70}$$

The fields given above have the same form as the electric dipole radiation fields previously found. Thus, for a point charge the dipole moment is $\mathbf{p}(t) = q\mathbf{r}_p(t)$ and therefore $\ddot{\mathbf{p}} = q\mathbf{a}$. There is one difference, since R and $\mathbf{n}$ here depend on the time dependent position of the charge. However, when sufficiently far from the charge this time dependence is less important. The lack of higher multipole contributions is clearly a consequence of the point-like charge distribution of the moving charge.

Poynting's vector is of the same form as for electric dipole radiation

$$\mathbf{S}(\mathbf{r}, t) = \frac{q^2\mu_0}{16\pi^2 r^2 c}[a^2 \sin^2\theta\,\mathbf{n}]_{ret}\,, \tag{12.71}$$

where θ is the angle between the direction of the acceleration $\mathbf{a}$ and the direction vector $\mathbf{n}$. The corresponding formula for the radiation power is

$$P = \frac{\mu_0 q^2}{6\pi c}\, a^2 \,. \tag{12.72}$$

This is called the Larmor radiation formula. Strictly speaking, as follows from our derivation, P in this equation should be interpreted as the radiation power measured in the far field, with a as the acceleration at the retarded time relative to the measurement. However, a reinterpretation of the equation is possible, with P regarded as the power transferred from the charge to the electromagnetic field at the instant of the acceleration a. This disregards the intermediate effects of the near field. It simply takes for granted that the power measured at time t in the far field is the same as the power transferred from the charge to the near field at the retarded time.

The expression given above thus gives a simple picture of the radiation process. At a given time along the space-time trajectory of the charge, the charge will radiate energy at a rate proportionally to the square of the acceleration of the charge. The energy emitted in a time interval dt will then propagate with the speed of light radially outwards from the charge, like an expanding spherical shell. The time delay when the shell moves outwards is the origin of the retardation effect. When the charge moves, the center of these shells of energy will continuously change, so that when viewed from a fixed point in space the radiation is at any time directed away from the position of the charge at the retarded time.

The derivation of the formula (12.72) is based on the non-relativistic approximation of the radiation fields (12.68). This implies that in a relativistic setting the formula is, strictly speaking, correct only in the instantaneous rest frame of the accelerated charge. However, we will as the final point discuss how it can be generalized to a relativistic formula. The derivation is based on the fact that the radiation power P is a Lorentz invariant quantity.

To give a simple argument for P to be Lorentz invariant, we consider the radiation emitted from the charge in a short time interval dt. We view this both in an inertial frame where the charge is moving, and in the instantaneous inertial rest frame of the charge, where we refer to the time interval as $d\tau$. The emitted energy is initially restricted to a small volume around the (point-like) charge. The emitted field energy and momentum, measured in the two reference frames, are related by the Lorentz transformation

$$dE = \gamma(dE_0 - \beta c\, dp_0)\,, \quad c\, dp = \gamma(c\, dp_0 - \beta dE)\,, \tag{12.73}$$

where dE_0 and $d\mathbf{p}_0$ are the energy and momentum in the instantaneous inertial rest frame. In the transformation formula only the component of the latter in the direction of the relative velocity of the two reference frames appears. However, the emitted field momentum in the given time interval vanishes, $d\mathbf{p}_0 = 0$, as follows from the symmetry of the emitted energy in electric dipole radiation, shown in Fig. 12.4. From this follows the simple relation $dE = \gamma dE_0$, and due to the time dilation formula $dt = \gamma d\tau$, the invariance of the radiation power follows

$$P = \frac{dE}{dt} = \frac{dE_0}{d\tau}\,. \tag{12.74}$$

The Larmor radiation formula, in relativistic form is now found simply by replacing the acceleration a with the proper acceleration a_0, and by further expressing a_0 in terms of the acceleration $\mathbf{a}$ and velocity $\mathbf{v}$ of the charge in an unspecified inertial frame (see (7.18)), we get

$$P = \frac{\mu_0 q^2}{6\pi c}\, a_0^2 = \frac{\mu_0 q^2}{6\pi c}\left(\gamma^4 a^2 + \gamma^6 \frac{(\mathbf{v}\cdot\mathbf{a})^2}{c^2}\right). \tag{12.75}$$

Example

Synchrotron radiation

We will here study *synchrotron radiation* from an accelerated charged particle. This is radiation where the acceleration of the particle is perpendicular to its velocity, for example for particles in cyclic accelerators. The acceleration is then caused by a strong magnetic field perpendicular to the plane of the particle orbit. To be more concrete we consider the situation in the Large Hadron Collider (LHC) at CERN, where protons circulate with ultrarelativistic velocity in the accelerator ring. The bending radius caused by strong magnets is $R = 2804$ m, and with particle momentum $p = 7.0$ TeV, the gamma factor of the circulating protons is $\gamma = 7463$.

The Larmor formula in this case gives for the radiation power

$$P = \gamma^4 \frac{\mu_0 e^2}{6\pi c} a^2 = \gamma^4 \frac{e^2 c}{6\pi\epsilon_0 R^2}\,, \tag{12.76}$$

where we have approximated the proton velocity by the speed of light. The radiation energy loss per particle, under one period of circulation, is then

$$\Delta\mathcal{E} = \frac{2\pi R}{c} P = \gamma^4 \frac{e^2}{3\epsilon_0 R}\,. \tag{12.77}$$

We calculate the following expression

$$\gamma^4 \frac{e}{3\epsilon_0 R} = 7463^4 \cdot \frac{1.6\cdot 10^{-19}}{3\cdot 8.85\cdot 10^{-12}\cdot 2804}\,\mathrm{V} = 6.7\cdot 10^3\,\mathrm{V}\,. \tag{12.78}$$

This gives the following value for the energy loss, expressed in electron volt (eV),

$$\Delta\mathcal{E} = 6.7\,\text{keV}\,. \tag{12.79}$$

When viewed in the instantaneous inertial rest frame S' of the particle, the angular distribution of the radiation energy is given by Poynting's vector, as shown in (12.71). The distribution is symmetric around the axis defined by the acceleration **a**, with a minimum along this axis and a maximum in the plane orthogonal to the symmetry axis. For the discussion to follow it is convenient to choose the z-axis in the instantaneous direction of motion of the particle and the x-axis in the plane of motion in the direction opposite to **a**. With this choice of coordinate axes the Poynting vector (12.71), now takes the form

$$\mathbf{S}'(\mathbf{r},t) = \frac{\mu_0 e^2}{16\pi^2 r^2 c} a_0^2 (1 - \cos^2\phi' \sin^2\theta')\,\mathbf{n}'\,, \tag{12.80}$$

where a_0 is the proper acceleration and $\mathbf{n}'$ is a unit vector with ϕ' and θ' as the corresponding polar angles. We will show that the picture this expression gives of the radiation will change in an essential way when viewed in the lab frame. Thus, the transformation from the rest frame to the lab frame will change the energy distribution to be concentrated to within a small angle in the forward direction of the moving particle.

To transform from the rest frame S' to the lab frame S we start with the Lorentz transformation, which here takes the form

$$ct = \gamma(ct' + \beta z') \quad z = \gamma(z' + \beta ct')\,, \quad x = x'\,, \quad y = y'\,. \tag{12.81}$$

We will apply this to the radiation that is emitted from the charge at the point corresponding to the origin of the two reference systems, $t' = t = 0$ and $r' = r = 0$. The light will propagate in radial directions, with coordinates given by $r' - ct' = r - ct = 0$. What is of interest is to relate the polar coordinates of light rays in the two reference frames. (Note that this transformation is essentially the same as we have earlier studied in Sect. 7.4, on *Doppler effect with photons*).

With the relation above between the time and radial coordinates included in the Lorentz transformation equations, these take the form

$$\begin{aligned} r &= \gamma r'(1 + \beta\cos\theta')\,, \\ r\cos\theta &= \gamma r'(\cos\theta' + \beta)\,, \\ r\cos\phi\sin\theta &= r'\cos\phi'\sin\theta'\,, \\ r\sin\phi\sin\theta &= r'\sin\phi'\sin\theta'\,. \end{aligned} \tag{12.82}$$

The two last equations show that $\tan\phi = \tan\phi'$, which implies $\phi = \phi'$, and from this follows

$$r\sin\theta = r'\sin\theta' \,. \tag{12.83}$$

We eliminate the radial variables, and find

$$\cos\theta = \frac{\cos\theta' + \beta}{1 + \beta\cos\theta'}\,, \quad \sin\theta = \frac{1}{\gamma}\frac{\sin\theta'}{1 + \beta\cos\theta'}\,. \tag{12.84}$$

The last equation shows that, due to the small value of the factor $1/\gamma$, the angle θ will be very small, unless θ' is extremely close to 180°, which corresponds to the direction opposite of the motion of the particle. This means that almost all the radiation will, in the lab frame be emitted in a small cone around the forward direction of the moving particle. Here the small angle approximation of θ is valid, which gives

$$\theta \approx \frac{1}{\gamma}\frac{\sin\theta'}{1 + \cos\theta'} = \frac{1}{\gamma}\tan(\theta'/2)\,. \tag{12.85}$$

Thus, almost all of the radiation is emitted within a very narrow cone with opening angle $\Delta\theta \approx 1/\gamma$.

To show this more explicitly, we consider the mapping of the radiation power from the instantaneous rest frame to the lab frame. In the rest frame the expression for the power per unit solid angle is

$$\frac{dP'}{d\Omega} = S'r^2 = \frac{\mu_0 e^2}{16\pi^2 c}a_0^2(1 - \cos^2\phi'\sin^2\theta')\,. \tag{12.86}$$

As previously discussed, the (integrated) power $P = \int \frac{dP}{d\Omega}d\Omega$ is a Lorentz invariant. For the transformation of the differential power this means

$$\frac{dP}{d\Omega}(\theta,\phi)d\Omega = \frac{dP'}{d\Omega'}(\theta',\phi')d\Omega'\,. \tag{12.87}$$

We write $d\Omega = \sin\theta d\theta d\phi = d(\cos\theta)d\phi$, and with the angle ϕ invariant under the transformation, the equation above can be re-written as

$$\frac{dP}{d\Omega}(\theta,\phi) = \frac{dP'}{d\Omega'}(\theta',\phi)\,\frac{d(\cos\theta')}{d(\cos\theta)}\,. \tag{12.88}$$

The right-hand side can be expressed as function of θ by use of the equations (12.84). Note that these equations can be inverted, $\theta \leftrightarrow \theta'$, simply by changing the sign $\beta \to -\beta$. We get from this

$$\sin^2\theta' = \frac{1}{\gamma^2}\frac{\sin^2\theta}{(1 - \beta\cos\theta)^2} \tag{12.89}$$

and

$$\frac{d(\cos\theta')}{d(\cos\theta)} = \frac{1}{1-\beta\cos\theta} + \beta\frac{1}{(1-\beta\cos\theta)^2}$$
$$= \frac{1}{\gamma^2(1-\beta\cos\theta)^2}. \tag{12.90}$$

Inserting this in (12.87), and using the fact that $\cos\phi = \cos\phi'$, we get for the differential power in the lab frame,

$$\frac{dP}{d\Omega}(\theta,\phi) = \frac{\mu_0 e^2}{16\pi^2 c}\gamma^4 a^2\left(1-\frac{1}{\gamma^2}\frac{\cos^2\phi\sin^2\theta}{(1-\beta\cos\theta)^2}\right)\frac{1}{\gamma^2}\frac{1}{(1-\beta\cos\theta)^2}$$
$$= \frac{\mu_0 e^2 a^2}{16\pi^2 c}\frac{\gamma^2}{(1-\beta\cos\theta)^2}\left(1-\frac{1}{\gamma^2}\frac{\cos^2\phi\sin^2\theta}{(1-\beta\cos\theta)^2}\right), \tag{12.91}$$

where a now is the acceleration in the lab frame. For large γ and small θ we make the approximations $\beta \approx 1 - \frac{1}{2\gamma^2}$, $\sin\theta \approx \theta$ and $\cos\theta \approx 1-\theta^2/2$. If we further average over the angle ϕ, which implies $\cos^2\phi \to 1/2$, the expression above can be simplified to

$$\frac{d\overline{P}}{d\Omega}(\theta) = \frac{\mu_0 e^2 a^2}{4\pi^2 c}\frac{\gamma^6(1+\gamma^4\theta^4)}{(1+\gamma^2\theta^2)^4}. \tag{12.92}$$

The form of the function, which is illustrated in Fig. 12.6a, has a sharply defined peak in the forward direction, with the width of the peak $\Delta\theta \approx 1/\gamma$, in accordance with what we have already shown.

These results show that the radiation from the accelerated proton is emitted in a narrowly defined cone, tangentially to the particle orbit. However, in the lab frame the proton moves essentially with the same speed as the emitted radiation. This implies that the radiation cone will rotate so rapidly that the radiation will be sprayed in a spiral-like curve from the particle's orbit. This is illustrated in Fig. 12.6b, where the radiation is restricted to a thin thread, which seems attached to the moving charge.

12.6 Exercises

Problem 12.1

A so-called *half-wave center-fed* antenna is formed by a linear conductor of length a. It is oriented along the z-axis as shown in the Fig. 12.7. An alternating current, which is running in the antenna, is of the form

$$I(z,t) = I_0\cos\left(\frac{\pi z}{a}\right)\cos\omega t\,, \quad -a/2 < z < a/2\,. \tag{12.93}$$

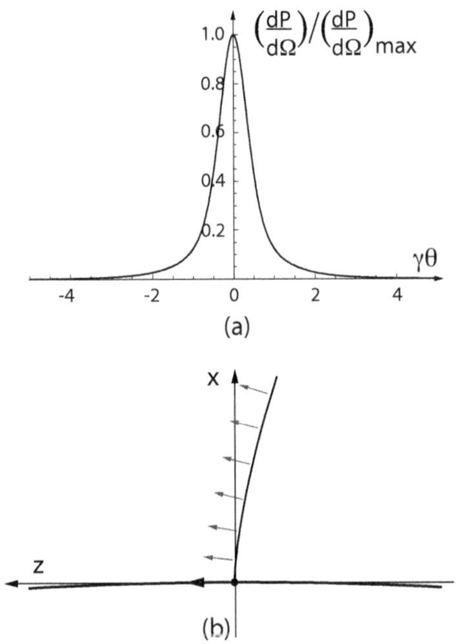

Fig. 12.6 Synchrotron radiation for an ultra-relativistic charged particle. Fig. 12.6a shows the power distribution of the radiation around the forward direction of the particle. Fig. 12.6b shows an instantaneous picture of the location of the radiation. The radiation is restricted to a thin, curved thread which emanates from the charge, which here is located at the origin. The direction of propagation of the radiation is shown by the arrows. A part of the curved trajectory of the particle is also shown, close to the z-axis in the figure.

In the following $\lambda(z,t)$ denotes the linear charge density of the antenna (charge per unit length). At time $t = 0$ the antenna is charge neutral, so that $\lambda(z,0) = 0$.

a) Show that the charge density and current satisfy the relation

$$\frac{\partial \lambda}{\partial t} + \frac{\partial I}{\partial z} = 0\,, \tag{12.94}$$

and find λ as a function of z and t.

b) Show that the electric dipole moment of the antenna has the form

$$\mathbf{p}(t) = p_0 \sin\omega t\,\mathbf{k}\,, \tag{12.95}$$

with $\mathbf{k}$ as the unit vector along the z-axis, and determine the constant p_0.

c) Use the expressions for electric dipole radiation to determine the electric and magnetic fields at a point with large distance r from the antenna, and

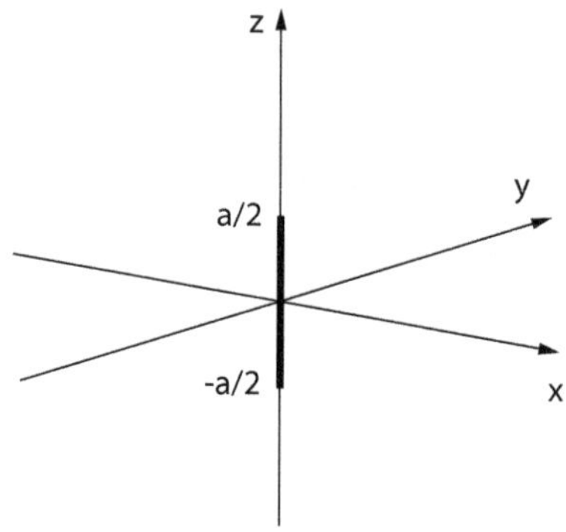

Fig. 12.7 The antenna in Problem 12.1.

localized on the x-axis. What is the type of polarization of the radiation from the antenna in this direction?

Problem 12.2

In a simple, classical model of the hydrogen atom, the negatively charged electron moves in a circular orbit around the positively charged atomic nucleus (a proton). The electron has a small mass $m_e = 9.1 \cdot 10^{-31}\,\text{kg}$ and a charge $e = -1.60 \cdot 10^{-19}\,\text{C}$. The nucleus is much heavier, and we can regard it as having, all the time, a fixed position. The charge of the nucleus is equally large as that of the electron, but with opposite sign. The radius of the electron's orbit we regard to be equal to the Bohr radius $a_0 = 0.53 \cdot 10^{-10}\,\text{m}$. We take the orbit plane to be the x, y-plane, with the atomic nucleus placed at the origin, as shown in Fig. 12.8. Both the electron and the nucleus we treat as point particles. We further disregard the intrinsic spin of the particles, and assume that the motion can be treated non-relativistically.

The vacuum permittivity has the value $\epsilon_0 = 8.85 \cdot 10^{-12}\,\text{C}^2\text{N}^{-1}\text{m}^{-2}$, the vacuum permeability is $\mu_0 = 4\pi \cdot 10^{-7}\,\text{N/A}^2$, and the speed of light is $c = 3.00 \cdot 10^8\,\text{m/s}$.

a) Disregard first the electromagnetic radiation from the electron. Find the angular velocity ω of the electron when it moves in the Coulomb potential of the atomic nucleus.

b) Find next, by use of Larmor's formula, the radiation power.

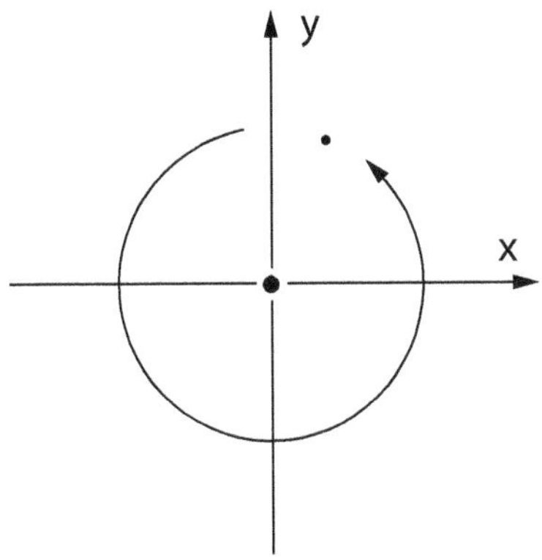

Fig. 12.8 A classical atomic model. The electron moves in a circular orbit around the atomic nucleus.

c) Energy conservation implies that the radiated energy has to be accompanied by a reduction of the (kinetic and potential) energy of the electron. Make the assumption that the energy loss gives rise to a slow reduction of the radius of the electron's orbit. Determine the rate at which the radius is reduced when $r = a_0$, and make from this an estimate of the life time of this classical atom.

Problem12.3

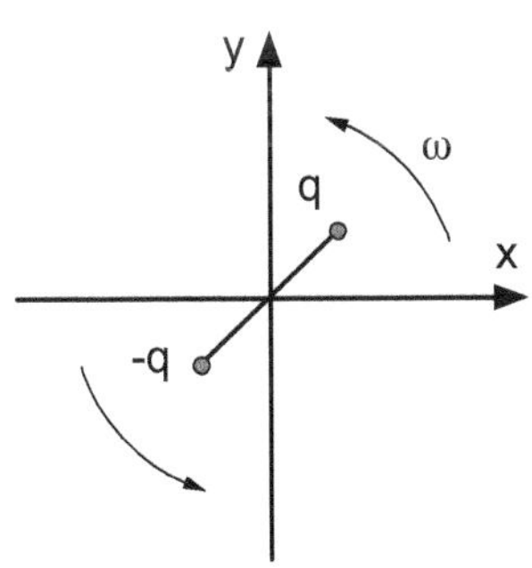

Fig. 12.9 A rotating electric dipole.

A thin rigid rod of length l rotates in a horizontal plane (the x, y-plane) as shown in Fig. 12.9. At the two end points there are fixed charges with opposite signs, $+q$ and $-q$. The rod is rotating with constant angular frequency ω. This gives rise to a time dependent electric dipole moment

$$\mathbf{p}(t) = ql(\cos\omega t\,\mathbf{i} + \sin\omega t\,\mathbf{j})\,. \tag{12.96}$$

a) Use the general expression for the radiation fields of an electric dipole to show that the magnetic field in the present case can be written as

$$\begin{aligned}\mathbf{B}(\mathbf{r},t) = B_0(r)\ \Big[&\cos\theta\sin\left(\omega\left(t-\frac{r}{c}\right)\right)\mathbf{i} - \cos\theta\cos\left(\omega\left(t-\frac{r}{c}\right)\right)\mathbf{j}\\ &- \sin\theta\sin\left(\omega\left(t-\frac{r}{c}\right)-\phi\right)\mathbf{k}\Big]\,,\end{aligned} \tag{12.97}$$

with (r,θ,ϕ) as the polar coordinates of $\mathbf{r}$. Find the expression for $B_0(r)$.

What is the general relation between the electric component $\mathbf{E}(\mathbf{r},t)$ and the magnetic component $\mathbf{B}(\mathbf{r},t)$ of the radiation field?

b) Show that radiation in the x-direction is linearly polarized. What is the polarization of the radiation in the z-direction?

c) Find the time-averaged expression for the energy density of the radiation. In what direction has the radiated energy its maximum?

Problem 12.4

In a circular loop of radius a an oscillating current of the form $I = I_0\cos\omega t$ is running. The current loop lies in the x,y-plane, with the center of the loop at the origin. The loop is at all times charge neutral. We assume $a\,\omega << c$.

a) Show that the magnetic dipole moment has the following time dependence, $\mathbf{m}(t) = m_0\cos\omega t\,\mathbf{e}_z$, with m_0 as a constant and $\mathbf{e}_z$ as a unit vector in the z-direction. Find m_0 expressed in terms of a and I_0. Far from the current loop, the radiation fields are dominated by the magnetic dipole radiation field. Explain why.

b) The magnetic dipole radiation fields have the general form

$$\mathbf{E}(\mathbf{r},t) = -\frac{\mu_0}{4\pi c r}\ddot{\mathbf{m}}_{ret}\times\mathbf{n}\,,\quad \mathbf{B}(\mathbf{r},t) = -\frac{1}{c}\mathbf{E}(\mathbf{r},t)\times\mathbf{n}\,, \tag{12.98}$$

with $\mathbf{m}_{ret} = \mathbf{m}(t-r/c)$, $\mathbf{n} = \mathbf{r}/r$, and $r >> a$. For the present case give the full expressions of the fields, as functions of r and t, and written in terms of the orthonormalized vectors $\{\mathbf{n} = \mathbf{e}_r, \mathbf{e}_\theta, \mathbf{e}_\phi\}$ of the polar coordinate system. Do they form, as expected, waves that propagate in the radial direction away from the loop? Explain. Characterize the polarization of the waves.

c) Use the general expression for Poynting's vector $\mathbf{S}$ to find the radiation power per unit solid angle $\frac{dP}{d\Omega}$, expressed as a function of the polar angle θ (angle relative to the z-axis). Find the total power, integrated over all directions, and averaged over time.

d) A second conducting loop, identical to the first one, is placed at a large distance r from the first loop, with the center of the loop in the x, y-plane. It is used as a receiver, with the radiation from the first loop inducing a current in the second loop. Let $\mathbf{e}$ be a unit vector orthogonal to the plane of the second loop. In what direction should $\mathbf{e}$ be oriented for the second loop to receive the maximal signal?

Problem 12.5

An electron with charge e and mass m is moving with constant speed in a circle under the influence of a constant magnetic field $\mathbf{B}_0$. The magnetic field is directed along the z-axis while the motion of the electron takes place in the x, y-plane. We assume the motion of the electron to be non-relativistic.

a) Since the electron is accelerated it will radiate electromagnetic energy and thereby lose kinetic energy when no energy is added to the particle. By use of Larmor's radiation formula, find an expression for the radiated energy per unit time expressed in terms of the radius r of the electron orbit and the cyclotron frequency $\omega = -eB_0/m$.

b) Show that the radius of the electron orbit is slowly reduced with an exponential time dependence, $r = r_0\, e^{-\lambda t}$, and determine λ.

Problem 12.6

An antenna is composed of two parts, as shown in Fig. 12.10. One part is a linear antenna along the z-axis, with end points $z = \pm a/2$. It carries the current

$$I_1 = I_0 \sin\omega t \cos\left(\frac{\pi z}{a}\right). \tag{12.99}$$

The other part is a circular antenna, which lies in the x, y-plane, and is centered at the origin of the coordinate system. It has radius $2a$ and carries the current

$$I_2 = I_0 \sin\omega t\,. \tag{12.100}$$

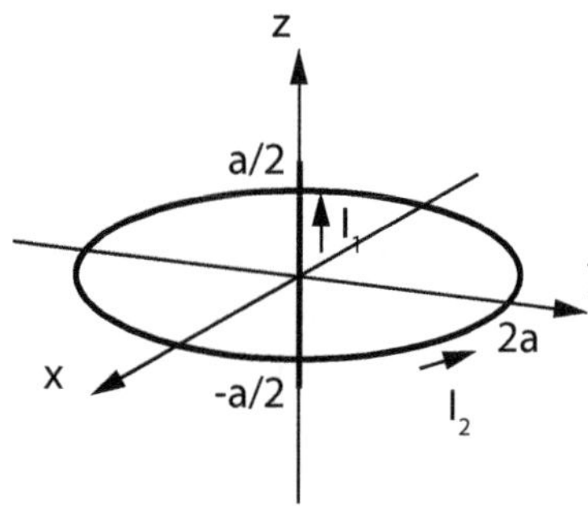

Fig. 12.10 A composite antenna, with a circular and a linear part.

The linear charge density $\lambda(z,t)$ of the linear antenna is determined by the continuity equation for charge

$$\frac{\partial \lambda}{\partial t} + \frac{\partial I_1}{\partial z} = 0\,, \qquad (12.101)$$

while the circular one is all the time charge neutral.

We assume that the radiation from the antenna is dominated by electric and magnetic dipole radiation.

a) Show that the time derivative of the electric dipole moment is given by

$$\dot{\mathbf{p}} = \frac{2}{\pi} a I_0 \sin\omega t\, \mathbf{k}\,, \qquad (12.102)$$

and the magnetic dipole moment is

$$\mathbf{m} = 4\pi a^2 I_0 \sin\omega t\, \mathbf{k}\,. \qquad (12.103)$$

As a reminder, the dipole contributions to the electric and magnetic radiation fields are

$$\begin{aligned} \mathbf{E}(\mathbf{r},t) &= \frac{\mu_0}{4\pi r}\left((\ddot{\mathbf{p}} \times \mathbf{n}) \times \mathbf{n} - \frac{1}{c}\ddot{\mathbf{m}} \times \mathbf{n} + ...\right)_{ret} \\ \mathbf{B}(\mathbf{r},t) &= -\frac{1}{c}\mathbf{E}(\mathbf{r},t) \times \mathbf{n}\,, \end{aligned} \qquad (12.104)$$

with $\mathbf{n} = \mathbf{r}/r$.

b) Assume the frequency ω is chosen so that the time average of the power of the electric and magnetic dipole radiation from the antenna are equal. What is then the angular frequency ω? Find for this case the radiation power per unit solid angle, $\frac{dP}{d\Omega}$, expressed as a function of the angle θ between the vector $\mathbf{n}$ and the z-axis.

c) What is in this case the polarization of the radiation? If the frequency ω changes so that the time average of the power of the electric and magnetic dipole radiation are no longer equal, how would that influence the polarization?

Problem 12.7

An electromagnetic plane wave has electric field components

$$E_x = E_0 \cos(kz - \omega t)\,, \quad E_y = E_z = 0\,. \tag{12.105}$$

An electron, with charge e and mass m, performs oscillations in the field.

a) Give the expressions for the components of the magnetic field $\mathbf{B}$ and determine Poynting's vector $\mathbf{S}$ of the plane wave.

b) Assume the motion of the electron, within a good approximation, is given by

$$\dot{x} = -\frac{eE_0}{m\omega}\sin(kz - \omega t)\,, \quad \dot{y} = \dot{z} = 0\,, \tag{12.106}$$

with (x, y, z) as the electron coordinates. Specify the conditions for this approximation to be valid.

c) The interaction cross section σ between the wave and the electron is defined as the power of the radiation energy from the electron, divided by the energy current density of the incoming wave, both averaged over time. Determine the cross section σ, with the electron motion given by the above approximation.

d) Determine the time averaged, differential power $\frac{dP}{d\Omega}$ of the radiation from the electron. Specify what are directions with maximal and minimal radiation. Characterize also the polarization of the radiation.

We give the following (non-relativistic) expressions for the radiation power (Larmor's formula),

$$P = \frac{\mu_0 e^2}{6\pi c}\mathbf{a}^2\,, \tag{12.107}$$

and for the radiation fields from the accelerated charge e,

$$\mathbf{B}_{rad}(\mathbf{r}, t) = \frac{\mu_0 e}{4\pi c}\left[\frac{\mathbf{a} \times \mathbf{n}}{R}\right]_{ret}\,, \quad \mathbf{E}_{rad}(\mathbf{r}, t) = c\mathbf{B}_{rad}(\mathbf{r}, t) \times \mathbf{n}_{ret}\,. \tag{12.108}$$

Here $\mathbf{a}$ is the acceleration of electron, $\mathbf{n} = \mathbf{R}/R$, with $\mathbf{R} = \mathbf{r} - \mathbf{r}(t)$, as the vector between the point $\mathbf{r}$ where the radiation fields are measured, and the position $\mathbf{r}(t)$ of the electron. Far from the charge we may use the approximation $\mathbf{R} = \mathbf{r}$.

Problem 12.8

Figure 12.11 shows a linear antenna of length $2a$ lying along the z-axis with its center at the origin. We assume that the charge of the antenna is at all times located at the endpoints. The current in the antenna (between the charged end points) is given by $I = I_0 \sin \omega t$, where ω and I_0 are constants. The antenna is electrical neutral at time $t = 0$. The point A where the field is evaluated is given by the position vector $\mathbf{r}$, expressed in spherical coordinates as (r, θ, ϕ). The corresponding orthonormal vector basis is $\{\mathbf{e}_r, \mathbf{e}_\theta, \mathbf{e}_\phi\}$. We will in the following assume that the point A is far away from the antenna, $r >> a$, and that the radiation from the antenna can be treated as electric dipole radiation.

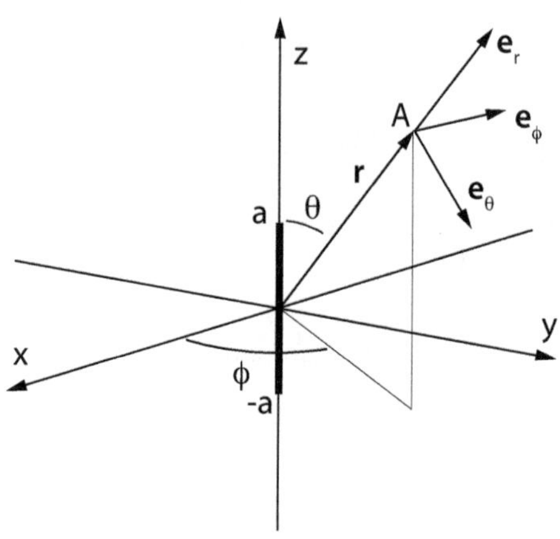

Fig. 12.11 The linear antenna in Problem 12.8.

a) Show that the antenna's electrical dipole moment at time t is given by

$$\mathbf{p}(t) = \frac{2aI_0}{\omega}(1 - \cos \omega t)\mathbf{k}\,, \tag{12.109}$$

where $\mathbf{k}$ is the unit vector in the z-direction.

b) Find the radiation fields $\mathbf{B}$ and $\mathbf{E}$ at point A, expressed in the spherical vector basis, and written as functions of time t. What is the polarization of the radiation field?

c) Find the expression for Poynting's vector. Show that the time average of the radiated power in all directions can be written as $\bar{P} = \frac{1}{2}RI_0^2$ and determine R (the radiation resistance). What is the time average of the total power consumed by the antenna if it has an 'ordinary' resistance R_0 as well?

d) Find the radiation resistance R for an antenna of length $2a = 5\,\mathrm{cm}$, which is conducting a current with frequency $f = 150\,\mathrm{MHz}$. What is the time average of the radiation power when $I_0 = 30\,\mathrm{A}$?

Problem 12.9

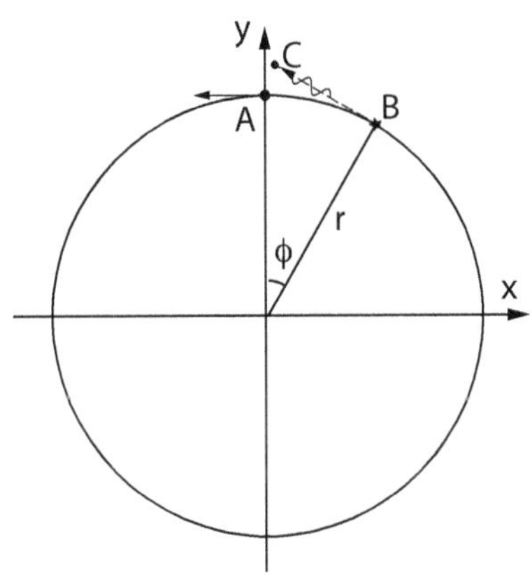

Fig. 12.12 Synchrotron radiation.

Assume a charged particle moves with ultrarelativistic speed in a circular orbit, as shown in Fig. 12.12. As discussed in the example on synchrotron radiation, the radiation which is caused by acceleration of the charge, will almost exclusively be sent in the forward direction, tangentially to the circular orbit. In the figure three different points are labelled by A, B, and C. Point A corresponds to the position of the particle at time $t = 0$, point B is the position of the particle at an earlier time $t_B < 0$, and point C is the position of the radiation emitted from this point, at the later time $t = 0$. The angle ϕ between A and B on the circle is given by $\phi = c|t_B|/r$, with r as the radius of the circle, when we approximate the speed of the particle with the speed of light.

a) Show that the coordinates of point C are given by

$$x = r(\sin\phi - \phi\cos\phi)\,, \quad y = r(\cos\phi + \phi\sin\phi)\,. \qquad (12.110)$$

b) The curve in the x, y-plane, formed by the position of the radiation at time $t = 0$, is determined by the coordinates of C with variable angle $\phi > 0$. Make a plot which shows this curve close to the position A of the charged particle at $t = 0$, and compare this to Fig. 12.6 in the example on synchrotron radiation. Indicate with an arrow the direction of propagation of the radiation.

c) Make a similar plot, which now shows the radiation curve at $t = 0$, for radiation emitted under a full period of the particle motion, $0 < \phi < 2\pi$. Assume the radiation can freely propagate outwards from the particle orbit. Also here indicate with arrows the direction of propagation of the radiation.

Summary

We have in this part studied how Maxwell's equations form the theoretical basis for understanding the variety of electromagnetic phenomena. Beginning from the four equations that constitute the set of Maxwell's equations, we have first seen how these can be compactified into two covariant field equations which involve the electromagnetic field tensor rather than the electric and magnetic field separately. This covariant form is attractive, not only because of its compactness and elegance, but also because the relativistic invariance of electromagnetic theory is made explicit in the covariant equations. Relativistic invariance and symmetry under Lorentz transformations are important properties of Maxwell's equations. When applied to the electromagnetic theory the relativistic invariance predicts the specific way in which the **E** and **B**-fields are mixed when changing from one inertial reference frame to another.

The main problem addressed here has been how to solve Maxwell's equation under different conditions. As a first step it is then of interest to simplify the equations by introducing the electromagnetic potentials. These are not uniquely determined by the **E** and **B**-fields, and we may therefore impose a gauge condition, here chosen as the Lorentz gauge condition, to simplify the equations. In these notes we have looked at three different situations. The first one has been to study solutions when the sources of the fields, *i.e.*, the charge and current distributions, vanish. The next one has been to study when the sources (in a given inertial frame) are time independent, and finally we have studied the general situation with space and time dependent distributions of charge and current. In all these cases we assume that there are no non-trivial boundary conditions, so we look for solutions in the open, infinite space.

In the first case, where the charge and current densities vanish, Maxwell's equations have solutions in the form of freely propagating waves. These are the electromagnetic waves, which appear in a wide variety of forms, depending on the frequency of the waves. In our somewhat brief discussion of electromagnetic waves we have focussed on the property of polarization, which characterizes all these different types of wave phenomena. The special cases of linear and circular polarization can be understood as depending on the phase and amplitude relations between two orthogonal components of the radiation, and that is also so for the general type of elliptic polarization.

When the charge and current distributions are time independent, the equations for the electric and magnetic fields decouple completely and they can be examined separately in the form of electrostatic and magnetostatic equations. It is easy to find solutions of these equations by applying the linearity of the equations. Thus the general solutions of the electrostatic problem can be found as a linear superposition of the Coulomb potentials of all the small parts of the charge distributions. The magnetostatic equations are of the same form and solutions can be found by the same method. In both cases the general solutions for the (scalar or vector) potentials can be written as integrals over the charge or current distributions.

Even if explicit solutions can be found for the field equations with general stationary sources, it is often of interest to make simplifications for the resulting integral in the form of approximations that are valid for points not too close to the charges and currents. The multipole expansion is based on the assumption that the distance from the source to the point where the potential should be determined is much larger than the extension of the source itself. For the electrostatic field the leading term is the Coulomb potential, the next is the electric dipole potential, then the electric quadrupole potential etc. For the magnetostatic potential there is a similar expansion, but here the leading term is the magnetic dipole potential. There is in fact a simple symmetry between the electrostatic and magnetostatic expansions, so that term by term the **E** and **B**-fields, for dipole, quadrupole etc., are of precisely the same form.

The method used with stationary sources can with some modifications be used also to solve Maxwell's equations with general time and space dependent sources. As a first step in finding the general solution we have introduced the Fourier transform in time and thereby brought the equations into a form similar to the static cases. The type of differential equation we then meet is not identical to that of the electrostatic case, but the same general method can be used. This means that we first look for solutions of the problem with a point source, which is now a modified Coulomb potential. We next extend this to the general case by making a linear superposition over contributions from all point-like parts of the charge and current densities. Finally the inverse Fourier transformation gives the solution in the form of an integral over the time and space dependent charge and current distributions. As we have seen the solution is strikingly similar to the corresponding solutions for the electrostatic and magnetostatic potentials. The main difference with time dependent sources is the retardation effect. Thus the integral over the charge and current distributions is not taken at a fixed time, but the integral is instead over the past light cone relative to the point where the potential should be determined.

The retardation effect one should clearly expect from the theory of relativity, where the influence of a source on the field at a distant point is delayed by the limit of propagation set by the speed of light. Even this delay may look innocent, it contains the important physical effect of radiation from a time dependent source. To see this we have focussed on the radiation fields, which fall off with distance as $1/r$, and therefore are the dominating components of the fields far from the sources. The derivation shows that these parts of the fields appear as a consequence of the position dependence of the retarded time. We have found explicit expressions for the first few terms of the multipole expansion of the radiation fields, where normally the electric dipole contribution is the most important one.

Obviously there are a lot of interesting further developments of the theory that are not covered here. That is true also for the other parts of the course, where the motivation has been to focus on some of the most important and simplest parts of the classical theory of mechanics and electrodynamics. One of the main objectives has been to show that the analytic approach applied in this part of physics gives the theory an attractive and elegant form, but also to show that these methods are important in solving the fundamental equations and revealing the underlying structure of the physical phenomena.

PART 4

Classical Field Theory

Introduction

The main aim of this part of the book is to extend the use of the Lagrange–Hamilton method from discrete physical systems to continuous fields. We begin with a demonstration of how a discrete set of interacting point particles in one dimension can approximate a continuous system, with the Lagrangian replaced by the Lagrangian density of the continuous system.

Hamilton's principle for the continuous system is then expressed in terms of derivatives of the Lagrangian density. This is first shown for cases where the variables are restricted to only two. This is to see more easily how Hamilton's principle gives rise to the Euler–Lagrange equation. As the next step, general fields are introduced, which are useful as a start for finding the Euler–Lagrange equations with more variables of the Lagrange densities.

The Lagrange formalism can easily be transformed between different coordinate sets, with a corresponding change of the Lagrangian density. This is shown in the general case, with demonstration in some concrete examples. We focus in particular on transformations where the Lagrangian densities are conserved under the transformation. These are referred to as *symmetry transformations*.

Noether's theorem shows in general terms how symmetries are connected to the presence of conserved quantities. In Part 1, this was demonstrated for discrete physical systems, and here, we continue this demonstration by showing that this is the case also for continuous fields. We give some examples, where energy and momentum are shown as the conserved quanta related to the symmetries.

In the final chapter, we focus on two different types of fields, the first one with Klein–Gordon fields and also with ϕ^4 added to the field. In the first cases, the general solutions are found as a superposition of harmonic oscillators. In the other cases, there are internal symmetries that can be used to find the minimum of the potential. With negative ϕ^4, this give interesting minimum potentials.

In the second case, Maxwell's theory is extended with Lagrange formulations. This implies the use of potentials A_μ and gauge transformations. The standard tensor $T^{\mu\nu}$ is in this case not enough to describe correctly the energy–momentum tensor, but with an added term $t^{\mu\nu}$ that is sufficient. There is however an alternative solution which is discussed in the last section of the book.

Chapter 13

Lagrangian and Hamiltonian formulations

13.1 From particles to fields

In Part 1, the Lagrangian formalism is restricted to discrete physical systems. We take such a system here as our starting point and show how the system, in principle, can be changed from a discrete to a continuous system within the Lagrangian formalism. The result gives an indication of how to apply such descriptions for continuous systems more generally.

As a concrete choice, we consider a one-dimensional set of identical, interacting particles connected by metal springs, as shown in Fig. 13.1. The idea is to consider the particles to change in size and mass and simultaneously change the strength between the neighbouring particles when they move. When this is done in a correct way, we can think of that as a way to change the system from a discrete toward a continuous system and use this result also for systems that are truly continuous.

We denote by q_i the displacement from equilibrium of the ith particle. We write the kinetic energy as

$$T = \frac{1}{2}\sum_i m\dot{q}_i^2, \tag{13.1}$$

and the potential energy as

$$V = \frac{1}{2}\sum_i k(q_{i+1} - q_i)^2, \tag{13.2}$$

with k as a constant. The potential energy V is then assumed to be well described by the first-order expansion of neighbouring positions $(q_{i+1}-q_i)^2$.

Fig. 13.1 A system of particles move with variations of the springs.

For the Lagrangian of the system, this implies that

$$L = T - V$$
$$= \frac{1}{2}\sum_i (m\dot{q_i}^2 - k(q_{i+1} - q_i)^2). \tag{13.3}$$

The Lagrange equation is

$$\frac{d}{dt}\frac{\partial L}{\partial \dot{q}_i} - \frac{\partial L}{\partial q_i} = 0, \tag{13.4}$$

which here means

$$m\ddot{q}_i - k(q_{i+1} - 2q_i + q_{i-1}) = 0. \tag{13.5}$$

A continuum form of the particle system can be introduced in the following way. The distance between neighbouring particles at equilibrium is denoted a, and we introduce the limit of reducing the distance $a \to 0$ in the following way:

$$\sum_i a \to \int dx,$$
$$m/a \to \mu,$$
$$ka \to \eta,$$
$$\frac{q_{i+1} - q_i}{a} \to \frac{\partial q}{\partial x}, \tag{13.6}$$

where μ is the linear mass density and η is known as elasticity modulus of the continuous string. In this limit, the variable $q_i(t)$ is replaced by the field variable $q(x, t)$.

For the Lagrangian, this gives the following expression:

$$lim_{a\to 0}\ L = \sqrt{\frac{1}{2}}\int dx \left\{\mu\left(\frac{\partial q}{\partial t}\right)^2 - \eta\left(\frac{\partial q}{\partial x}\right)^2\right\}, \tag{13.7}$$

with the corresponding field equation,

$$\mu\frac{\partial^2 q}{\partial t^2} - \eta\frac{\partial^2 q}{\partial x^2} = 0 \quad \Rightarrow \quad \frac{\partial^2 q}{\partial t^2} - v^2\frac{\partial^2 q}{\partial x^2} = 0. \tag{13.8}$$

The variable $v = \sqrt{\eta/\mu}$, in the last expression, is interpreted as the velocity of the moving density.

The Lagrangian density is defined by $L = \int \mathcal{L} dx$, which in the present case is

$$\mathcal{L} = \frac{1}{2}(\mu\, \dot{q}^2 - \eta\, q'^2), \tag{13.9}$$

where $\dot{q} = \partial q/\partial t$ and $q' = \partial q/\partial x$.

Hamilton's principle and the Euler–Lagrange equation

In the system introduced above, the Lagrangian density depends only on the two variables $\dot{q}$ and q'. This is a consequence of the symmetries of the system with respect to space and time. We now continue with a related system, which however has no similar symmetries. In such a case, $\mathcal{L}$ will depend on additional variables:

$$\mathcal{L} = \mathcal{L}(q, \dot{q}, q'; x, t). \tag{13.10}$$

We note in particular the dependence of time, which earlier, in Section 2.1, has been referred to as *explicit* time dependence. In this case, there are two explicit dependences corresponding to t and x. This differs from the implicit dependence of t and x, which are shown by $\dot{q}$ and q'.

Let us start with Hamilton's principle, which in the discrete case was studied in Section 3.6. We will show from this how to derive the equation of motion, which is referred to as the Euler–Lagrange equation. The starting point is the *action integral*, written in terms of Lagrangian density,

$$S[q] = \iint \mathcal{L} dx dt, \tag{13.11}$$

which to first order in variations gives

$$\begin{aligned} \delta S &= \delta \iint \mathcal{L} dx dt \\ &= \iint \left(\frac{\partial \mathcal{L}}{\partial q}\delta q + \frac{\partial \mathcal{L}}{\partial \dot{q}}\delta \dot{q} + \frac{\partial \mathcal{L}}{\partial q'}\delta q' \right) dx dt, \end{aligned} \tag{13.12}$$

In the derivation, q, $\dot{q}$, and q' are treated as independent variables. Note that the variation is applied to q and its derivatives, but not to the coordinates, x and t, which are kept fixed.

The action is assumed to be restricted to a region in the (x, t)-plane, and for simplicity, we take the region to have a quadratic form, as shown in Fig. 13.2.

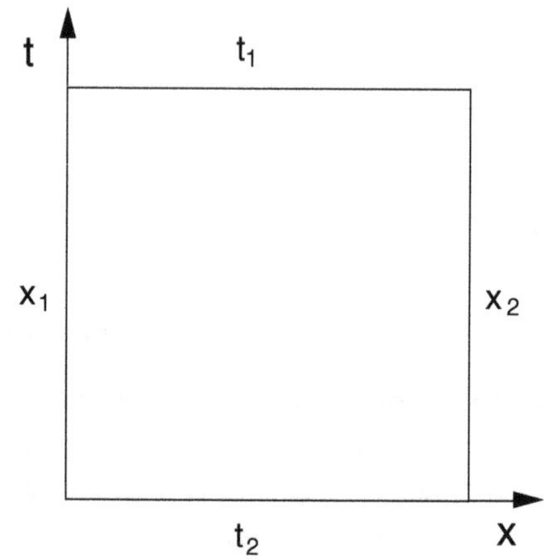

Fig. 13.2 Region with a quadratic form.

With integration by parts, we get

$$\int_{t_1}^{t_2} dt \frac{\partial \mathcal{L}}{\partial \dot{q}} \delta \dot{q} = - \int_{t_1}^{t_2} dt \frac{d}{dt} \left(\frac{\partial \mathcal{L}}{\partial \dot{q}} \right) \delta q + \left[\frac{\partial \mathcal{L}}{\partial \dot{q}} \delta q \right]_{t_1}^{t_2}, \tag{13.13}$$

and similarly,

$$\int_{x_1}^{x_2} dx \frac{\partial \mathcal{L}}{\partial q'} \delta q' = - \int_{x_1}^{x_2} dx \frac{d}{dx} \left(\frac{\partial \mathcal{L}}{\partial q'} \right) \delta q + \left[\frac{\partial \mathcal{L}}{\partial q'} \delta q \right]_{x_1}^{x_2}, \tag{13.14}$$

where the last contributions vanish, in each equation, due to the restriction $\delta q = 0$ at the boundary of the region. The variation in $S[q]$ can therefore be written as

$$\delta S[q] = \int_{t_1}^{t_2} \int_{x_1}^{x_2} \left(\frac{\partial \mathcal{L}}{\partial q} - \frac{d}{dt} \left(\frac{\partial \mathcal{L}}{\partial \dot{q}} \right) - \frac{d}{dx} \left(\frac{\partial \mathcal{L}}{\partial q'} \right) \right) \delta q \; dx dt. \tag{13.15}$$

Note that the operators d/dt and d/dx in these expressions are total derivatives of the form

$$\begin{aligned} \frac{d}{dt} &= \frac{\partial}{\partial t} + \dot{q} \frac{\partial}{\partial q} + \ddot{q} \frac{\partial}{\partial \dot{q}}, \\ \frac{d}{dx} &= \frac{\partial}{\partial x} + q' \frac{\partial}{\partial q} + q'' \frac{\partial}{\partial q'}. \end{aligned} \tag{13.16}$$

Hamilton's principle defines $\delta S[q]$ to vanish for arbitrary infinitesimal variations $\delta q(x, t)$, as shown above. This gives the following form of the

Euler–Lagrange equation:

$$\frac{\partial \mathcal{L}}{\partial q} - \frac{d}{dt}\left(\frac{\partial \mathcal{L}}{\partial \dot{q}}\right) - \frac{d}{dx}\left(\frac{\partial \mathcal{L}}{\partial q'}\right) = 0, \tag{13.17}$$

which leads to the corresponding field equation when the Lagrangian density is specified. In the following example, we illustrate this in a simple case.

Example: Transverse waves on a string

We consider here vibrations in a string in the (x, y) plane, which we describe as a smooth function $y(x, t)$. x is then the horizontal and y the vertical coordinate, as shown in Fig. 13.3. The string has mass density μ and tension τ, and we assume that deviations from constant values of these are negligible. The field variable is identical to the vertical displacement, $y = y(x, t)$. The Lagrangian density is

$$\mathcal{L}(\dot{y}, y') = \frac{1}{2}\mu\, \dot{y}^2 - \frac{1}{2}\tau\, y'^2, \tag{13.18}$$

and we note that there is no dependence of y, x or t here. As a starting point, we treat the two variables $\dot{y}$ and y' as independent.

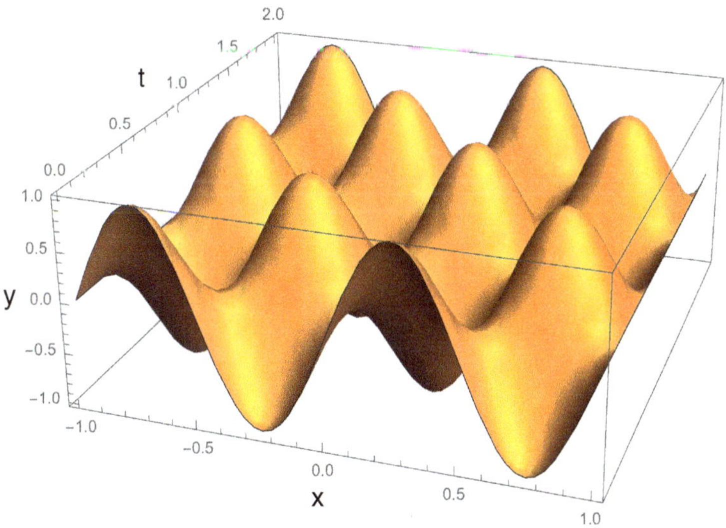

Fig. 13.3 A vibrating string. The string is suspended on two points along the x-direction with vibrations restricted to the y-direction. In the figure, variations in time are shown in the t-direction.

The action of the string is the double integral,

$$S[y] = \int_{x_1}^{x_2} dx \int_{t_1}^{t_2} dt \; \mathcal{L}(\dot{y}, y'), \tag{13.19}$$

and the (infinitesimal) variation of S is

$$\delta S = -\int_{x_1}^{x_2} dx \int_{t_1}^{t_2} dt \left(\frac{d}{dx}\left(\frac{\partial \mathcal{L}}{\partial y'}\right) + \frac{d}{dt}\left(\frac{\partial \mathcal{L}}{\partial \dot{y}}\right)\right) \delta y, \tag{13.20}$$

where the boundary contributions of δS have been set to zero. This is correct for any choice of the boundary, which introduces the Euler–Lagrange equation in the following form:

$$\frac{d}{dx}\left(\frac{\partial \mathcal{L}}{\partial y'}\right) + \frac{d}{dt}\left(\frac{\partial \mathcal{L}}{\partial \dot{y}}\right) = 0. \tag{13.21}$$

The Lagrangian density, shown in (13.18), and the expressions for the derivatives shown in (13.16), with q replaced by y, gives the following relations:

$$\frac{d}{dt}\frac{\partial \mathcal{L}}{\partial \dot{y}} = \mu \ddot{y}, \quad \frac{d}{dx}\frac{\partial \mathcal{L}}{\partial y'} = -\tau y''. \tag{13.22}$$

By use of these results, the Euler–Lagrange equation (13.21) can be written as

$$\mu \frac{\partial^2 y}{\partial t^2} - \tau \frac{\partial^2 y}{\partial x^2} = 0, \tag{13.23}$$

which we write as

$$\frac{\partial^2 y}{\partial t^2} - v^2 \frac{\partial^2 y}{\partial x^2} = 0, \tag{13.24}$$

where $v = \sqrt{\tau/\mu}$ describes the velocity of waves on the string.

The general solution for this equation is

$$y(x,t) = f(x - vt) + g(x + vt), \tag{13.25}$$

where f and g are unspecified functions of $x - vt$ and $x + vt$. We assume further that the string is restricted to the interval $x \epsilon [0, L]$, with $y(0,t) = y(L,t) = 0$. This gives

$$g(vt) = -f(-vt), \quad g(L - vt) = -f(L - vt), \tag{13.26}$$

and since it is valid for any value of vt, this implies the identity $g(vt - x) = -f(x - vt)$. Therefore, we have

$$y(x,t) = g(vt + x) - g(vt - x). \tag{13.27}$$

With $x = L$, this gives

$$g(vt + L) - g(vt - L) = 0, \tag{13.28}$$

which shows that the function $g(\xi)$ is periodic, with period $2L$. A Fourier expansion can then be used to examine the function,

$$g(\xi) = \sum_{n=1}^{\infty} \left\{ a_n \cos\left(\frac{n\pi\xi}{L}\right) + b_n \sin\left(\frac{n\pi\xi}{L}\right) \right\}, \tag{13.29}$$

with a_n and b_n as unspecified coefficients. The full solution for $y(x, t)$ is then

$$y(x,t) = \sum_{n=1}^{\infty} \sin\left(\frac{n\pi x}{L}\right) \left\{ A_n \cos\left(\frac{n\pi vt}{L}\right) + B_n \sin\left(\frac{n\pi vt}{L}\right) \right\}, \tag{13.30}$$

and a simple solution is

$$y(x,t) = A \sin\left(\frac{n\pi x}{L}\right) \cos\left(\frac{n\pi vt}{L}\right), \tag{13.31}$$

where n is a chosen integer. Such a vibrating string, in a simplified form, is shown in Fig. 13.3.

13.2 General Fields

The results in the previous section can easily be extended to include fields also of higher dimensions both in the relativistic and non-relativistic cases. To describe fields more generally, we introduce $\phi^i(x)$, where $i = 1, 2, ..., n$ runs over the field components and x indicates the full set of space–time variables x^μ. The Lagrangian density is written in a condensed way as

$$\mathcal{L} = \mathcal{L}(\phi^i,\ \phi^i_{,\mu}\ ; x), \tag{13.32}$$

where $\phi^i_{,\mu}$ denotes the derivative of ϕ^i in the direction of x^μ. Thus, the three following expressions have the same meaning:

$$\phi^i_{,\mu} \equiv \frac{\partial \phi^i}{\partial x^\mu} \equiv \partial_\mu \phi^i. \tag{13.33}$$

However, in the same way as the variables $q, \dot{q}, q'$, in the Lagrangian density (13.10), are considered to be independent, this is also the case for the variables $\phi^i, \phi^i_{,\mu}$ in the Lagrangian density (13.32).

The action is then written as

$$S = \int_{\Omega} \mathcal{L}(\phi^i,\ \phi^i_{,\mu}\ ; x)dx, \tag{13.34}$$

where the integration describes a volume integral with dimension equal to the number of coordinates x^μ, with Ω referring to the integration region. (For comparison, see Eqs. (13.10), (13.11) and (13.17)). A compact form of the Euler–Lagrange equation is then

$$\frac{\partial \mathcal{L}}{\partial \phi^i} - \frac{d}{dx^\mu}\left(\frac{\partial \mathcal{L}}{\partial {\phi^i}_{,\mu}}\right) = 0. \tag{13.35}$$

Note that the Euler–Lagrange equation can also be expressed as

$$\frac{\partial \mathcal{L}}{\partial \phi^i} - \left(\frac{\partial \mathcal{L}}{\partial {\phi^i}_{,\mu}}\right)_{,\mu} = 0. \tag{13.36}$$

Even if the notation x^μ is inspired by relativity theory, this does not mean that the Lagrangian density is restricted to relativistic space–time cases. Thus, both in relativistic and non-relativistic cases, the presence of upper and lower μ indicates the use of (Einstein) summation convention. This notation is only a convenient way to give compact descriptions in the use of the space–time coordinates. These we may write as (ct, x, y, z) or (x^0, x^1, x^2, x^3).

An explicit expression of equation (13.35) (or (13.37)), for a field in four dimensions, will then take the inconvenient form,

$$\frac{\partial \mathcal{L}}{\partial \phi^i} - \frac{d}{dt}\left(\frac{\partial \mathcal{L}}{\partial {\phi^i}_{,t}}\right) - \frac{d}{dx}\left(\frac{\partial \mathcal{L}}{\partial {\phi^i}_{,x}}\right) - \frac{d}{dy}\left(\frac{\partial \mathcal{L}}{\partial {\phi^i}_{,y}}\right) - \frac{d}{dz}\left(\frac{\partial \mathcal{L}}{\partial {\phi^i}_{,z}}\right) = 0. \tag{13.37}$$

For this reason, the compact description of the Euler–Lagrange equations will be used independently of the number of space–time coordinates. This number will instead be explicitly defined.

Equation (13.35) is often written as

$$\frac{\delta \mathcal{L}}{\delta \phi^i} - \frac{d}{dt}\left(\frac{\partial \mathcal{L}}{\partial {\phi^i}_{,t}}\right) = 0, \tag{13.38}$$

where

$$\frac{\delta \mathcal{L}}{\delta \phi^i} = \frac{\partial \mathcal{L}}{\partial \phi^i} - \frac{d}{dx^k}\left(\frac{\partial \mathcal{L}}{\partial \phi^i_{,k}}\right) \tag{13.39}$$

is referred to as a *functional derivative* (also called *variational derivative*). The coordinate x^k represents the space variables, with $1/\delta\phi^i$ as the variational derivatives in the space dimensions.

Example: Complex scalar field

We study here a scalar field written in covariant form,

$$\mathcal{L}(\phi, \phi^*, \phi_{,\mu}, \phi^*_{,\mu}) = -(\phi^*_{,\mu}\phi^{,\mu} + \mu^2\phi^*\phi), \tag{13.40}$$

with $x^\mu = (ct, \mathbf{x})$ and $x_\mu = (-ct, \mathbf{x})$. The index i in the Lagrangian density (13.32) corresponds here to the two fields ϕ and ϕ^*. The variables in the expression of $\mathcal{L}$ are treated as independent, and in the same way as the string example, there is no explicit dependence on the coordinates x_μ. This simplifies the way of finding the field equations defined by the Lagrangian density (13.40).

The first step is to find the derivatives of the Lagrangian densities, which in the present case are the following:

$$\frac{\partial \mathcal{L}}{\partial \phi} = -\mu^2\phi^*, \quad \frac{\partial \mathcal{L}}{\partial \phi_{,\mu}} = -\frac{\partial \phi^*}{\partial x_\mu}, \tag{13.41}$$

where the Euler–Lagrange equation is found as

$$\frac{\partial \mathcal{L}}{\partial \phi} - \frac{d}{dx^\mu}\left(\frac{\partial \mathcal{L}}{\partial \phi_{,\mu}}\right) = 0$$

$$\Rightarrow \; -\mu^2\phi^* + \frac{\partial^2 \phi^*}{\partial x^\mu \partial x_\mu} = 0$$

$$\Rightarrow \; (\Box - \mu^2)\,\phi^* = 0. \tag{13.42}$$

As a complex conjugate equation, we have

$$(\Box - \mu^2)\,\phi = 0. \tag{13.43}$$

where we remind you of the definition,

$$\Box = \partial_\mu \partial^\mu = \nabla^2 - \frac{1}{c^2}\frac{\partial^2}{\partial t^2}. \tag{13.44}$$

Equations (13.42) and (13.43) are Klein–Gordon equations, which can be interpreted as a relativistic wave equation for a particle with charge, but without spin. A further discussion of the Klein–Gordon field is found in the final chapter.

Different cases

The general Lagrangian density (13.32) has an unspecified number of field components i and also an unspecified set of indeces x^μ. However, in the latter case, the number of space–time variables is maximally four. As an example, in the case of the vibrating string, we have $\{i = 1\}$ and $\{\boldsymbol{x} = (t, x)\}$, i.e. there is a time coordinate and one space coordinate.

This shows that the Euler–Lagrange equation (13.35) can appear in many different forms. Special cases are found earlier in this book. Thus, in Chapter 2, there are cases where only time coordinates $\{\boldsymbol{x} = t\}$ are included. This brings us back to the particle dynamics, where $q_i(t)$ describe how particles move. In Chapter 3, there are fields where only a single space coordinate x is included. In these cases, there is no time change, but there is a field that varies only with x.

In the first case, the first-order variations have the form,

$$\delta S = \int_{t_1}^{t_2} \sum_k \left[\frac{\partial L}{\partial q_k} - \frac{d}{dt}\frac{\partial L}{\partial \dot{q}_k} \right] \delta q_k \, dt. \tag{13.45}$$

This shows that the Lagrangian is present rather than the Lagrangian *density*. In the second case, the same thing happens where the time coordinate is exchanged with a single space coordinate.

Equivalent Lagrangian densities

In Section 2.4, we have seen that two Lagrangians are equivalent if they differ by a total time derivative,

$$L'(q, \dot{q}, t) = L(q, \dot{q}, t) + \frac{d}{dt} f(q, t). \tag{13.46}$$

A similar situation in the case of Lagrangian densities has the form

$$\mathcal{L}'(\phi^i,\ \phi^i_{,\mu},\ x) = \mathcal{L}(\phi^i,\ \phi^i_{,\mu},\ x) + \frac{d}{dx^\mu} f^\mu(\phi, x), \tag{13.47}$$

where f^μ is a function which depends only on ϕ and x^μ. We show in the following the equivalence between $\mathcal{L}$ and $\mathcal{L}'$ in essentially the same way as shown with L and L' in Section 2.4.

In order to do so, we introduce the function

$$\begin{aligned} g(\phi^i, \phi^i_{,\mu}, x) &= \frac{d}{dx^\mu} f^\mu(\phi, x) \\ &= \frac{\partial f^\mu}{\partial x^\mu} + \frac{\partial f^\mu}{\partial \phi^i} \phi^i_{,\mu} \end{aligned} \tag{13.48}$$

and focus on the Euler–Lagrange equation in the two cases, $\mathcal{L}'$ and $\mathcal{L}$. We have the following identity:

$$\frac{\partial \mathcal{L}'}{\partial \phi^i} - \frac{d}{dx^\mu}\left(\frac{\partial \mathcal{L}'}{\partial \phi^i{}_{,\mu}}\right) = \frac{\partial \mathcal{L}}{\partial \phi^i} - \frac{d}{dx^\mu}\left(\frac{\partial \mathcal{L}}{\partial \phi^i{}_{,\mu}}\right) - \frac{\partial g}{\partial \phi^i} + \frac{d}{dx^\mu}\left(\frac{\partial g}{\partial \phi^i_{,\mu}}\right). \tag{13.49}$$

The two last contributions can be rewritten as

$$\frac{\partial g}{\partial \phi^i} = \frac{\partial}{\partial \phi^i}\left(\frac{\partial f^\mu}{\partial x^\mu} + \frac{\partial f^\mu}{\partial \phi^j}\phi^j_{,\mu}\right) = \frac{\partial^2 f^\mu}{\partial \phi^i\, \partial x^\mu} + \frac{\partial^2 f^\mu}{\partial \phi^i \partial \phi^j}\phi^j_{,\mu}, \tag{13.50}$$

and

$$\frac{d}{dx^\mu}\left(\frac{\partial g}{\partial \phi^i_{,\mu}}\right) = \frac{d}{dx^\mu}\frac{\partial f^\mu}{\partial \phi^i} = \frac{\partial^2 f^\mu}{\partial x^\mu \partial \phi^i} + \frac{\partial^2 f^\mu}{\partial \phi^i \partial \phi^j}\phi^j_{,\mu}. \tag{13.51}$$

The results from (13.50) and (13.51) is

$$\frac{\partial g}{\partial \phi^i} - \frac{d}{dx^\mu}\left(\frac{\partial g}{\partial \phi^i_{,\mu}}\right) = 0, \tag{13.52}$$

which implies that the Euler–Lagrange equations of $\mathcal{L}$ and $\mathcal{L}'$ are identical:

$$\frac{\partial \mathcal{L}'}{\partial \phi^i} - \frac{d}{dx^\mu}\left(\frac{\partial \mathcal{L}'}{\partial \phi^i{}_{,\mu}}\right) = \frac{\partial \mathcal{L}}{\partial \phi^i} - \frac{d}{dx^\mu}\left(\frac{\partial \mathcal{L}}{\partial \phi^i{}_{,\mu}}\right). \tag{13.53}$$

13.3 Hamilton formulation

In Chapter 3, we have discussed how the Lagrangian of a particle system can be transformed to the Hamiltonian, thereby changing the set of generalized coordinates and velocities from $(q, \dot{q})$ to the set of generalized coordinates and conjugate momenta (q, p). Even if the classical fields transformation from Lagrangian to Hamiltonian formulation can be done, there is a difference, since the Hamiltonian formulation does not handle space and time in the same way. This makes the Hamiltonian formulation more complicated, and even if we study some of the basics of this formulation, we will soon go back to the formulation with Lagrangian densities.

We first look at the Hamiltonian where there is a single space coordinate x and a set of particles with positions q_i along the coordinate. The Hamiltonian can then be expressed in the following way:

$$H = p_i\dot{q}_i - L = \frac{\partial L}{\partial \dot{q}_i}\dot{q}_i - L, \tag{13.54}$$

where p_i is the momentum of particle i and L is Lagrangian of the particles. In the continuum limit, this gives

$$H = \int dx \left(\frac{\partial \mathcal{L}}{\partial \dot{\phi}} \dot{\phi} - \mathcal{L} \right), \tag{13.55}$$

where the particle positions q_i are replaced by the field $\phi(x)$ and the particle momentum $p_i = \partial L / \partial \dot{q}_i$ is replaced by the *momentum density*,

$$\pi(x) = \frac{\partial \mathcal{L}}{\partial \dot{\phi}}. \tag{13.56}$$

Hamiltonian (13.55) is the x-integral of the *Hamiltonian density* defined by

$$\mathcal{H} = \pi \dot{\phi} - \mathcal{L}. \tag{13.57}$$

The generalization of (13.56) to a three-dimensional field, with the field quantities $\phi^i(x^k)$, $k = 1, 2, 3$, is

$$\pi_i(x) = \frac{\partial \mathcal{L}}{\partial \dot{\phi}^i}. \tag{13.58}$$

The full expressions of the variables in the Hamiltonian formulation are then $\phi^i(x^k, t)$ and $\pi_i(x^k, t)$. They define an infinite-dimensional space, which corresponds to the finite-dimensional *phase spaces* earlier discussed in Chapter 3.

Generalization of the Hamilton density (13.57) to the three-dimensional case is then

$$\mathcal{H}(\phi^i, \phi^i_{,k}, \pi_i, x^k, t) = \pi_i \dot{\phi}^i - \mathcal{L}. \tag{13.59}$$

From (13.56) and (13.59), we find

$$\frac{\partial \mathcal{H}}{\partial \pi^i} = \dot{\phi}^i + \pi_j \frac{\partial \dot{\phi}^j}{\partial \pi_i} - \frac{\partial \mathcal{L}}{\partial \dot{\phi}^j} \frac{\partial \dot{\phi}^j}{\partial \pi_i} = \dot{\phi}_i, \tag{13.60}$$

and the related derivative with π_i replaced by ϕ^i is

$$\frac{\partial \mathcal{H}}{\partial \phi^i} = \pi_i + \pi_j \frac{\partial \dot{\phi}^j}{\partial \pi_i} - \frac{\partial \mathcal{L}}{\partial \dot{\phi}^j} \frac{\partial \dot{\phi}^j}{\partial \phi^i} - \frac{\partial \mathcal{L}}{\partial \phi^i} = -\frac{\partial \mathcal{L}}{\partial \phi^i}. \tag{13.61}$$

Using the Euler–Lagrange equation,

$$\begin{aligned} \frac{\partial \mathcal{L}}{\partial \phi^i} &= \frac{\partial}{\partial x^\mu} \left(\frac{\partial \mathcal{L}}{\partial \phi^i_{,\mu}} \right) \\ &= \frac{\partial}{\partial x^k} \left(\frac{\partial \mathcal{L}}{\partial \phi^i_{,k}} \right) - \dot{\pi}_i, \end{aligned} \tag{13.62}$$

we can get rid of the reference to the Lagrangian density $\mathcal{L}$ and rewrite (13.61) as

$$\frac{\partial \mathcal{H}}{\partial \phi^i} - \frac{\partial}{\partial x^k}\left(\frac{\partial \mathcal{H}}{\partial \phi^i_{,k}}\right) = -\dot{\pi}_i. \tag{13.63}$$

Equations (13.60) and (13.63) can be regarded as the canonical field equations in the Hamiltonian formulation. A simpler and more equal form of the two can be found if the equations are expressed in the form of functional derivatives. In the two cases, we then have equations

$$\dot{\phi}^i = \frac{\delta \mathcal{H}}{\delta \pi_i}, \qquad \dot{\pi}_i = -\frac{\delta \mathcal{H}}{\delta \phi^i}. \tag{13.64}$$

A general observable of the field will be a functional, $F = F[\phi, \pi, t]$, with the possibility of an explicit time included. The time development of ϕ and π, according to Hamilton's equations, gives the following time derivative of F:

$$\begin{aligned}\dot{F} &= \int_{\Omega_3} d^3x \left(\frac{\delta F}{\delta \phi^i}\dot{\phi}^i + \frac{\delta F}{\delta \pi^i}\dot{\pi}^i\right) + \frac{\partial F}{\partial t}\\ &= \int_{\Omega_3} d^3x \left(\frac{\delta F}{\delta \phi^i}\frac{\delta H}{\delta \pi^i} - \frac{\delta F}{\delta \pi^i}\frac{\delta H}{\delta \phi^i}\right) + \frac{\partial F}{\partial t},\end{aligned} \tag{13.65}$$

which we write as

$$\dot{F} = \{F, H\} + \frac{dF}{dt}, \tag{13.66}$$

where $\{F, H\}$ is referred to as the Poisson bracket for classical fields of the two observables $F[\phi, \pi, t]$ and $H[\phi, \pi, t]$. The fundamental Poisson brackets are

$$\begin{aligned}&\{\phi^i(\mathbf{r}), \phi^j(\mathbf{s})\} = 0, \quad \{\pi_i(\mathbf{r}), \pi_j(\mathbf{s})\} = 0,\\ &\{\phi^i(\mathbf{r}), \pi_j(\mathbf{s})\} = \delta^i_j\, \delta^3(\mathbf{r} - \mathbf{s}).\end{aligned} \tag{13.67}$$

The Poisson brackets give a prescription for *quantizing* the classical field. The variables ϕ^i and π_i then are identified as operators in a Hilbert space, and the following so-called canonical commutation relations correspond to the classical Poisson brackets:

$$\begin{aligned}&[\phi^i(\mathbf{r}), \phi^j(\mathbf{s})] = 0, \quad [\pi_i(\mathbf{r}), \pi_j(\mathbf{s})] = 0,\\ &[\phi^i(\mathbf{r}), \pi_j(\mathbf{s})] = \delta^i_j\, \delta^3(\mathbf{r} - \mathbf{s}).\end{aligned} \tag{13.68}$$

Energy density and energy flux on the string

We return to the example with waves on a string. The Hamiltonian of the string is

$$\mathcal{H} = \pi \dot{y} - \mathcal{L}, \tag{13.69}$$

with π as the momentum density,

$$\pi = \frac{\partial \mathcal{L}}{\partial \dot{y}} = \mu \, \dot{y}. \tag{13.70}$$

This gives

$$\mathcal{H} = \frac{\pi^2}{2\mu} + \frac{1}{2}\tau \, y'^2. \tag{13.71}$$

As used in the string, μ is the mass density and τ is the tension, and we take these as constants also here.

The energy density $\mathcal{E}$ can then be written as

$$\mathcal{E} = \frac{1}{2}\mu \, \dot{y}^2 + \frac{1}{2}\tau \, y'^2, \tag{13.72}$$

with the time derivative as

$$\frac{\partial \mathcal{E}}{\partial t} = \mu \, \frac{\partial y}{\partial t}\frac{\partial^2 y}{\partial t^2} + \tau \, \frac{\partial y}{\partial x}\frac{\partial^2 y}{\partial t \partial x}. \tag{13.73}$$

This can be simplified by use of the equation of motion, (13.24),

$$\begin{aligned}
\frac{\partial \mathcal{E}}{\partial t} &= \tau \, \frac{\partial y}{\partial t}\frac{\partial^2 y}{\partial x^2} + \tau \, \frac{\partial y}{\partial x}\frac{\partial^2 y}{\partial t \partial x} \\
&= \tau \frac{\partial}{\partial x}\left[\frac{\partial y}{\partial x}\frac{\partial y}{\partial t}\right] \\
&\equiv -\frac{\partial j}{\partial x}.
\end{aligned} \tag{13.74}$$

This can be identified as the *continuity equation*,

$$\frac{\partial \mathcal{E}}{\partial t} + \frac{\partial j}{\partial x} = 0, \tag{13.75}$$

with

$$j = -\tau \frac{\partial y}{\partial x}\frac{\partial y}{\partial t}, \tag{13.76}$$

as the *energy flux* along the string.

13.4 Exercises

Problem 13.1

This is an electrostatic situation, where ρ is the electric charge distribution and ϕ is the electric potential. The Lagrangian density has the form,

$$\mathcal{L} = \frac{1}{2}(\nabla\phi)^2 - \frac{\rho}{\epsilon}\,\phi, \tag{13.77}$$

and we use the definition,

$$\phi_{,i} \equiv \frac{\partial\phi}{\partial x_i}, \quad i = \{x, y, z\}. \tag{13.78}$$

The problem is to determine the derivatives

$$\frac{\partial\mathcal{L}}{\partial\phi}, \quad \frac{\partial\mathcal{L}}{\partial\phi_{,i}}, \tag{13.79}$$

and find the corresponding Euler–Lagrange equation.

Check that the result is in accordance with Poisson's equation,

$$\nabla^2\phi = -\frac{\rho}{\epsilon}. \tag{13.80}$$

Problem 13.2

The Lagrangian density of the so-called ϕ^4-theory in one space and one time dimension is

$$\mathcal{L} = \frac{1}{2}(\partial_t\phi)^2 - \frac{1}{2}(\partial_x\phi)^2 - \frac{1}{4}\lambda\left(\phi^2 - \rho_0\right)^2, \tag{13.81}$$

where $c = 1$, and λ and ρ_0 are constants.

(a) Find the corresponding Euler–Lagrange equation and show that if ϕ is a solution, so is $-\phi$.

(b) Find the expression for the energy density.

(c) Find the classical vacuum states (i.e. the lowest energy states).

(d) Show that the static expressions,

$$\phi_\pm(x) = \pm\sqrt{\rho_0}\tanh\left[\sqrt{\frac{\lambda\rho_0}{2}}(x-a)\right], \tag{13.82}$$

with a as an integration constant, are solutions to the Euler–Lagrange equation.

(e) Find the energy density of the "kink" solution ϕ_+ as function of x.
(f) Show that the total energy is

$$E_{kink} = \frac{2\sqrt{2}}{3}\sqrt{\lambda\rho_0^3}. \tag{13.83}$$

(g) Find the solution for a *kink* which moves with the velocity v. What is the energy in this case?

Problem 13.3

A string has a constant mass density μ (density per length). It vibrates around the x-axis with end points, $x = 0$ and $x = a$. Between these points, the string vibrates freely, with displacements which vary with position and time along the string, $y = y(x,t)$ and $z(x,t)$. We assume small variations, which means that the length of the string, l, vary little. The string tension, σ, is therefore close to constant, and the same is the case for the potential energy, $V = \sigma l$.

(a) Show that for small variations, the Lagrangian is

$$L = \int_0^a dx\left[\frac{1}{2}\mu\left\{\left(\frac{\partial y}{\partial t}\right)^2 + \left(\frac{\partial z}{\partial t}\right)^2\right\} - \frac{1}{2}\sigma\left\{\left(\frac{\partial y}{\partial x}\right)^2 + \left(\frac{\partial z}{\partial x}\right)^2\right\}\right]. \tag{13.84}$$

(b) Determine the Euler–Lagrange equations and solve them (find the most general solution).

(c) The functions $y(x,t)$ and $z(x,t)$ can be Fourier-expanded with coefficients that are functions of t. Express the Lagrangian L in terms of the Fourier components of y and z.

(d) Express the Euler–Lagrange equations with the Fourier components, and compare this with the general solutions in point (b).

Chapter 14

Symmetry transformations

14.1 Transformations

The Lagrangian formalism has the advantage that the equations of motion can easily be transferred from one set of coordinates to another. This can be seen already in the case of discrete systems. Assume $L(q, \dot{q}, t)$ is a Lagrange function, with q representing a coordinate set q_i of coordinates, and let us assume that we wish to change this to another coordinate set Q, with a transformation

$$q_i = f_i(Q, t), \tag{14.1}$$

where both sets include the same time t. The Lagrangians in the two cases are then the same, in the sense

$$L'(Q, \dot{Q}, t) = L(q, \dot{q}, t), \tag{14.2}$$

where the relations between the two sets $(q, \dot{q}, t)$ and $(Q, \dot{Q}, t)$ are determined by the relation in (14.1).

A similar transformation rule can be introduced in the Lagrange formalism for fields, but we will do that more generally by including in the transformation also the field components.

Let us begin with the following Lagrangian density:

$$\mathcal{L} = \mathcal{L}(\phi^i(x), \phi^i(x)_{,\mu}, x), \tag{14.3}$$

where we initially assume cartesian coordinates and assume that the initial fields are transformed from ϕ to ψ, combined with a transformation of the coordinates x to y,

$$\phi^i(x) = F^i(\psi(y), y). \tag{14.4}$$

The y coordinates are then functions of the x coordinates,

$$y^\mu = y^\mu(x), \tag{14.5}$$

with the transformation assumed to be invertible.

To find the differential equation for the function $\psi(y)$, we begin with the action expressed with $\phi(x)$:

$$S = \int_\Omega \mathcal{L}(\phi, \phi_{,\mu}; x) dx. \tag{14.6}$$

We next change the coordinates from x to y. This introduces the functional determinant $|\partial x/\partial y|$ and changes the description of the integration region from Ω to Ω'. The action is then expressed as

$$S = \int_{\Omega'} \mathcal{L}'\, dy = \int_{\Omega'} \mathcal{L}(\phi, \phi_{,\mu}; x) \left|\frac{\partial x}{\partial y}\right| dy, \tag{14.7}$$

with $\mathcal{L}'$ taking the same value as $\mathcal{L}|\partial x/\partial y|$. However, $\mathcal{L}'$ is expressed in terms of the variables ψ and y, which implies

$$\mathcal{L}'\left(\psi(y), \frac{\partial\psi(y)}{\partial y^\mu}; y\right) = \mathcal{L}\left(\phi(x), \frac{\partial\phi(x)}{\partial x^\mu}; x\right) \left|\frac{\partial x}{\partial y}\right|. \tag{14.8}$$

The action S shown in (14.6) is stationary under variations of ϕ, which go to zero on the boundary of Ω. However, this implies that the same action, shown in (14.7) as an integral over $\mathcal{L}'$, is stationary with respect to variations of ψ that go to zero on the boundary of Ω'.

Furthermore, with variation of $\int_{\Omega'} \mathcal{L}'\, dy$ means zero for variations of ψ which vanishes on the boundary of Ω', and this implies that $\psi(y)$ satisfies the Euler–Lagrange equation,

$$\frac{\partial \mathcal{L}'}{\partial \psi^i} - \frac{d}{dy^\mu}\left(\frac{\partial \mathcal{L}'}{\partial {\psi^i}_{,\mu}}\right) = 0. \tag{14.9}$$

This has the same form as the Euler–Lagrange equation, expressed with the Lagrangian density $\mathcal{L}$:

$$\frac{\partial \mathcal{L}}{\partial \phi^i} - \frac{d}{dx^\mu}\left(\frac{\partial \mathcal{L}}{\partial {\phi^i}_{,\mu}}\right) = 0. \tag{14.10}$$

Example of a transformation

The Lagrangian density of field $\phi(x)$ is

$$\mathcal{L} = \mathcal{L}(\phi(x), \phi(x)_{,\mu}; x), \quad x = (x^0, \mathbf{x}), \tag{14.11}$$

A transformed field ψ is introduced as a scale transformation,

$$\phi(x) = \frac{1}{b}\psi(y), \quad y^\mu = ax^\mu, \tag{14.12}$$

with a and b as constants. This gives

$$\frac{d\phi}{dx^\mu} = \frac{1}{b}\frac{d\psi}{dy^\nu}\frac{\partial y^\nu}{\partial x^\mu} = \frac{1}{b}\frac{d\psi}{dy^\nu}a\delta^\nu_\mu = \frac{a}{b}\frac{d\psi}{dy^\mu}, \tag{14.13}$$

which together with the determinant

$$\left|\frac{\partial x}{\partial y}\right| = \frac{1}{a^4}, \tag{14.14}$$

gives the following Lagrangian density for the field $\psi(y)$:

$$\mathcal{L}' = \frac{1}{a^4}\,\mathcal{L}\left(\frac{1}{b}\psi(y), \frac{a}{b}\frac{d\psi}{dy^\mu}, \frac{1}{a}y\right). \tag{14.15}$$

This is a transformation of the Lagrangian density $\mathcal{L}$ but not a symmetry transformation.

Symmetry transformation

A transformation, with the Lagrangian density $\mathcal{L}$ being changed to $\mathcal{L}'$, is called symmetric if the transformation leaves the system invariant. This means that a symmetry transformation is of the type

$$\phi^i(x) = f^i(\psi(y), y), \quad y^\mu = y^\mu(x), \tag{14.16}$$

with the property

$$\mathcal{L}\left(\psi(y), \frac{d\psi(y)}{dy^\mu}, y\right) = \mathcal{L}\left(\phi(x), \frac{d\phi(x)}{dx_\mu}, x\right)\left|\frac{\partial x}{\partial y}\right|. \tag{14.17}$$

The two Lagrangian densities here, $\mathcal{L}(\psi(y), ...)$ and $\mathcal{L}(\phi(x), ...)$, are the same, in the sense that the corresponding Euler–Lagrange equations are the same for ϕ and ψ. The only difference is that the functions $\psi(y)$ and $\phi(x)$ have different names.

Displacements in space and time

We study the symmetry transformation in a case where $\mathcal{L}$ does not depend explicitly on coordinates x^μ. A simple case is the real scalar field,

$$\mathcal{L}\left(\phi(x), \frac{\partial\phi}{\partial x^\mu}\right) = -\frac{1}{2}\left(\frac{\partial\phi}{\partial x^\mu}\frac{\partial\phi}{\partial x_\mu} + \mu^2\phi^2\right). \tag{14.18}$$

The equation of motion is

$$\frac{\partial^2\phi(x)}{\partial x^\mu \partial x_\mu} - \mu^2\phi(x) = 0, \tag{14.19}$$

which can be interpreted as the (quantum) Klein–Gordon equation, for a free particle without spin and charge, and with mass $m = \mu\hbar/c$.

We introduce now the transformation

$$\phi(x) = \psi(y), \quad y^\mu = x^\mu + b^\mu, \tag{14.20}$$

with b^μ as constants. This is a symmetry transformation, as follows from (14.17), since the functional determinant, in this case, is 1 and $\partial\phi/\partial x^\mu = \partial\psi/\partial y^\mu$. This means that ψ satisfies the same equation as ϕ:

$$\frac{\partial^2\psi(y)}{\partial y^\mu \partial y_\mu} - \mu^2\psi(y) = 0, \tag{14.21}$$

which after change of coordinates, $y \to x$, is

$$\frac{\partial^2\psi(x)}{\partial x^\mu \partial x_\mu} - \mu^2\psi(x) = 0. \tag{14.22}$$

The transformation $\phi \to \psi$ can be understood as an active change, in the form of a displacement of the field in the four-dimensional space–time,

$$\phi(x) \to \psi(x) = \phi(x - b). \tag{14.23}$$

Quite generally, if the Lagrangian density does not have any explicit dependence of the coordinates t and $\mathbf{x}$, then new solutions of the field equation can be introduced from a solution already found by translating the solution in space and time (see Fig. 14.1).

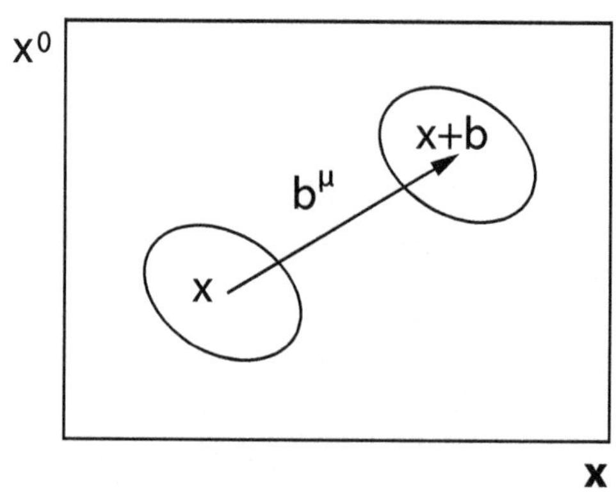

Fig. 14.1 The symmetry is here illustrated by showing how a local field keeps its form under a translation between two places.

14.2 Noether's theorem

The central content of Noether's theorem is that symmetry transformations give rise to conservation laws. A condition for this to be the case is that the symmetry transformation is a continuous function, such as the one shown in (14.17). The starting point here is an infinitesimal symmetry transformation,

$$x \to y: \quad y^\nu = x^\nu + \delta x^\nu, \quad \delta x^\nu = \delta x^\nu(x), \tag{14.24}$$

$$\phi \to \psi: \quad \psi^i(y) = \phi^i(x) + \delta\phi^i, \quad \delta\phi^i = \delta\phi^i(\phi, x), \tag{14.25}$$

where δx^ν is a function of x and $\delta\phi^i$ is a function of ϕ and x. The expressions for y and $\psi(y)$ can be taken from equation (14.17), which implies that it is a symmetry transformation. We take this to be a first-order Taylor expansion in the infinitesimal variations $\delta x^\nu(x)$ and $\delta\phi^i(\phi, x)$.

Derivation of (14.24) with respect to y^μ gives

$$\delta^\nu_\mu = \frac{\partial x^\nu}{\partial y^\mu} + \frac{\partial(\delta x^\nu)}{\partial y^\mu} \to \frac{\partial x^\nu}{\partial y^\mu} + \frac{\partial(\delta x^\nu)}{\partial x^\mu}, \tag{14.26}$$

where the last expression is equal to the previous one, up to first order in the infinitesimal variables. By reorganizing the terms, we get

$$\frac{\partial x^\nu}{\partial y^\mu} = \delta^\nu_\mu - \delta x^\nu_{,\mu}. \tag{14.27}$$

Note that in the following, we use the identities

$$\delta x^\nu_{,\mu} = \frac{\partial(\delta x^\nu)}{\partial x^\mu}, \quad \phi^i_{,\mu} = \frac{\partial\phi^i}{\partial x^\mu}, \quad \mathcal{L}_\mu = \frac{d\mathcal{L}}{dx^\mu}. \tag{14.28}$$

Derivation of (14.25) with respect to y^μ gives to first order

$$\begin{aligned}\frac{\partial\psi^i}{\partial y^\mu} = \frac{\partial\psi^i}{\partial x^\nu}\frac{\partial x^\nu}{\partial y^\mu} &= \left(\frac{\partial\phi^i}{\partial x^\nu} + \frac{\partial(\delta\phi^i)}{\partial x^\nu}\right)(\delta^\nu_\mu - \delta x^\nu_{,\mu}) \\ &= \phi^i_{,\mu} + \delta\phi^i_{,\mu} - \phi^i_{,\nu}\,\delta x^\nu_{,\mu}.\end{aligned} \tag{14.29}$$

Since the difference between x^μ and y^μ is infinitesimal, we can write the functional determinant in (14.17), to first order, as

$$|\partial x/\partial y| = 1 - \delta x^\mu_{,\mu}. \tag{14.30}$$

We continue to find the Taylor expansion of the full equation (14.17) to first order. The left part is

$$\begin{aligned}
\mathcal{L}\left(\psi^i, \frac{\partial \psi^i}{\partial y^\mu}, y\right) &= \mathcal{L}(\phi^i + \delta\phi^i,\ \phi^i_{,\mu} + \delta\phi^i_{,\mu} - \phi^i_{,\nu}\ \delta x^\nu_{\ \mu},\ x + \delta x) \\
&= \mathcal{L} + \frac{\partial \mathcal{L}}{\partial \phi^i}\delta\phi^i + \frac{\partial \mathcal{L}}{\partial \phi^i_{,\mu}}\delta\phi^i_{,\mu} - \frac{\partial \mathcal{L}}{\partial \phi^i_{,\mu}}\phi^i_{,\nu}\ \delta x^\nu_{\ ,\mu} + \frac{\partial \mathcal{L}}{\partial x^\nu}\delta x^\nu \\
&= \mathcal{L} + \left(\frac{\partial \mathcal{L}}{\partial \phi^i_{,\mu}}\delta\phi^i - \frac{\partial \mathcal{L}}{\partial \phi^i_{,\mu}}\phi^i_{,\nu}\ \delta x^\nu\right)_{,\mu} + \mathcal{L}_\nu \delta x^\nu, \qquad (14.31)
\end{aligned}$$

where the final expression has been modified by use of the Euler–Lagrange equation,

$$\frac{\partial \mathcal{L}}{\partial \phi^i} - \left(\frac{\partial \mathcal{L}}{\partial \phi^i_{,\mu}}\right)_{,\mu} = 0. \qquad (14.32)$$

The right part of equation (14.17) is

$$\mathcal{L}(\phi, \phi_{,\mu}, x)\ |\partial x/\partial y| = \mathcal{L} - \mathcal{L}\ \delta x^\mu_{\ ,\mu}, \qquad (14.33)$$

and combining this with the result in (14.31), we find

$$\left(\frac{\partial \mathcal{L}}{\partial \phi^i_{,\mu}}\delta\phi^i - \frac{\partial \mathcal{L}}{\partial \phi^i_{,\mu}}\phi^i_{,\nu}\ \delta x^\nu + \mathcal{L}\ \delta x^\mu\right)_{,\mu} = 0, \qquad (14.34)$$

or, more compact,

$$\frac{d}{dx^\mu}\left(\frac{\partial \mathcal{L}}{\partial \phi^i_{,\mu}}(\delta\phi^i - \phi^i_{,\nu}\ \delta x^\nu) + \mathcal{L}\ \delta x^\mu\right) = 0. \qquad (14.35)$$

This equation describes the conservation laws of Noether's theorem.

Conservation laws

The equation (14.35), which is named the Noether current, can be written as

$$\frac{dN^\mu}{dx^\mu} = 0, \qquad (14.36)$$

with

$$N^\mu \equiv \frac{\partial \mathcal{L}}{\partial \phi^i_{,\mu}}\ \delta\phi^i + T^\mu_{\ \nu}\ \delta x^\nu, \qquad (14.37)$$

$$T^\mu_{\ \nu} \equiv \mathcal{L}\ \delta^\mu_\nu - \frac{\partial \mathcal{L}}{\partial \phi^i_{,\mu}}\phi^i_{,\nu}. \qquad (14.38)$$

We take x^0 to be identical to ct, with x^i denoting the three space-like cartesian coordinates. Equation (14.36) is then

$$\frac{dN^0}{dx^0} + \frac{dN^i}{dx^i} = 0 \quad \Rightarrow \quad \frac{dN^0}{dt} + \boldsymbol{\nabla} \cdot (c\mathbf{N}) = 0. \tag{14.39}$$

We can recognise this as an equation similar to the continuity equation for electric charges:

$$\frac{\partial \rho}{\partial t} + \boldsymbol{\nabla} \cdot \mathbf{j} = 0. \tag{14.40}$$

Integration of (14.39) over an arbitrary spacelike volume V implies

$$\int_V \left(\frac{dN^0}{dt} + \boldsymbol{\nabla} \cdot (c\mathbf{N}) \right) dV = 0, \tag{14.41}$$

$$\Rightarrow \quad -\frac{d}{dt} \int_V N^0 dV = \int_V \boldsymbol{\nabla} \cdot (c\mathbf{N}) dV = \int_\sigma c\mathbf{N} \cdot d\sigma. \tag{14.42}$$

In general terms, this can be understood as the following: The integral of N^0 represents the quantity of an unspecified "substance" (for example: charge) in the volume V. The contribution to the left in (14.42) shows how much of this substance decreases per unit time. The contribution to the right in (14.42) compensate for this loss. Thus, the interpretations of N^0 and $c\mathbf{N}$ are

N^0 the density of the substance,

$c\mathbf{N}$ the flux density of the same. (14.43)

Assume now that the surface of the volume σ is expanded to infinity while the field goes rapidly to zero. The result is the conservation law

$$\frac{d}{dt} \int N^0 dV = 0 \quad \Rightarrow \quad \int N^0 dV = \text{const.} \tag{14.44}$$

This shows the essential content of Noether's theorem, that an infinitesimal symmetry transformation gives rise to a continuity equation.

When the variables δx^μ and $\delta \phi^i$, in the transformation, imply several (independent) parameters, there will be one continuity equation for each of these. As an example, a homogenous Lorentz transformation includes six independent equations.

14.3 Applications

Translation invariance

We assume in this case that the Lagrangian density $\mathcal{L}$ does not have explicit dependence of time t and position $\boldsymbol{x}$. The transformation then has the form

$$\psi^i(y) = \phi^i(x), \quad y^\mu = x^\mu + b^\mu, \tag{14.45}$$

which shows that it is a symmetry transformation. It also shows that the functional determinant is $|\partial x/\partial y| = 1$, which implies that equation (14.17) is simplified as

$$\mathcal{L}\left(\psi(y), \frac{\partial\psi(y)}{\partial y^\mu}, y\right) = \mathcal{L}\left(\phi(x), \frac{\partial\phi(x)}{\partial x_\mu}, x\right). \tag{14.46}$$

The infinitesimal variations are $\delta\phi^i = 0$ and $\delta x_\nu = b_\nu$. From (14.37), we find

$$N^\mu = T^{\mu\nu} b_\nu. \tag{14.47}$$

Since the four parameters b_ν are independent, this gives four continuity equations:

$$\frac{dT^{\mu\nu}}{dx^\mu} = 0, \quad (\nu = 0, 1, 2, 3). \tag{14.48}$$

The corresponding preserved quantities are interpreted as energy and momentum:

$$\begin{aligned}
\text{Energy:}\quad E &= \int T^{00} dV \\
&= \int \left(\mathcal{L}\, g^{00} - \frac{\partial\mathcal{L}}{\partial\phi^i_{,0}}\phi^{i,0}\right) dV \\
&= \int \left(-\mathcal{L} + \frac{\partial\mathcal{L}}{\partial\dot{\phi}^i}\dot{\phi}^i\right) dV, \\
\text{Momentum:}\quad P^k &= \frac{1}{c}\int T^{0k} dV \\
&= -\frac{1}{c}\int \frac{\partial\mathcal{L}}{\partial\phi^i_{,0}}\,\phi^{i,k}\, dV \\
&= -\int \frac{\partial\mathcal{L}}{\partial\dot{\phi}^i}\,\phi^{i,k}\, dV.
\end{aligned} \tag{14.49}$$

Concerning the final result of energy E, we can note an analogy with the expression for the Hamiltonian in the case of a point description of the energy

$$H = -L + \frac{\partial L}{\partial \dot{q}_i}\dot{q}_i. \tag{14.50}$$

Energy can thus be interpreted as a quantity connected to translation in time, while momentum is being interpreted as translation in space. There is also a relation with quantum mechanics, where energy and time have an uncertainty relation, and so has momentum and position.

Back to classical physics. In a relativistic invariant theory, we have

$$P^\mu = \frac{1}{c}\int T^{0\mu}\, dV, \tag{14.51}$$

where P^μ is interpreted as a four vector for energy and momentum, while $T^{\mu\nu}$ is interpreted as an energy–momentum (density) tensor. As follows from the interpretation of N^0 and $c\mathbf{N}$ in (14.43), we have the following interpretation of the different parts of $T^{\mu\nu}$:

$$\begin{aligned} T^{00} &= \text{energy density}, \\ cT^{i0} &= \text{ flux density of energy in the direction of } x^i, \\ \frac{1}{c}T^{0k} &= \text{component of a k-momentum density}, \\ T^{ik} &= \text{flux density of k-momentum in the direction } x^i. \end{aligned} \tag{14.52}$$

Lorentz transformation

In a relativistic field theory, the Lagrangian density is form invariant under Lorentz transformations, which means that equation (14.46) is satisfied. This secures that the relativity principle is satisfied, which means that energy, momentum, charge, etc. are expressed in the same way in all the basic reference frames.

We have already studied the inhomogeneous transformation, $x^\mu \to x^\mu + b^\mu$, and we will now study a homogeneous Lorentz transformation. The infinitesimal change is then

$$x^\mu \to y^\mu = x^\mu + \omega^{\mu\lambda}x_\lambda, \quad \omega^{\mu\lambda} = -\omega^{\lambda\mu}, \tag{14.53}$$

with $\omega^{\mu\lambda}$ as an infinitesimal constant. This implies

$$\delta x_\mu = \omega_{\mu\lambda}\, x^\lambda. \tag{14.54}$$

Which to the first order gives the identity $y^\mu y_\mu = x^\mu x_\mu$.

Due to the antisymmetry in the indices, $\omega_{\mu\lambda}$ includes six independent parameters. Physically, $\omega_{\mu\lambda}$ can be interpreted as an angular velocity and a linear velocity, both infinitesimal. Thus, y is produced by rotating x an angle

$$\boldsymbol{\Omega} = (\omega_{32}, \omega_{12}, \omega_{21}), \tag{14.55}$$

and boosted with the velocity $c\boldsymbol{\beta}$, where

$$\boldsymbol{\beta} = (\omega_{10}, \omega_{20}, \omega_{30}). \tag{14.56}$$

A scalar field

We continue with including a scalar field ϕ, with $\mathcal{L}$ as the Lagrangian density. The Lorentz transformation of the scalar field is a symmetry transformation, expressed by

$$x \to y = x + \delta x, \quad \phi(x) \to \psi(y) = \phi(x),$$

$$\delta x_\mu = \omega_{\mu\lambda} x^\lambda, \quad \delta\phi = 0. \tag{14.57}$$

As follows from (14.37), we have

$$N^\mu = T^{\mu\nu} x^\lambda \; \omega_{\nu\lambda}, \tag{14.58}$$

and by use of the antisymmetry of $\omega_{\nu\lambda}$, this can be rewritten as

$$\begin{aligned} N^\mu &= \frac{1}{2}(T^{\mu\nu} x^\lambda - T^{\mu\lambda} x^\nu) \; \omega_{\nu\lambda} \\ &= \frac{1}{2} L^{\mu\nu\lambda} \; \omega_{\nu\lambda}. \end{aligned} \tag{14.59}$$

We have here introduced the rank three tensor $L^{\mu\nu\lambda}$,

$$L^{\mu\nu\lambda} \equiv T^{\mu\nu} x^\lambda - T^{\mu\lambda} x^\nu. \tag{14.60}$$

It is antisymmetric in the two last indices, ν and λ.

The continuity equation (14.36) in this case can be written as

$$\frac{dL^{\mu\nu\lambda}}{dx^\mu} = 0, \tag{14.61}$$

where there is one equation for each choice of index pair $\omega_{\nu\lambda}$ in (14.59).

Constants of motion

The constants of motion are found in (14.60) by setting $\mu = 0$ and integrating over space. A factor $-1/c$ is also added:

$$\begin{aligned} L^{\nu\lambda} &= -\frac{1}{c} \int L^{0\nu\lambda} \; dV \\ &= \frac{1}{c} \int (x^\nu T^{0\lambda} - x^\lambda T^{0\nu}) dV. \end{aligned} \tag{14.62}$$

Due to the antisymmetry, $L^{\nu\lambda} = -L^{\lambda\nu}$, there are six independent constants of motion. We will look at how they are interpreted.

1. We first consider the situation where both ν and λ are being space indexes:

$$L^{jk} = \frac{1}{c}\int (x^j T^{0k} - x^k T^{0j})\, dV. \tag{14.63}$$

Since T^{0k}/c is the k component of the momentum density g, this can also be written as

$$L^{jk} = \int (x^j g^k - x^k g^j) dV, \tag{14.64}$$

which are the components of the moment of momentum around the origin,

$$\mathbf{L} = (L^{23}, L^{31}, L^{12}) = \int \boldsymbol{r} \times \boldsymbol{g}\; dV. \tag{14.65}$$

2. Next, we look at how L^{0k} is explained. It depends explicitly on time, but nevertheless, it is a constant of motion,

$$\begin{aligned} L^{0k} &= \frac{1}{c}\int (x^0 T^{0k} - x^k T^{00})\, dV \\ &= x^0 \int \frac{1}{c} T^{0k}\, dV - \frac{1}{c}\int x^k T^{00}\, dV \\ &= ctP^k - (E/c)\langle x^k\rangle \end{aligned} \tag{14.66}$$

Here is P^k the momentum and E the energy:

$$P^k = \int \frac{1}{c} T^{0k}\, dV, \quad E = \int T^{00}\, dV, \tag{14.67}$$

where both of these are constants of motion. $\langle x^k\rangle$ can be interpreted as coordinate for the centre of the energy (or mass). In vector form,

$$E\langle \boldsymbol{r}\rangle = \int T^{00} \boldsymbol{r} dV, \tag{14.68}$$

with T^{00} as the energy density.

The fact that L^{0k} is a constant of motion can therefore be explained by the equation

$$\langle \boldsymbol{r}\rangle = (c^2 \boldsymbol{P}/E)\, t + \text{const.} \tag{14.69}$$

This equation is the same as the one which is relevant for a free particle. The velocity of the particle has the well-known description in form of $\boldsymbol{p}$ and E,

$$c^2 \mathbf{p}/E = c^2 \gamma m_0 \boldsymbol{v}/(\gamma m_0 c^2) = \boldsymbol{v}. \tag{14.70}$$

In addition to the continuity equation (14.61), there is the continuity equation (14.48), $dT^{\mu\nu}/dx^{\mu} = 0$, which follows from the symmetry under translation in space and time. When introduced in (14.61), this gives

$$\begin{aligned} \frac{dL^{\mu\nu\lambda}}{dx^{\mu}} &= \frac{\partial}{\partial x^{\mu}}(T^{\mu\nu}x^{\lambda} - T^{\mu\lambda}x^{\nu}) \\ &= T^{\mu\nu}\delta^{\lambda}_{\mu} - T^{\mu\lambda}\delta^{\nu}_{\mu} \\ &= T^{\lambda\nu} - T^{\nu\lambda} \\ &= 0. \end{aligned} \tag{14.71}$$

The symmetry $T^{\lambda\nu} = T^{\nu\lambda}$, which applies to energy–momentum tensors of scalar fields, is hidden in the fact that $\mathcal{L}$ is form invariant under Lorentz transformations. As an example with the Lagrangian density,

$$\mathcal{L} = -(\phi^{*,\nu}\phi_{,\nu} + \mu^2\phi^*\phi), \tag{14.72}$$

the symmetry of $T^{\mu\nu}$ is

$$\begin{aligned} T^{\mu\nu} &= g^{\mu\nu}\mathcal{L} - \frac{\partial\mathcal{L}}{\partial\phi_{,\mu}}\phi^{,\nu} - \frac{\partial\mathcal{L}}{\partial\phi^*_{,\mu}}\phi^{*,\nu} \\ &= g^{\mu\nu}\mathcal{L} + \phi^{*,\mu}\phi^{,\nu} + \phi^{*,\nu}\phi^{,\mu}, \end{aligned} \tag{14.73}$$

which shows explicitly how $T^{\mu\nu}$ is symmetric.

14.4 Exercises

Problem 14.1

The Lagrangian density of a string with mass density μ and string tension σ is

$$\mathcal{L} = \frac{1}{2}\mu\left\{\left(\frac{\partial y}{\partial t}\right)^2 + \left(\frac{\partial z}{\partial t}\right)^2\right\} - \frac{1}{2}\sigma\left\{\left(\frac{\partial y}{\partial x}\right)^2 + \left(\frac{\partial z}{\partial x}\right)^2\right\}, \tag{14.74}$$

as has already been used, see (13.84). The end points of the string are $x = 0$ and $x = a$.

(a) Find the energy–momentum tensor. Is energy and momentum conserved?

(b) Show that for an arbitrary constant α, the following transformation is symmetry transformation:

$$\begin{pmatrix} y \\ z \end{pmatrix} \to \begin{pmatrix} y' \\ z' \end{pmatrix} = \begin{pmatrix} \cos\alpha & -\sin\alpha \\ \sin\alpha & \cos\alpha \end{pmatrix}\begin{pmatrix} y \\ z \end{pmatrix}. \tag{14.75}$$

What is the physical meaning of the symmetry and of the conservation law?

Problem 14.2

We consider a complex scalar field with Lagrangian density

$$\mathcal{L} = -\phi^{*,\nu}\phi_{,\nu} - \mu^2\, \phi^*\phi. \tag{14.76}$$

A transformation

$$\phi \to \psi = \phi\, e^{i\lambda}, \tag{14.77}$$

where λ is a real-valued constant, is a local symmetry transformation. It is an example of a *gauge transformation*, which changes variables ϕ without changing the coordinates x^μ. This is called a *global* gauge transformation, since λ does not change with the space–time positions x^μ.

(a) The parameter λ can in principle take any real value and therefore also an infinitesimal one. This introduces an infinitesimal change in the field ϕ:

$$\delta\phi = i\lambda\phi, \quad \delta\phi^* = -i\lambda\phi^*. \tag{14.78}$$

The function N^μ is defined as

$$N^\mu = \frac{\partial\mathcal{L}}{\partial\phi_{,\mu}}\delta\phi + \frac{\partial\mathcal{L}}{\partial\phi^*_{,\mu}}\delta\phi^*. \tag{14.79}$$

Find the expression of this field in terms of ϕ and ϕ^*.

(b) By use of this expression, find if the following continuity equation is satisfied:

$$\frac{dN^\mu}{dx^\mu} = 0. \tag{14.80}$$

(c) Find, in a similar way, the expression for the momentum–energy tensor $T^{\mu\nu}$, which follows from the translation invariance.

Problem 14.3

(a) Show that the Lagrangian density

$$\mathcal{L} = \frac{1}{2}i\hbar(\phi^*\dot{\phi} - \phi\,\dot{\phi}^*) - \frac{\hbar^2}{2m}\nabla\phi^*\cdot\nabla\phi - V\phi^*\dot{\phi} \tag{14.81}$$

gives the Schrödinger equation, and the complex conjugate, as the Euler–Lagrange equations.

(b) What is the condition for symmetry under time translation? Find the corresponding conserved quantity and give its interpretation.

What is the condition for symmetry under space translation. Find the corresponding conserved quantity also here and give its interpretation.

Chapter 15

Relativistic fields

15.1 Klein–Gordon field

The Klein–Gordon field is the simplest of relativistic classical fields, and for this reason, we begin with studying this field here. It is a scalar field, which can be a real Klein–Gordon field or a complex one. With ϕ as the Klein–Gordon field, the Klein–Gordon equation takes the form

$$(\Box - \mu^2)\phi = 0, \tag{15.1}$$

where μ can be interpreted as the Compton wave length, $\mu = mc/\hbar$, and m as the mass of quantum spinless particles.

The equation can be understood as a relation between energy E, momentum $\mathbf{p}$ and rest mass m,

$$E^2 - \mathbf{p}^2 c^2 = m^2 c^4, \tag{15.2}$$

with the substitutions

$$E \to i\hbar\frac{\partial}{\partial t}, \quad \mathbf{p} \to i\hbar\nabla. \tag{15.3}$$

Originally, the KG equation was meant to give a relativistic generalization of the Schrödinger equation. However, there is a problem with this interpretation since the wave equation, referred to as a first quantization of the equation, leads to problems with negative energies. But the so-called *second-quantization* of the field avoids this problem, and there are particles referred to as *pions* or *pi mesons*, which satisfy this criteria, although the short life time of these particles makes this correspondence not so important. *Higgs bosons* are also elementary scalar particles which are very unstable, and we will later have a look at this as a generalization of the Klein–Gordon equations. However, for classical KG fields, which are our focus here, the problem with negative energies are avoided.

Lagrange formulation

Real Klein–Gordon field

The Lagrangian density for the real Klein–Gordon field is

$$\begin{aligned}\mathcal{L} &= -\frac{1}{2}(\phi_{,\nu}\phi^{,\nu} + \mu^2\phi^2) \\ &= \frac{1}{2}\left(\frac{1}{c^2}\left(\frac{\partial\phi}{\partial t}\right)^2 - (\nabla\phi)^2 - \mu^2\phi^2\right). \end{aligned} \tag{15.4}$$

This gives an Euler–Lagrange equation which is consistent with the Klein–Gordon equation,

$$\frac{\partial\mathcal{L}}{\partial\phi} - \frac{d}{dx^\mu}\left(\frac{\partial\mathcal{L}}{\partial\phi_{,\mu}}\right) = (\Box - \mu^2)\,\phi = 0. \tag{15.5}$$

From the above Lagrangian density, the energy–momentum tensor gets the following form (see (14.37)):

$$\begin{aligned}T^{\mu\nu} &= -\frac{\partial\mathcal{L}}{\partial\phi_{,\nu}}\phi^{,\mu} + g^{\mu\nu}\mathcal{L} \\ &= \phi^{,\mu}\phi^{,\nu} + g^{\mu\nu}\mathcal{L}. \end{aligned} \tag{15.6}$$

The energy density is

$$T^{00} = (\phi^{,0})^2 - \mathcal{L} = \frac{1}{2}\left(\frac{1}{c^2}\left(\frac{\partial\phi}{\partial t}\right)^2 + (\nabla\phi)^2 + \mu^2\phi^2\right), \tag{15.7}$$

which has non-negative values. If we compare this with quantized Klein–Gordon fields, only the second-quantized Klein–Gordon fields have non-negative energies.

Complex Klein–Gordon field

The real Klein–Gordon field can in a simple way be changed to a complex field by changing the real field ϕ to a set of complex conjugate fields ϕ and ϕ^*. The Lagrangian density satisfies the equation

$$\mathcal{L} = -(\phi^*_{,\nu}\phi^{,\nu} + \mu^2\phi^*\phi). \tag{15.8}$$

The Euler–Lagrange equation for ϕ is found by variation of ϕ^*, and vice versa, and the results are

$$\begin{aligned}&\frac{\partial\mathcal{L}}{\partial\phi^*} - \frac{d}{dx^\mu}\left(\frac{\partial\mathcal{L}}{\partial\phi^*_{,\mu}}\right) = (\Box - \mu^2)\,\phi = 0, \\ &\frac{\partial\mathcal{L}}{\partial\phi} - \frac{d}{dx^\mu}\left(\frac{\partial\mathcal{L}}{\partial\phi_{,\mu}}\right) = (\Box - \mu^2)\,\phi^* = 0. \end{aligned} \tag{15.9}$$

This means that both ϕ and ϕ^* satisfy the same Klein–Gordon equation as the real Klein–Gordon field. In fact, the complex fields can be be expressed in terms of two real fields, ϕ^1 and ϕ^2 in the following way:

$$\phi = \frac{1}{2}(\phi^1 + i\phi^2), \quad \phi^* = \frac{1}{2}(\phi^1 - i\phi^2). \tag{15.10}$$

The energy–momentum tensor for the complex Klein–Gordon field is

$$\begin{aligned} T^{\mu\nu} &= -\left(\frac{\partial\mathcal{L}}{\partial\phi_{,\nu}}\phi^{,\mu} + \frac{\partial\mathcal{L}}{\partial\phi^*_{,\nu}}\phi^{*,\mu}\right) + g^{\mu\nu}\mathcal{L} \\ &= \phi^{*,\mu}\phi^{,\nu} + \phi^{,\mu}\phi^{*,\nu} + g^{\mu\nu}\mathcal{L}. \end{aligned} \tag{15.11}$$

It is symmetric, and the energy here is also non-negative:

$$T^{00} = 2|\phi^{,0}|^2 - \mathcal{L} = \frac{1}{c^2}\left(\frac{\partial\phi}{\partial t}\right)^2 + |\nabla\phi|^2 + \mu^2|\phi|^2. \tag{15.12}$$

Solutions to the Klein–Gordon equations

In order to find the solutions, we introduce the Fourier transform of a real Klein–Gordon field ϕ,

$$\phi(x) = \int d^4k\ e^{-i\underline{\mathbf{k}}\cdot\underline{\mathbf{x}}}\ \tilde{\phi}(k). \tag{15.13}$$

where $\underline{\mathbf{k}} = (k_0, \mathbf{k})$, $\underline{\mathbf{x}} = (ct, \mathbf{x})$, $\underline{\mathbf{k}}\cdot\underline{\mathbf{x}} = k_\mu x^\mu$ and $\underline{\mathbf{k}}^2 = k_\mu k^\mu$. We use this in the Klein–Gordon equation,

$$(\Box - \mu^2)\phi = -\int d^4k\ e^{-i\underline{\mathbf{k}}\cdot\underline{\mathbf{x}}}(\underline{\mathbf{k}}^2 + \mu^2)\tilde{\phi}(k) = 0, \tag{15.14}$$

which implies

$$(\underline{\mathbf{k}}^2 + \mu^2)\tilde{\phi}(k) = 0. \tag{15.15}$$

From this equation, we deduce that $\tilde{\phi}(k)$ can be written as

$$\tilde{\phi}(k) = \delta(\underline{\mathbf{k}}^2 + \mu^2)\ \tilde{\varphi}(k), \tag{15.16}$$

with $\tilde{\varphi}(k)$ as an unspecified function. The reason for this follows from (15.16), where $\tilde{\phi}(k) \neq 0$ only when $\underline{\mathbf{k}}^2 + \mu^2 = 0$. In (15.15), this is expressed by including an infinitesimal change from the Dirac delta function $\delta(\underline{\mathbf{k}}^2 + \mu^2)$. The result $\underline{\mathbf{k}}^2 + \mu^2 = 0$ also shows that the components of $\underline{\mathbf{k}}^2$ are not all independent.

In order to study this further, we make use of how the following Dirac delta function can be simplified (see Appendix A):

$$\delta(g(x)) = \sum_i \frac{1}{|g'(x_i)|}\delta(x - x_i), \tag{15.17}$$

where x_i is a point where $g(x_i) = 0$. In our case, we have the correspondence

$$g(x) \to (\underline{\mathbf{k}}^2 + \mu^2) = k_0^2 - \omega_{\mathbf{k}}^2/c^2, \tag{15.18}$$

where $\omega_{\mathbf{k}}/c = \sqrt{\mathbf{k}^2 + \mu^2}$, and we introduce $\pm x_0$ as the two points where g vanishes, $g(\pm x_0) = 0$. The result is

$$\delta(g(x)) = \sum_i \frac{1}{|g'(x_0)|}(\delta(x + x_0) + \delta(x - x_0)). \tag{15.19}$$

Following the correspondence (15.18), this gives

$$\delta(\underline{\mathbf{k}}^2 + \mu^2) = \frac{c}{2\omega_{\mathbf{k}}}(\delta(k_0 - \omega_{\mathbf{k}}) + \delta(k_0 + \omega_{\mathbf{k}})), \tag{15.20}$$

and this can be used to integrate over k_0,

$$\phi(x) = \int \frac{d^3k}{(2\pi)^3} \frac{c}{2\omega_{\mathbf{k}}}(a(k)e^{-i\underline{\mathbf{k}}\cdot\underline{\mathbf{x}}} + a^*(k)e^{i\underline{\mathbf{k}}\cdot\underline{\mathbf{x}}}), \tag{15.21}$$

where $d^3k/((2\pi)^3 2\omega_{\mathbf{k}})$ is the Lorentz invariant, as follows from the identity (15.20), and $\underline{\mathbf{k}}$ is now to be understood as $(\omega_{\mathbf{k}}/c, \mathbf{k})$.

This expression shows that the general solution of the Klein–Gordon equation has the simple form of a superposition of harmonic oscillators.

Generalizations of the Klein–Gordon equation

Internal symmetries

We consider first the real Klein–Gordon field, where we add a ϕ^4 term. The Lagrangian density can be written as

$$\mathcal{L} = -\frac{1}{2}g^{\mu\nu}\phi_{,\mu}\phi_{,\nu} - V(\phi), \tag{15.22}$$

with

$$V(\phi) = \frac{1}{2}\mu^2\phi^2 + \frac{\lambda}{4}\phi^4. \tag{15.23}$$

The internal symmetry acts as

$$\phi(x) \to -\phi(x), \tag{15.24}$$

where the transformation leaves (15.22) invariant since the scalar potential V contains only terms with even number of ϕ.

The Euler–Lagrange equation gives rise to the following equation of motion:

$$\Box\phi - V'(\phi) = 0. \tag{15.25}$$

This shows that, if $V'(v) = 0$ for some v, there are solutions where $\phi(x)$ takes the constant value v, independent of its position in space–time. If the constant λ is negative, this implies that the quartic part of $V(\phi)$ will be unbounded from below, with negative energy as $\phi \to \infty$. To avoid this, there is a restriction $\lambda > 0$, which implies that there is a single minimum at $\phi = v = 0$.

However, an interesting change appears when we consider the situation where we change $\mu^2 \to -\mu^2$,

$$V(\phi) = -\frac{1}{2}\mu^2\phi^2 + \frac{\lambda}{4}\phi^4. \tag{15.26}$$

This does not imply that $V(\phi)$, in this case, will be unbounded from below, since the quartic part of V will prevent that from happening.

What happens is that $\phi = 0$ defines a maximum of the potential, and there are two minima (see Fig. 15.1):

$$\phi_\pm = v_\pm = \pm\sqrt{\frac{\mu^2}{\lambda}}. \tag{15.27}$$

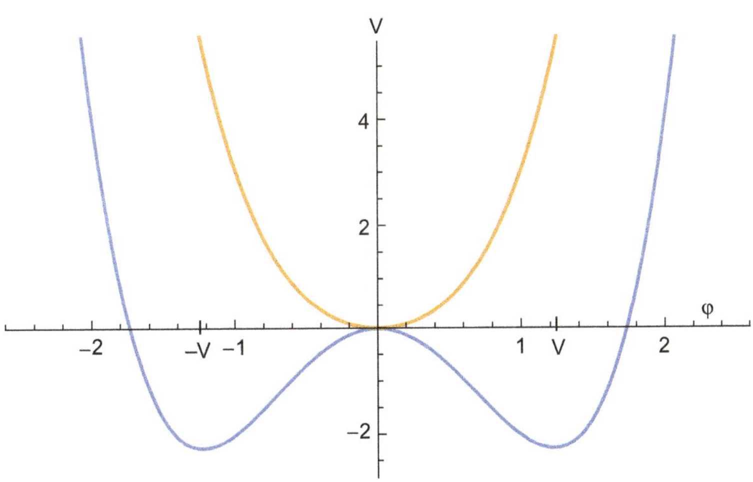

Fig. 15.1 Two types of Klein–Gordon fields with a term proportional to ϕ^4 added. The yellow curve shows results with $V(\phi)$ being positive and with a minimum at $\phi = 0$. The blue curve shows results where $V(\phi)$ takes both positive and negative values. In this case, there are two minima at positions $\phi = \pm v$. In both cases, the Euler–Lagrange solutions correspond to constant values of ϕ^2.

An interesting interpretation of $V(\phi)$ is the so-called *Higgs potential.* One of the minimal points $v_\pm$ is taken as what is known as the vacuum of the *Higgs field.* In the quantum description, Higgs boson particles[1] can be introduced by perturbations from the minimum $V(v_\pm)$,

$$V(v_\pm + \epsilon) = V(v_\pm) + \left(-\frac{1}{2}\mu^2 + \frac{3}{2}\lambda v^2\right)\epsilon^2. \tag{15.28}$$

As follows from (15.27), we have $\lambda = \mu^2/v^2$, and by use of (15.28) and $\mu = mc/\hbar^2$, we get

$$m_H^2 = -\frac{1}{2}m^2 + \frac{3}{2}\lambda v^2\hbar^2/c^2 = m^2. \tag{15.29}$$

The result is therefore that the Higgs boson mass m_H in this case is the same as the Klein–Gordon mass m.

We will next examine how a complex Klein–Gordon field with a ϕ^4 term added will change the results of the real Klein–Gordon field. The Lagrangian now takes the form,

$$\mathcal{L} = -\partial_\mu\phi^*\partial^\mu\phi - V(\phi^*\phi), \tag{15.30}$$

with

$$V(\phi^*\phi) = V_0 - \mu^2\phi^*\phi + \frac{1}{4}\lambda(\phi^*\phi)^2, \tag{15.31}$$

where we consider m^2 to give a negative contribution to V.

At the circle with minimum value of V, we may introduce the angular variable θ with

$$\phi(\theta) = ve^{i\theta} = \sqrt{\frac{2\mu^2}{\lambda}}e^{i\theta}, \tag{15.32}$$

which gives

$$\lambda = \frac{2\mu^2}{v^2} = \frac{2m^2\hbar^2}{v^2c^2}. \tag{15.33}$$

Similar to the Higgs boson discussed above, we can introduce boson particles also here (see Fig. 15.2). In fact, there are two types, one of them being massless. The first one, which has a mass, is found by making

[1]Peter Higgs (1929–2024) was a British theoretical physicist. He was a Professor at the University of Edinburgh and was a Nobel laureate in Physics in 2013 for his work which predicted the existence of a new particle: the Higgs boson.

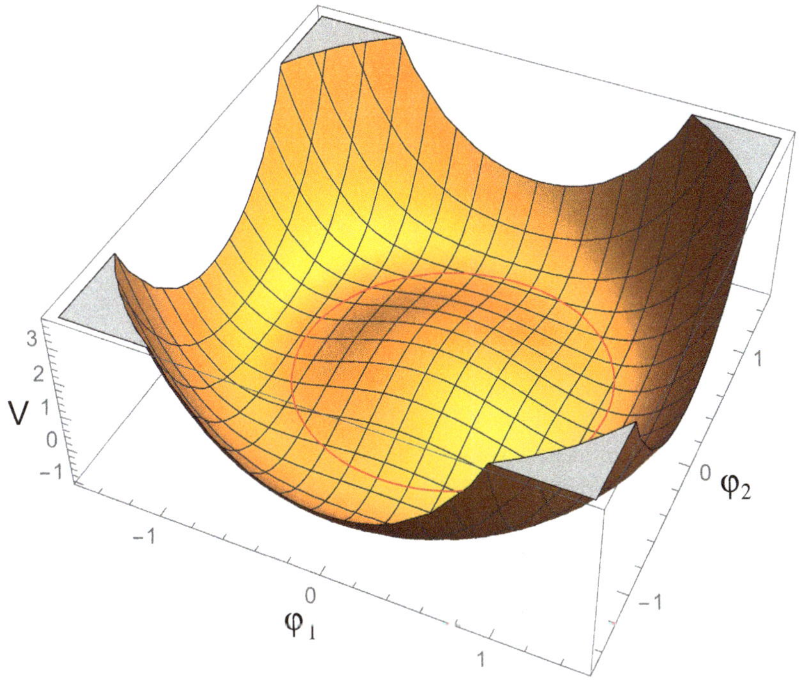

Fig. 15.2 A complex Klein–Gordon field with a term proportional to ϕ^4 included. The real variables in the figure, ϕ_1 and ϕ_2, are related to the two complex ones by $\phi_1 = (1/\sqrt{2})(\phi + \phi^*)$ and $\phi_2 = (1/\sqrt{2})(-\phi + \phi^*)$. The function $V(\phi^*\phi)$ is two-dimensional and rotationally invariant, with a local maximum at the centre $\phi = 0$. The red circle shows the position of the minimum of V.

a perturbation ϵ from the minimum circle in the orthogonal direction. We write this as

$$V(v + \epsilon) = V_0 - \mu^2(v + \epsilon)^2 + \frac{1}{4}\lambda(v + \epsilon)^4$$

$$= V(v) + \mu^2\epsilon^2 + 0(\epsilon^3), \tag{15.34}$$

where we make use of the rotational symmetry and set $\theta = 0$.

This result implies, in the same way as for the Higgs boson, that the present boson has mass $m_G^2 = m^2$. The mass of the other boson particle, in this case, is extracted from the infinitesimal change of $V(v)$ *along* the circle. But due to the rotational symmetry, this is zero, indicating that the boson is massless.

Sine-Gordon equation

The name “sine-Gordon equation” is a pun on the well-known Klein–Gordon equation in physics. The two equations, first Klein–Gordon,

$$\partial_x^2\,\phi - \partial_t^2\,\phi + \phi = 0, \tag{15.35}$$

and then sine-Gordon,

$$\partial_x^2\,\phi - \partial_t^2\,\phi - \sin\phi = 0 \tag{15.36}$$

(both with $c = 1$), in fact seem to be similar.

But the physics of the two are rather different. The sine-Gordon equation includes the interesting *solitons*. These are localized excitations, and there are two types, where the *kink* represents a positive twist, $\phi \to \phi + 2\pi$, and the *antikink* makes an opposite twist, $\phi \to \phi - 2\pi$.

The Lagrange density of the sine-Gordon equation is

$$\mathcal{L} = -\frac{1}{2}\partial_\mu\phi\partial^\mu\phi + (\cos\phi - 1), \tag{15.37}$$

and the corresponding Euler–Lagrange equations is

$$\partial_\mu\partial^\mu\phi - \sin\phi = 0, \tag{15.38}$$

where μ is restricted to the two coordinates, x and t. We will first see what the one-kink soliton looks like as a function $\phi(x)$, and afterward, see what happens with a collision of two-kink solitons. The solution of the sine-Gordon equation with a one-kink soliton at rest is

$$\phi_s(x) = 4\ \arctan\left(e^{x-x_0}\right), \tag{15.39}$$

which is illustrated in Fig. 15.3. The corresponding energy density is shown in the following equation:

$$\mathcal{E}(x) = \left(\frac{\partial\phi}{\partial x}\right)^2 = \left(\frac{4e^x}{1+e^{2x}}\right)^2, \tag{15.40}$$

and a following figure of the energy density is shown in Fig. 15.4.

A two-soliton system shows some of the characteristic features of interactions between solitons. We will study what happens in the case of a collision of two-kink solitons. For this situation, we have the following solution:

$$\phi(x,t) = 4\arctan\left(\frac{v\sinh(\gamma x)}{\cosh(\gamma v t)}\right), \tag{15.41}$$

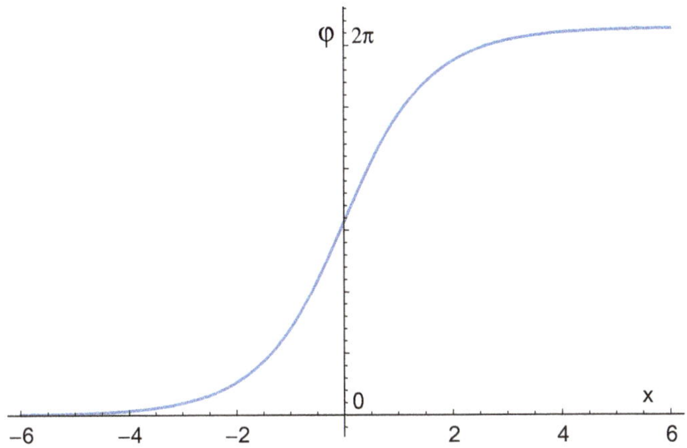

Fig. 15.3 A one-kink soliton with ϕ as a function of the coordinate x.

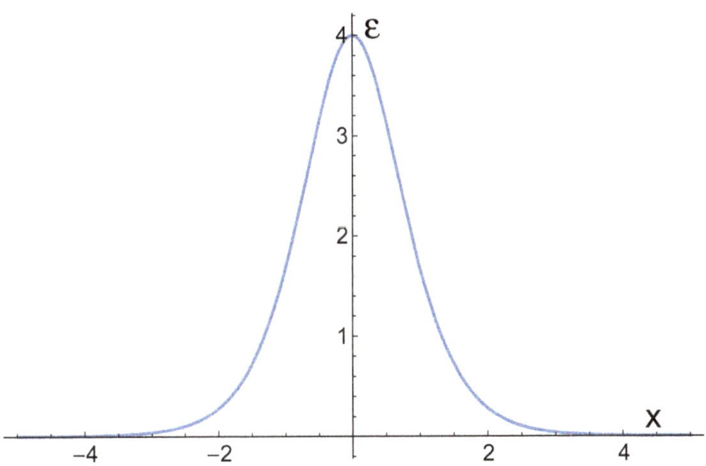

Fig. 15.4 The energy density of a soliton at rest is shown as a function of x.

with $\gamma = 1/\sqrt{1-v^2}$. It describes how the solitons collide and how they recover their velocity and shape. The interaction is referred to as an elastic collision, and the only observed effect is a phase shift, which can be interpreted as a time shift of the two solitons during their overlap.

In Fig. 15.5, a sequence of times with the corresponding position of the solitons are shown.

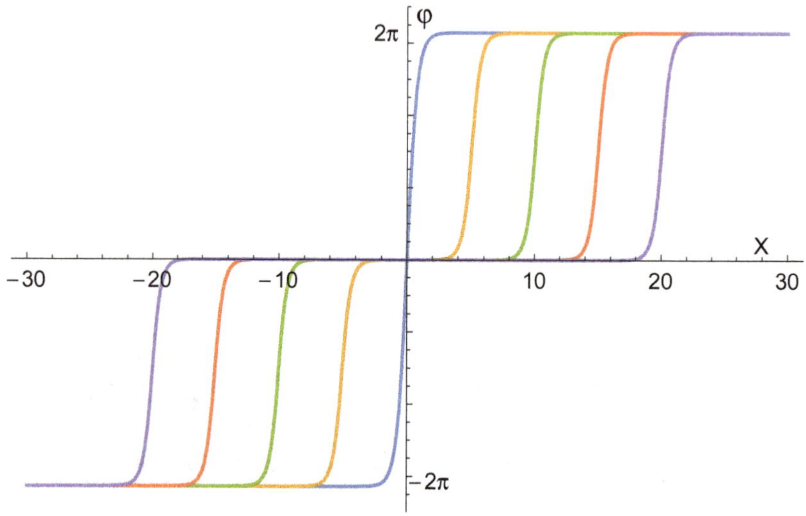

Fig. 15.5 A collision of two-kink solitons. The different colored lines show a sequence of positions of the solitons at increasing times. Since the solitons before and after the collision are precisely the same, the figure can be considered as showing both the situations before and after the collision.

15.2 Maxwell theory

In Part 3, we have studied many aspects of Maxwell's equations. The intention here is to examine how the Lagrangian formulation can add something to what we have already discussed.

Electromagnetic fields

We make a brief repetition of Maxwell's equations in vacuum:

$$\nabla \cdot \mathbf{E} = \frac{\rho}{\epsilon_0}, \quad \nabla \times \mathbf{B} - \frac{1}{c^2}\frac{\partial}{\partial t}\mathbf{E} = \mu_0 \mathbf{j}, \tag{15.42}$$

$$\nabla \cdot \mathbf{B} = 0, \quad \nabla \times \mathbf{E} + \frac{\partial}{\partial t}\mathbf{B} = 0, \tag{15.43}$$

where the source-free equations, (15.43), are referred to as homogenous and the equations (15.42), with sources, are referred to as inhomogenous equations.

Maxwell's equations have the same form in all inertial systems, and they can be expressed in covariant form, with the field variables written in the form of a 4×4 matrix tensor $F^{\mu\nu}$,

$$(F^{\mu\nu}) = \begin{pmatrix} 0 & E_1/c & E_2/c & E_3/c \\ -E_1/c & 0 & B_3 & -B_2 \\ -E_2/c & -B_3 & 0 & B_1 \\ -E_3/c & B_2 & -B_1 & 0 \end{pmatrix}. \tag{15.44}$$

The corresponding, *dual* field tensor $\tilde{F}^{\mu\nu}$, is in matrix form,

$$(\tilde{F}^{\mu\nu}) = \begin{pmatrix} 0 & -B_1 & -B_2 & -B_3 \\ B_1 & 0 & E_3/c & -E_2/c \\ B_2 & -E_3/c & 0 & E_1/c \\ B_3 & E_2/c & -E_1/c & 0 \end{pmatrix}, \tag{15.45}$$

and Maxwell's equations, when expressed in terms of the field tensors, have the compact form,

$$\partial_\nu F^{\mu\nu} = \mu_0 j^\mu, \tag{15.46}$$

$$\partial_\nu \tilde{F}^{\mu\nu} = 0. \tag{15.47}$$

The four-vector current density is composed of the charge and current densities in the following way:

$$(j^\mu) = (c\rho, \mathbf{j}). \tag{15.48}$$

Since the field tensor $F^{\mu\nu}$ is antisymmetric in the variables μ and ν, we note that equation (15.42) implies the *continuity equation*

$$\partial_\mu j^\mu = 0. \tag{15.49}$$

Potentials and gauge transformations

The homogeneous equations (15.43) are reduced to identities in the case when we introduce the potentials ϕ and $\mathbf{A}$, defined by

$$\mathbf{B} = \boldsymbol{\nabla} \times \mathbf{A}, \quad \mathbf{E} = -\boldsymbol{\nabla}\phi - \frac{\partial \mathbf{A}}{\partial t}, \tag{15.50}$$

and the tensor $F^{\mu\nu}$ gets the following expression as a function of A^μ:

$$F^{\mu\nu} = \partial^\mu A^\nu - \partial^\nu A^\mu. \tag{15.51}$$

with $(A^\mu) = (\phi/c, \mathbf{A})$. This expression for $F^{\mu\nu}$ is a general solution to (15.43), and installed in the inhomogeneous equation (15.42), it gives the following equation for the potential A^μ:

$$\partial_\nu \partial^\mu A^\nu - \partial_\nu \partial^\nu A^\mu = \mu_0 j^\mu. \tag{15.52}$$

Another way to describe this is

$$\Box A^\mu - \frac{\partial}{\partial x_\mu}\left(\frac{\partial A^\nu}{\partial x^\nu}\right) = -\mu_0 j^\mu. \tag{15.53}$$

The relations between the field tensors and the potentials are not unique, with an infinity of possible potentials A^μ being associated with the same field tensor $F^{\mu\nu}$. If $\bar{A}^\mu$ and A^μ are two such potentials, there is a relation between the two of the form where $\chi(x)$ is an arbitrary function of space and time,

$$\bar{A}^\mu = A^\mu + \partial^\mu \chi(x). \tag{15.54}$$

This gives the following relation between the field tensors that correspond to the two potentials:

$$\begin{aligned}\bar{F}^{\mu\nu} &= \partial^\mu \bar{A}^\nu - \partial^\nu \bar{A}^\mu \\ &= \partial^\mu (A^\nu + \partial^\nu \chi) - \partial^\nu (A^\mu + \partial^\mu \chi) \\ &= \partial^\mu A^\nu - \partial^\nu A^\mu + (\partial^\nu \partial^\mu - \partial^\mu \partial^\nu)\chi \\ &= F^{\mu\nu}.\end{aligned} \tag{15.55}$$

This shows that under the transformation (15.54), which is called a *gauge transformation*, there is no change in the field tensor, which is therefore gauge-invariant. The interpretation is that whereas $F^{\mu\nu}$ is a physical variable, this is not the case with the potentials A^μ.

The use of potentials is nevertheless useful since they can often simplify the mathematics in the solution of Maxwell's equations by choosing a gauge which is adapted to the problem to be solved. However, when using the Lagrange formulation as we do here, choosing potentials rather than field tensors as variables is simply a necessity.

Gauge fixing

The gauge invariance can be used to further make restrictions on A^μ. Such a restriction is

$$\partial_\mu A^\mu = 0, \tag{15.56}$$

which is called the Lorenz gauge. This gives the simpler wave equation,

$$\Box A^\mu = -\mu_0 j^\mu. \tag{15.57}$$

However, the Lorenz gauge is incomplete in some sense. There remains a subspace of gauge transformations which can also preserve the constraint.

These degrees of freedom correspond to gauge functions which satisfy the wave equation,

$$\partial_\mu \partial^\mu \psi = 0, \tag{15.58}$$

and propagate with the speed of light.

Another choice for the gauge invariance is the Coulomb gauge,

$$\nabla \cdot \mathbf{A} = 0, \tag{15.59}$$

which implies that

$$\Box A^\mu - \partial^\nu(\partial_0 A^0) = -\mu_0 j^\mu. \tag{15.60}$$

The Coulomb gauge has a number of properties, where the potentials can be expressed in terms of instantaneous values of the fields and densities,[2]

$$\begin{aligned} \mathbf{A}(\mathbf{r},t) &= \boldsymbol{\nabla} \times \int \frac{\mathbf{B}(\mathbf{r}',t)}{4\pi R} d^3\mathbf{r}', \\ A^0(\mathbf{r},t) &= \frac{1}{4\pi\epsilon_0} \int \frac{\rho(\mathbf{r}',t)}{R} d^3\mathbf{r}', \end{aligned} \tag{15.61}$$

where $\mathbf{R} = \mathbf{r} - \mathbf{r}'$, $R = |\mathbf{R}|$. $\mathbf{r}$ is any position vector in space, and $\mathbf{r}'$ is a point in the charge or current distribution, while $\boldsymbol{\nabla}$ operates on $\mathbf{r}$.

In the case where $j^\mu = 0$, we can further set $A^0 = 0$, which implies that

$$\Box \mathbf{A} - 0 \tag{15.62}$$

in addition to $\nabla \cdot \mathbf{A} = 0$. There are solutions in this case which are plane waves of the form

$$\mathbf{A}(\underline{\mathbf{x}}) = \int \frac{d^3k}{(2\pi)^3\, 2|\mathbf{k}|} (\boldsymbol{a}(\mathbf{k})\ \mathrm{e}^{-\mathrm{i}\ \underline{\mathbf{k}}\cdot\underline{\mathbf{x}}} + \boldsymbol{a}^*(\mathbf{k})\ \mathrm{e}^{\mathrm{i}\ \underline{\mathbf{k}}\cdot\underline{\mathbf{x}}}), \tag{15.63}$$

where $\underline{\mathbf{x}} = (ct, \mathbf{x})$, $\underline{\mathbf{k}} = (|\mathbf{k}|, \mathbf{k})$, $\underline{\mathbf{k}} \cdot \underline{\mathbf{x}} = k_\mu x^\mu$, and $\mathbf{k} \cdot \boldsymbol{a} = \mathbf{0}$. (See also related comments in Section 10.1, *Electromagnetic waves.*)

Lagrange formulation

By use of potentials, Maxwell's equations can be rewritten in the form of an Euler–Lagrange equation. In order to show this, we introduce the

[2] A description of this is found in A.M. Stewart, *Eur. J. Phys.* **24**, 519 (2003).

Lagrangian density in the following way:

$$\begin{aligned}\mathcal{L} &= \mathcal{L}(A^\mu, A^\mu{}_{,\nu}\ ; x) \\ &= -\frac{1}{4\mu_0} F_{\alpha\beta}F^{\alpha\beta} + j_\alpha A^\alpha,\end{aligned} \tag{15.64}$$

with $F^{\alpha\beta}$ expressed in terms of potentials,

$$\begin{aligned}F^{\alpha\beta} &= \partial^\alpha A^\beta - \partial^\beta A^\alpha \\ &\equiv A^{\beta,\alpha} - A^{\alpha,\beta},\end{aligned} \tag{15.65}$$

This gives the following result:

$$\frac{\partial}{\partial A^\nu_{,\mu}}(F_{\alpha\beta}F^{\alpha\beta}) = 4F^\mu{}_\nu, \tag{15.66}$$

and the Euler–Lagrange equation

$$\begin{aligned}&\frac{\partial \mathcal{L}}{\partial A^\nu} - \frac{d}{dx^\mu}\left(\frac{\partial \mathcal{L}}{\partial A^\nu_{,\mu}}\right) = 0 \\ \Rightarrow \quad &\frac{\partial}{\partial x^\mu} F^\mu{}_\nu = \mu_0 j_\nu.\end{aligned} \tag{15.67}$$

This is consistent with what we have previously found in (3.46) and (15.52).

In the expression we have used for the Lagrangian density, we have included the coupling of the field to the charge and current, j_α. With this as a contribution to the Lagrangian density, it will be included as an external, *explicit* dependence of the coordinates x_μ. The contribution $j_\alpha A^\alpha$ is an expression for the interaction between the Maxwell field and the charge and current. When j_α is taken as an explicit contribution, this means that we neglect the back reaction from the Maxwell field on the charge and current source. A more complete Lagrange formulation will therefore contain both the field variable, the charge variables and the combination of the two.

The free Maxwell field and the energy–momentum tensor

With a Maxwell field without sources, the Lagrangian density has the simple form

$$\mathcal{L} = -\frac{1}{4\mu_0} F_{\alpha\beta}F^{\alpha\beta}. \tag{15.68}$$

Since it does not depend, explicitly, on points in space and time, there exists an energy–momentum tensor $T^{\mu\nu}$ which respects the continuity condition, $\partial T^{\mu\nu}/\partial x^\mu = 0$. According to standard theory, the tensor $T^{\mu\nu}$ is defined by the following expression, see (14.38):

$$\begin{aligned} T^{\mu\nu} &= \mathcal{L}g^{\mu\nu} - \frac{\partial \mathcal{L}}{\partial A^\sigma_{,\mu}} A^{\sigma,\nu} \\ &= \mathcal{L}g^{\mu\nu} + \frac{1}{\mu_0} F^\mu_{\ \sigma} A^{\sigma,\nu} \\ &= \mathcal{L}g^{\mu\nu} + \frac{1}{\mu_0} F^{\mu\sigma}\, \partial^\nu A_\sigma. \end{aligned} \tag{15.69}$$

We have earlier, in (14.52), given an interpretation of $T^{\mu\nu}$, but this does not apply in the present case due to the lack of gauge invariance. We also note that here $T^{\mu\nu}$ is not symmetric. As a consequence, it has no obvious physical interpretation.

To compensate for this, Belinfante in 1939 introduced a modification to the energy–momentum tensor, which makes it symmetric and gauge-invariant. The idea is to add a term $t^{\mu\nu}$, which keeps the continuity equation intact and also keeps intact the constants of motion, $\int T^{0\nu} dV$. The term is

$$\begin{aligned} t^{\mu\nu} &= -\frac{1}{\mu_0} \partial_\sigma (F^{\mu\sigma} A^\nu) \\ &= -\frac{1}{\mu_0} F^{\mu\sigma} \partial_\sigma A^\nu, \end{aligned} \tag{15.70}$$

where the identity $\partial_\sigma F^{\mu\sigma} = 0$ has been used. It has the properties,

$$\frac{\partial t^{\mu\nu}}{\partial x^\mu} = -\frac{1}{4\mu_0} \partial_\mu \partial_\sigma (F^{\mu\sigma} A^\nu) = 0, \tag{15.71}$$

where the result is due to the antisymmetry of field tensor. Further more, we have

$$\begin{aligned} \int t^{0\nu} dV &= -\frac{1}{\mu_0} \int \partial_\mu (F^{0\mu} A^\nu) dV \\ &= -\frac{1}{\mu_0} \int \partial_i (F^{0i} A^\nu) dV \\ &= -\frac{1}{\mu_0} \int_\sigma (F^{0i} A^\nu) d\sigma_i \\ &\to 0, \end{aligned} \tag{15.72}$$

where the boundary of the integration region is taken to infinity.

The modified tensor is then

$$\begin{aligned}\theta^{\mu\nu} &= T^{\mu\nu} + t^{\mu\nu} \\ &= \mathcal{L}g^{\mu\nu} + \frac{1}{\mu_0}F^{\mu\sigma}\partial^\nu A_\sigma - \frac{1}{\mu_0}F^{\mu\sigma}\partial_\sigma A^\nu \\ &= \mathcal{L}g^{\mu\nu} + \frac{1}{\mu_0}F^{\mu\sigma}F^\nu{}_\sigma \\ &= \frac{\epsilon_0}{2}(E^2 - c^2B^2)g^{\mu\nu} + \frac{1}{\mu_0}F^{\mu\sigma}F^\nu{}_\sigma, \end{aligned} \tag{15.73}$$

where we have also made use of $\partial_\sigma F^{\mu\sigma} = 0$. This tensor is both gauge-invariant and symmetric.

Special cases of $\theta^{\mu\nu}$are

$$\begin{aligned}\theta^{00} &= \frac{\epsilon_0}{2}(E^2 + c^2B^2) \quad \text{energy density} \\ c\,\theta^{j0} &= \epsilon_0 c^2\ (\mathbf{E}\times\mathbf{B})_j \quad \text{Poynting's vector} \\ \frac{1}{c}\,\theta^{0k} &= \epsilon_0\ (\mathbf{E}\times\mathbf{B})_k \quad \text{momentum density.}\end{aligned} \tag{15.74}$$

$T^{\mu\nu}$ is called the *canonical tensor*, whereas $\theta^{\mu\nu}$ is referred to as the *symmetric energy–momentum tensor.*

Noether's theorem and the gauge-invariant transformation

As an interesting comment, we will in the final section show that the modified field tensor $\theta^{\mu\nu}$ can be related directly to the validity of Noether's theorem. This does not work, however, when $\theta^{\mu\nu} = T^{\mu\nu} + t^{\mu\nu}$, where each of these include A_μ on four-vector form. What we do instead is to lead A_μ to gauge-invariant transformations.[3]

Let us briefly consider how the background is. We assume that there is no external interference with the electromagnetic system, so the Lagrangian density has the form

$$\mathcal{L} = -\frac{1}{4\mu_0}F_{\alpha\beta}F^{\alpha\beta}; \quad F_{\mu\nu} = \partial_\mu A_\nu - \partial_\nu A_\mu. \tag{15.75}$$

$F_{\mu\nu}$ transforms as a tensor, and with an infinitesimal Lorentz transformation, there is a symmetry transformation of the Lagrangian. As follows

[3]A more complete description is found in E. Eriksen and J. M. Leinaas, *Phys. Scr.* **22**, 199 (1980).

from Noether's theorem, this symmetry transformation implies the continuity equation,

$$\partial_\mu N^\mu = 0, \quad N^\mu = \frac{\partial \mathcal{L}}{\partial A_{\sigma,\mu}} \delta A_\sigma + \mathcal{L} \delta x^\mu, \tag{15.76}$$

which makes the volume integral $\int N^0 d^3x$ a conserved quantity.

The variation of the vector potential, δA_μ, is usually determined by assuming A_μ to transform as a vector field. Conservation of energy and momentum is related to $\mathcal{L}$ under these transformations, and a direct application of (15.76) gives the continuity equation,

$$\partial_\mu T^{\mu\nu} = 0, \tag{15.77}$$

where $T^{\mu\nu}$ is the canonical energy–momentum tensor,

$$T^{\mu\nu} = \frac{\partial \mathcal{L}}{\partial A_{\sigma,\mu}} \partial^\nu A_\sigma + g^{\mu\nu} \mathcal{L}. \tag{15.78}$$

However, this is a gauge-dependent and non-symmetric quantity. This defect is usually repaired by means of Belinfante's method, where $T^{\mu\nu}$ is replaced by

$$\theta^{\mu\nu} = T^{\mu\nu} - \frac{1}{\mu_0} F^{\mu\sigma} \partial_\sigma A^\nu. \tag{15.79}$$

It is directly related to the energy and momentum densities of the electromagnetic field, and it gives a fully acceptable way of deriving these quantities. However, there is no longer a direct contact of $\theta^{\mu\nu}$ with Noether's theorem. This is the reason to seek for a different transformation scheme.

The alternative approach is, rather than having A_μ as a vector field, to have A_μ gauge-independent under the transformation. For an infinitesimal Lorentz transformation, we have

$$x'^\mu = x^\mu + \delta x^\mu, \;\; \delta x^\mu = \omega^{\mu\nu} x_\nu + a^\mu, \tag{15.80}$$

and

$$A'_\mu(x) = A_\mu(x) + \delta A_\mu, \tag{15.81}$$

with the following expression for the variation of A_μ:

$$\delta A_\mu = -\delta x^\nu \; F_{\nu\mu}. \tag{15.82}$$

δA_μ is now gauge-invariant since $F_{\nu\mu}$ is gauge-invariant, while the standard transformation of A_μ is

$$\delta A_\mu = -\delta x^\nu \; F_{\nu\mu} - \partial_\mu(\delta x^\nu A_\nu), \tag{15.83}$$

which is gauge-dependent.

The continuity equation can now be written as

$$\partial_\mu N^\mu = 0, \quad N^\mu = \left(-\frac{\partial \mathcal{L}}{\partial A_{\sigma,\mu}} F^\nu{}_\sigma + g^{\mu\nu} \mathcal{L} \right) \delta x_\nu, \tag{15.84}$$

It is gauge-invariant since we have $\partial \mathcal{L}/\partial A_{\sigma\mu} = -F^{\mu\sigma}$.

The explicit form of N^μ in the case of the Lorentz transformation is

$$N^\mu = -\theta^{\mu\nu} \delta x_\nu, \quad \delta x_\nu = \omega_{\nu\lambda} x^\lambda + a_\nu, \tag{15.85}$$

where $\theta^{\mu\nu}$ now is the symmetric energy–momentum tensor,

$$\theta^{\mu\nu} = \frac{1}{\mu_0} F^{\sigma\mu} F^\nu{}_\sigma + g^{\mu\nu} \mathcal{L}. \tag{15.86}$$

It follows from Eq. (15.84) that the continuity equation of $\theta^{\mu\nu}$,

$$\partial_\mu \theta^{\mu\nu} = 0, \tag{15.87}$$

is now *a direct* consequence of the symmetry under translations. Similarly, the continuity equation for the angular momentum density follows directly from (15.84),

$$\partial_\mu J^{\mu\nu\lambda} = 0, \quad J^{\mu\nu\lambda} = \theta^{\mu\nu} x^\lambda - \theta^{\mu\lambda}, x^\mu, \tag{15.88}$$

as a consequence of the symmetry of $\mathcal{L}$ under homogeneous Lorentz transformations.

15.3 Exercises

Problem 15.1

We have the following Lagrangian density:

$$\mathcal{L} = -\frac{1}{2} \partial_\mu \phi \, \partial^\mu \phi + \alpha(\cos\phi - 1), \tag{15.89}$$

with α as a real constant.

(a) Find the corresponding Euler–Lagrange equation, referred to as a sine-Gordon equation, and show that the following function is a solution:

$$\phi_{ss}(x,t) = 4 \arctan \left[\frac{\beta \sinh(\sqrt{\alpha}\gamma x)}{\cosh(\sqrt{\alpha}\gamma v t)} \right], \tag{15.90}$$

with $\beta = v/c$, $\gamma = 1/\sqrt{1-\beta^2}$.

(b) Study the time evolution of a collision process by plotting ϕ_{ss} as a function of x for a series of t. Choose a small value of v, ($\beta \ll 1$) and a large value ($\beta \simeq 1$). The situation can be understood as an interaction between two parts (solitons) of the total field. How can one see that the interaction between the solitons is repulsive?

Problem 15.2

A Lagrangian density for the electromagnetic field is given by the relativistic covariant form

$$\mathcal{L} = -\frac{1}{4\mu_0}\left(F^{\mu\nu}F_{\mu\nu} - 2\partial^\mu A_\mu\, \partial^\nu A_\nu\right) + j_\mu A^\mu, \tag{15.91}$$

where $F^{\mu\nu} = \partial^\mu A^\nu - \partial^\nu A^\mu$ and A_μ is the potential four-vector. j_μ is a four-vector with components $(c\rho, \mathbf{j})$.

(a) Show that the Lagrangian density leads directly to the wave equation of the vector potential,

$$\Box A_\mu = \mu_0\, j_\mu. \tag{15.92}$$

(b) Show also that this Lagrangian density is identical with that given in the text (15.60) except for the middle term. Explain why this extra term does not have any influence on the physics of the system.

Problem 15.3

The field equations to a complex Klein–Gordon field, which interacts with a Maxwell field, can be derived from the Lagrangian density,

$$\mathcal{L} = \frac{1}{2\mu_0}F_{\mu\nu}F^{\mu\nu} - \left(\left[(\partial_\nu - i\frac{q}{\hbar c}A_\nu)\phi\right]^*\left[(\partial^\nu - i\frac{q}{\hbar c}A^\nu)\phi\right] + \mu^2\phi^*\phi\right). \tag{15.93}$$

(a) Find the corresponding Euler–Lagrange equations (field equations).

(b) Find the charge current (Noether current) in the same way as in the earlier problem, see (14.77).

(c) Show that the field equations are invariant under the (local) gauge transformations,

$$\begin{aligned}
A_\mu(x) &\to A'_\mu(x') = A_\mu + \partial_\mu\theta(x),\\
\phi(x) &\to \phi'(x') = \exp\left[i\frac{q}{\hbar c}\theta(x)\right]\phi(x),\\
\phi(x)^* &\to \phi'(x')^* = \exp\left[-i\frac{q}{\hbar c}\theta(x)\right]\phi(x)^*.
\end{aligned} \tag{15.94}$$

Summary

We have in the beginning of the new part of the book focussed on Hamilton's principle and the Euler–Lagrange equation, which is simplified by a small number of variables. A simple case is included *Transverse waves on a string*, where vibrations in the string are restricted to the two-dimensional plane. From the Lagrangian density, the Euler–Lagrange equation has been found, and the general solutions have been determined. This concrete description has had the meaning of showing what the basics of Hamilton's principle are when this is used in more complex cases.

In the earlier Chapter 14, it was shown how the Lagrangian of a system could be transformed to a Hamiltonian, $L(q, \dot{q}, t) \rightarrow H(q, p, t)$. These two are equivalent but different when the system consists of particles. In cases where the system consists of fields, Hamilton formulations can be defined, but in general with results which are more complicated than the Lagrangian formulation. We have also noted that in the case of *waves on the string*, the use of the Hamiltonian formulation works well.

An important part of the classical field theory is Noether's theorem. This theorem shows how symmetry transformations give rise to conservation laws. This is shown here, with a formulation referred to as *Noether currents* and with general continuity equations describing these currents. To understand this better, several concrete applications with translational invariance have been discussed, and the physical quantities, such as energy and momentum, have been connected to the related Noether currents.

In the final chapter of the book, which is titled "Relativistic fields", the Klein–Gordon field and the Maxwell theory have both been studied by how the Lagrangian densities can influence these fields. The real and complex KG fields have been shown, with respect to the energy–momentum tensor, and solutions of real Klein–Gordon field have been introduced with Fourier transform. We have also studied generalized Klein–Gordon equations with internal symmetries. These are interesting, since one of them have been, in its quantum formulation, identified as real physical particles, *Higgs bosons*.

In the second case of Maxwell theory with Lagrangian density formulation, these have to use A_μ as variables. It is interesting to see, as shown in the final section, that Noether's theorem can lead directly to a definition of A_μ. However, in this case, $\boldsymbol{A}$ is *not* a vector field.

Appendix A

A.1 Dirac's delta function

Dirac's delta function in one dimension is formally defined by the properties

$$a) \quad \delta(x) = \begin{cases} 0 & x \neq 0 \\ \infty & x = 0 \end{cases}, \qquad b) \quad \int_{-\infty}^{\infty} dx\, \delta(x) = 1\,. \tag{A.1}$$

This implies that, for any well-behaved function $f(x)$, we have

$$\int_{-\infty}^{\infty} dx\, \delta(x) f(x) = f(0)\,. \tag{A.2}$$

Clearly, with this definition, $\delta(x)$ cannot be a function in the usual sense. However, it may be useful to consider $\delta(x)$ as the limit $\epsilon \to 0$ of a smooth function $\delta_\epsilon(x)$, where ϵ characterizes the width of the function around its peak at $x = 0$. The actual form of the function is not important, but a specific choice is the Gaussian

$$\delta_\epsilon(x) = \sqrt{\frac{1}{\epsilon\pi}}\, e^{-x^2/\epsilon}, \qquad \epsilon > 0\,, \tag{A.3}$$

which is normalized to satisfy condition b) in (A.1) for any ϵ, and condition a) in the limit $\epsilon \to 0^+$. In principle this limit should be taken at the end of computations of expressions where $\delta(x)$ appears, but the use of the properties (A.1) simplifies this limiting procedure.

The use of the regularized function $\delta_\epsilon(x)$ also makes it possible to give meaning to the derivative $\delta'(x)$, when this appears in integrals, simply by treating $\delta(x)$ as a differentiable function. For example, by partial integration we find the following identity

$$\int_{-\infty}^{\infty} dx\, \delta'(x) f(x) = -f'(0)\,. \tag{A.4}$$

We sometimes meet an expression of the form $\delta(g(x))$ and want to express it in terms of $\delta(x)$. Clearly $\delta(g(x))$ vanishes in all points, except where $g(x)$ vanishes. Let us assume first that there is only one such point, x_0, $g(x_0) = 0$, with $g'(x_0) \neq 0$. We consider the integral

$$\int_{x_0-\epsilon}^{x_0+\epsilon} \delta(g(x))dx = \int_{-\epsilon|g'(x_0)|}^{\epsilon|g'(x)|} \delta(g)\frac{1}{|g'(x)|}dg = \frac{1}{|g'(x_0)|}\,, \tag{A.5}$$

where a change of integration variable $x \to g$ has been performed. The integral over $\delta(g)$ only gives a contribution at the point $g = 0 \Leftrightarrow x = x_0$. The result for the integral of $\delta(g(x)$ is consistent with the following identity

$$\delta(g(x)) = \frac{1}{|g'(x_0)|}\delta(x - x_0)\,. \tag{A.6}$$

It is straight forward to generalize the above result to the case where the function $g(x)$ has several zeros $g(x_i) = 0, i = 1, 2, \ldots$. In this case the result is

$$\delta(g(x)) = \sum_i \frac{1}{|g'(x_i)|}\delta(x - x_i)\,. \tag{A.7}$$

The three-dimensional delta function $\delta(\mathbf{r})$ can be defined in a similar way as the one-dimensional one, by the conditions

$$a)\quad \delta(\mathbf{r}) = \begin{cases} 0 & \mathbf{r} \neq 0 \\ \infty & \mathbf{r} = 0 \end{cases}\,, \qquad b)\quad \int d^3r\, \delta(\mathbf{r}) = 1\,. \tag{A.8}$$

This implies for any well-behaved function $f(\mathbf{r})$,

$$\int d^3r\, \delta(\mathbf{r}) f(\mathbf{r}) = f(0)\,, \tag{A.9}$$

and for the derivative of the delta function,

$$\int d^3r\, (\partial_k \delta(\mathbf{r})) f(\mathbf{r}) = -\partial_k f(0)\,. \tag{A.10}$$

For a change of integration variable $\mathbf{r} \to \mathbf{g}(\mathbf{r})$, we find, in a similar way as in one dimension,

$$\delta(\mathbf{g}(\mathbf{r})) = \sum_i \left(\left| \frac{\partial \mathbf{g}}{\partial \mathbf{r}} \right|_{\mathbf{r}=\mathbf{r}_i} \right)^{-1} \delta(\mathbf{r} - \mathbf{r}_i)\,, \tag{A.11}$$

where $\left|\frac{\partial \mathbf{g}}{\partial \mathbf{r}}\right|$ represents the Jacobian of the mapping $\mathbf{r} \to \mathbf{g}(\mathbf{r})$, and $\mathbf{r}_i, i = 1, 2, \ldots$ are the zeros of the function $\mathbf{g}(\mathbf{r})$.

A.2 The Levi-Civita symbol

The Levi-Civita symbol in three dimensions is a three-index symbol, ϵ_{ijk}, where each of the indices takes the values $1, 2, 3$. Depending on the values of the indices the symbol takes the values -1, 0 or 1.

The Levi-Civita symbol is fully antisymmetric under permutations of the indices, which means that it changes sign under transposition of any pair of indices. This implies that it vanishes unless the values of all indices are different, and the definition $\epsilon_{123} = 1$ determines the nonvanishing values of ϵ_{ijk}, with (ijk) as permutations of the sequence (123). This may be written as

$$\epsilon_{ijk} = (-1)^p \epsilon_{123} \tag{A.12}$$

with $(-1)^p$ as the signature of the permutation, $\{ijk\} = p\{123\}$, or explicitly

$$\epsilon_{ijk} = \begin{cases} +1 & (ijk) = (123), (231), (312)\,, \quad \text{(even permutations)} \\ -1 & (ijk) = (321), (132), (213)\,, \quad \text{(odd permutations)}. \\ 0 & \qquad\qquad i = j \text{ or } j = k \text{ or } k = i \end{cases} \tag{A.13}$$

The product of two Levi-Civita symbols, with summation of one or more pairs of indices is related to the Kronecker delta in the following way (*note the use of Einstein's summation convention*),

$$\begin{aligned} \epsilon_{ijk}\epsilon_{imn} &= \delta_{jm}\delta_{kn} - \delta_{jn}\delta_{km}, \\ \epsilon_{ijk}\epsilon_{ijn} &= 2\delta_k n, \\ \epsilon_{ijk}\epsilon_{ijk} &= 6. \end{aligned} \tag{A.14}$$

The application of the Levi-Civita symbol is useful for many expressions of vector and matrix identities. We mention some of these here, first the expression for the determinant of a 3×3 matrix $\mathbf{A}$ with matrix elements a_{ij},

$$det\mathbf{A} = \epsilon_{ijk} a_{1i} a_{2j} a_{3k}. \tag{A.15}$$

The matrix elements of a cross product of vectors is

$$(\mathbf{a} \times \mathbf{b})_i = \epsilon_{ijk} a_j b_k \tag{A.16}$$

and the triple product

$$(\mathbf{a} \times \mathbf{b}) \cdot \mathbf{c} = \epsilon_{ijk} a_i b_j c_k. \tag{A.17}$$

The Levi-Civita symbol in four-dimensional Minkowski space is a four index symbol $\epsilon_{\mu\nu\sigma\rho}$, each of the indices being a relativistic four-vector index. The symbol is antisymmetric under transposition of any pair of indices, and therefore vanishes in all cases where two of the indices take the same value. The non-vanishing components are determined by the antisymmetry from the definition $\epsilon_{0123} = 1$, so that all components $\epsilon_{\mu\nu\sigma\rho}$, where the sequence of indices $(\mu\nu\sigma\rho)$ is obtained from (0123) by an *even* permutation, take the value $+1$, and those obtained by an odd permutation take the value -1.

The symbol $\epsilon^{\mu\nu\sigma\rho}$, with raised indices is obtained from $\epsilon_{\mu\nu\sigma\rho}$, in the standard way, by using the (inversed) metric tensor $g^{\mu\nu}$ as a raising operator. This will introduce a sign change, so that $\epsilon^{0123} = -\epsilon_{0123} = -1$. The same sign change applies to all the other (non-vanishing) components. Products of the two, with contraction of upper and lower indices, can be expressed in terms of the Kronecker delta, in much the same way as in the three dimensional case. We list the identities here

$$
\begin{aligned}
\epsilon_{\mu\nu\rho\sigma}\epsilon^{\mu\alpha\beta\gamma} &= -\delta^\alpha_\nu\delta^\beta_\rho\delta^\gamma_\sigma - \delta^\beta_\nu\delta^\gamma_\rho\delta^\alpha_\sigma - \delta^\gamma_\nu\delta^\alpha_\rho\delta^\beta_\sigma + \delta^\alpha_\nu\delta^\gamma_\rho\delta^\beta_\sigma + \delta^\gamma_\nu\delta^\beta_\sigma\delta^\alpha_\rho + \delta^\beta_\nu\delta^\alpha_\rho\delta^\gamma_\sigma,\\
\epsilon_{\mu\nu\rho\sigma}\epsilon^{\mu\nu\beta\gamma} &= -2(\delta^\beta_\rho\delta^\gamma_\sigma - \delta^\gamma_\rho\delta^\beta_\sigma),\\
\epsilon_{\mu\nu\rho\sigma}\epsilon^{\mu\nu\rho\gamma} &= -6\delta^\gamma_\sigma,\\
\epsilon_{\mu\nu\rho\sigma}\epsilon^{\mu\nu\rho\sigma} &= -24.
\end{aligned}
\tag{A.18}
$$

Appendix B

B.1 Physical units, SI system

Name	Symbol	Quantity	In SI base units	In other SI units
meter	m	length	base unit	
kilogram	kg	mass	base unit	
second	s	time	base unit	
newton	N	force	$kg \cdot m \cdot s^{-2}$	
hertz	Hz	frequency	s^{-1}	
joule	J	energy	$kg \cdot m^2 \cdot s^{-2}$	$N \cdot m$ $(C \cdot V)$
watt	W	power	$kg \cdot m^2 \cdot s^{-3}$	J/s
ampere	A	electric current	base unit	
coulomb	C	electric charge	$s \cdot A$	
volt	V	electric potential	$kg \cdot m^2 \cdot s^{-3} \cdot A^{-1}$	W/A
tesla	T	magnetic field (B-field)	$kg \cdot s^{-2} \cdot A^{-1}$	
farad	F	capacitance	$kg^{-1} \cdot m^{-2} \cdot s^4 \cdot A^2$	C/V

B.2 Physical constants

Quantity	Symbol	Value
speed of light in vacuum	c	$299\,792\,458\ \text{m/s}$
elementary charge	e	$1.602\,176\,621 \cdot 10^{-19}\ \text{C}$
gravitational acceleration (standard value)	g_n	$9.806\,65\ \text{m/s}^2$
magnetic constant (vacuum permeability)	μ_0	$4\pi \cdot 10^{-7}\text{N/A}^2$
electric constant (vacuum permittivity)	ϵ_0	$8.854\,187\,817 \cdot 10^{-12}\ \text{F/m}$
classical electron radius	r_e	$2.817\,940\,323 \cdot 10^{-15}\ \text{m}$
Bohr radius	a_0	$5.291\,772\,107 \cdot 10^{-11}\ \text{m}$
electron mass	m_e	$0.510\,998\,946\ \text{MeV}/c^2$
proton mass	m_p	$938.272\,081\ \text{MeV}/c^2$

The energy unit *electron volt* is defined as
$1\text{eV} = e \cdot 1\text{V} = 1.602\,176\,621 \cdot 10^{-19}$ J, with e as the elementary charge.

Appendix C

C.1 Integration theorems

Helmholtz decomposition

Let $\mathbf{F}$ be a vector field defined on $\mathbb{R}^3$, where $\mathbf{F}$ is twice continuously differentiable. Then $\mathbf{F}$ can be decomposed into a curl-free component and a divergence-free component,

$$\mathbf{F} = -\boldsymbol{\nabla}\phi + \boldsymbol{\nabla} \times \mathbf{A}\,. \tag{C.1}$$

If $\mathbf{F}$ vanishes faster than $1/r$ as $r \to \infty$, the two components are

$$\phi(\mathbf{r}) = \frac{1}{4\pi}\int \frac{\boldsymbol{\nabla}' \cdot \mathbf{F}(\mathbf{r}')}{|\mathbf{r}-\mathbf{r}'|}dV'\,, \quad \mathbf{A}(\mathbf{r}) = \frac{1}{4\pi}\int \frac{\boldsymbol{\nabla}' \times \mathbf{F}(\mathbf{r}')}{|\mathbf{r}-\mathbf{r}'|}dV'\,. \tag{C.2}$$

Gauss' theorem

Let $\mathbf{F}$ be a continuously differentiable vector field defined on a compact volume V in $\mathbb{R}^3$, with a smooth boundary S. We then have

$$\int_V (\boldsymbol{\nabla} \cdot \mathbf{F})dV = \oint_S \mathbf{F} \cdot d\mathbf{S} \tag{C.3}$$

where $d\mathbf{S}$ is the area element directed outwards and perpendicular to the integration surface S .

Stokes' theorem

Let $\mathbf{F}$ be a continuously differentiable vector field defined on a two-dimensional surface S in $\mathbb{R}^3$, with C as its boundary. We then have

$$\int_S (\boldsymbol{\nabla} \times \mathbf{F}) \cdot d\mathbf{S} = \oint_C \mathbf{F} \cdot d\mathbf{r}. \tag{C.4}$$

The direction of line integral around C has positive orientation relative to the orientation of the surface S.

C.2 Vector identities in $\mathbb{R}^3$

Vector products

$$
\begin{aligned}
\mathbf{a}\cdot(\mathbf{b}\times\mathbf{c}) &= \mathbf{b}\cdot(\mathbf{c}\times\mathbf{a}) = \mathbf{c}\cdot(\mathbf{a}\times\mathbf{b}) \\
\mathbf{a}\times(\mathbf{b}\times\mathbf{c}) &= (\mathbf{a}\cdot\mathbf{c})\mathbf{b} - (\mathbf{a}\cdot\mathbf{b})\mathbf{c} \\
(\mathbf{a}\times\mathbf{b})\cdot(\mathbf{c}\times\mathbf{d}) &= (\mathbf{a}\cdot\mathbf{c})(\mathbf{b}\cdot\mathbf{d}) - (\mathbf{a}\cdot\mathbf{d})(\mathbf{b}\cdot\mathbf{c})
\end{aligned}
\tag{C.5}
$$

Derivatives

$$
\begin{aligned}
\boldsymbol{\nabla}\cdot(\boldsymbol{\nabla}\times\mathbf{a}) &= 0 \\
\boldsymbol{\nabla}\times(\boldsymbol{\nabla} f) &= 0 \\
\boldsymbol{\nabla}\cdot(\mathbf{a}f) &= f\boldsymbol{\nabla}\cdot\mathbf{a} + \mathbf{a}\cdot\boldsymbol{\nabla} f \\
\boldsymbol{\nabla}\times(\mathbf{a}f) &= f\boldsymbol{\nabla}\times\mathbf{a} - \mathbf{a}\times\boldsymbol{\nabla} f \\
\boldsymbol{\nabla}\cdot(\mathbf{a}\times\mathbf{b}) &= \mathbf{b}\cdot(\boldsymbol{\nabla}\times\mathbf{a}) - \mathbf{a}\cdot(\boldsymbol{\nabla}\times\mathbf{b}) \\
\boldsymbol{\nabla}\times(\mathbf{a}\times\mathbf{b}) &= (\mathbf{b}\cdot\boldsymbol{\nabla})\mathbf{a} - (\mathbf{a}\cdot\boldsymbol{\nabla})\mathbf{b} + (\boldsymbol{\nabla}\cdot\mathbf{b})\mathbf{a} - (\boldsymbol{\nabla}\cdot\mathbf{a})\mathbf{b} \\
\mathbf{a}\times(\boldsymbol{\nabla}\times\mathbf{a}) &= \frac{1}{2}\boldsymbol{\nabla}\mathbf{a}^2 - (\mathbf{a}\cdot\boldsymbol{\nabla})\mathbf{a} \\
&= \mathbf{a}(\boldsymbol{\nabla}\cdot\mathbf{a}) - \sum_i \mathbf{e}_i\boldsymbol{\nabla}\cdot\left(a_i\mathbf{a} - \frac{1}{2}\mathbf{e}_i\mathbf{a}^2\right) \\
\boldsymbol{\nabla}\times(\boldsymbol{\nabla}\times\mathbf{a}) &= \boldsymbol{\nabla}(\boldsymbol{\nabla}\cdot\mathbf{a}) - \boldsymbol{\nabla}^2\mathbf{a}
\end{aligned}
\tag{C.6}
$$

C.3 Curvilinear coordinate systems in $\mathbb{R}^3$

I Cylindrical coordinates (ρ, ϕ, z)

With (ρ, ϕ, z) as cylindrical coordinates and (x, y, z) as Cartesian coordinates, where z is a common coordinate, the remaining coordinates are related as for plane polar coordinates,

$$x = \rho \cos\phi\,, \quad y = \rho \cos\phi. \tag{C.7}$$

Length and volume elements

$$ds^2 = d\rho^2 + \rho^2 d\phi^2 + dz^2\,, \quad dV = \rho\, d\rho\, d\phi\, dz \tag{C.8}$$

Differential operations

Gradient

$$(\boldsymbol{\nabla} f)_\rho = \frac{\partial f}{\partial \rho}\,, \quad (\boldsymbol{\nabla} f)_\phi = \frac{1}{\phi}\frac{\partial f}{\partial \phi}\,, \quad (\boldsymbol{\nabla} f)_z = \frac{\partial f}{\partial z} \tag{C.9}$$

Divergence

$$\boldsymbol{\nabla} \cdot \mathbf{a} = \frac{1}{\rho}\frac{\partial}{\partial \rho}(\rho a_\rho) + \frac{1}{\rho}\frac{\partial a_\phi}{\partial \phi} + \frac{\partial a_z}{\partial z} \tag{C.10}$$

Curl

$$(\boldsymbol{\nabla} \times \mathbf{a})_\rho = \frac{1}{\rho}\frac{\partial a_z}{\partial \phi} - \frac{\partial a_\phi}{\partial z}$$

$$(\boldsymbol{\nabla} \times \mathbf{a})_\phi = \frac{\partial a_\rho}{\partial z} - \frac{\partial a_z}{\partial \rho}$$

$$(\boldsymbol{\nabla} \times \mathbf{a})_z = \frac{1}{\rho}\frac{\partial}{\partial \phi}(\rho a_\phi) - \frac{1}{\rho}\frac{\partial a_\phi}{\partial z} \tag{C.11}$$

Laplacian

$$\nabla^2 f = \frac{1}{r}\frac{\partial}{\partial r}\left(r\frac{\partial f}{\partial r}\right) + \frac{1}{r^2}\frac{\partial^2 f}{\partial \phi^2} + \frac{\partial^2 f}{\partial z^2} \tag{C.12}$$

II Spherical coordinates (r, θ, ϕ)

$$x = r\cos\phi\sin\theta\,, \quad y = r\cos\phi\sin\theta\,, \quad z = r\cos\theta \tag{C.13}$$

Length and volume elements

$$ds^2 = dr^2 + r^2 d\theta^2 + r^2\sin^2\theta\, d\phi^2\,, \quad dV = r^2\sin\theta\, dr\, d\theta\, d\phi \tag{C.14}$$

Differential operations

Gradient

$$(\boldsymbol{\nabla} f)_r = \frac{\partial f}{\partial r}\,, \quad (\boldsymbol{\nabla} f)_\theta = \frac{1}{r}\frac{\partial f}{\partial \theta}\,, \quad (\boldsymbol{\nabla} f)_\phi = \frac{1}{r\sin\theta}\frac{\partial f}{\partial \phi} \tag{C.15}$$

Divergence

$$\boldsymbol{\nabla}\cdot\mathbf{a} = \frac{1}{r^2}\frac{\partial}{\partial r}(r^2 a_\rho) + \frac{1}{r\sin\theta}\frac{\partial}{\partial\theta}(\sin\theta\, a_\theta) + \frac{1}{r\sin\theta}\frac{\partial a_\phi}{\partial\phi} \tag{C.16}$$

Curl

$$(\boldsymbol{\nabla}\times\mathbf{a})_r = \frac{1}{r\sin\theta}\left[\frac{\partial}{\partial\theta}(\sin\theta\, a_\phi) - \frac{\partial a_\theta}{\partial\phi}\right]$$

$$(\boldsymbol{\nabla}\times\mathbf{a})_\theta = \frac{1}{r\sin\theta}\left[\frac{\partial a_r}{\partial\phi} - \sin\theta\frac{\partial}{\partial r}(r a_\phi)\right]$$

$$(\boldsymbol{\nabla}\times\mathbf{a})_\phi = \frac{1}{r}\left[\frac{\partial}{\partial r}(r a_\theta) - \frac{\partial a_r}{\partial\theta}\right] \tag{C.17}$$

Laplacian

$$\nabla^2 f = \frac{1}{r^2}\frac{\partial}{\partial r}\left(r^2\frac{\partial f}{\partial r}\right) + \frac{1}{r^2\sin\theta}\frac{\partial}{\partial\theta}\left(\sin\theta\frac{\partial f}{\partial\theta}\right) + \frac{1}{r^2\sin^2\theta}\frac{\partial^2 f}{\partial\phi^2} \tag{C.18}$$

Index

www.ingramcontent.com/pod-product-compliance
Lightning Source LLC
Chambersburg PA
CBHW071247180926
51613CB00002B/155
9789811298448